HUMAN FACTORS IN LIGHTING

HUMAN FACTORS IN LIGHTING

P. R. BOYCE
B.Sc., Ph.D., M.C.I.B.S., M.I.S.

The Electricity Council Research Centre,
Capenhurst, Chester CH1 6ES, UK

MACMILLAN PUBLISHING CO., INC.
NEW YORK

COLLIER MACMILLAN CANADA, LTD.
TORONTO

Published in the USA by
SCIENTIFIC & TECHNICAL BOOKS
MACMILLAN PUBLISHING CO., INC.
866 Third Avenue, New York, N.Y. 10022

Distributed in Canada by
COLLIER MACMILLAN CANADA, LTD.

British Library Cataloging in Publication Data

Boyce, P R
   Human factors in lighting.
   1. Lighting—Human factors
   I. Title
   621.32'2   TH7725

Library of Congress Catalog Card Number 81-80912
ISBN 0-02-949250-5

WITH 30 TABLES AND 143 ILLUSTRATIONS

© APPLIED SCIENCE PUBLISHERS LTD 1981

Printed in Great Britain by Galliard (Printers) Ltd, Great Yarmouth

Dedication

*To my Mother and Father, for their encouragement.
To Susan and Anna, for their tolerance.*

Acknowledgements

The help and co-operation of the following authors and publishers in giving permission for the reproduction of copyright material is gratefully acknowledged.

The American Conference of Governmental Industrial Hygienists Inc., for Table 8.4 taken from *The threshold limit values for physical agents*, 1979.

Bell & Hyman Limited, for Figs 1.3 and 2.4 taken from *The Perception of Light and Colour* by C. A. Padgham and J. E. Saunders.

Mr B. R. Bridges, for Fig. 4.26.

Dr D. A. Canter, for Table 8.3 taken from *Environmental Interaction* published by Surrey University Press.

Mr D. A. Castle, for Figs 4.25, 8.2 and 8.3.

The Chartered Institution of Building Services, for Fig. 7.9 and Tables 9.4, 9.5 and 9.6, all taken from *The IES code for interior lighting*, 1977; and for Fig. 4.39 taken from 'Vehicle headlights' by J. S. Yerrell, *Lighting Research and Technology*, **8**, 69, 1976.

The Commission Internationale de l'Eclairage, for Tables 9.2 and 9.3 taken from CIE Publication 29, *Guide on interior lighting*; and for Tables 9.8 and 9.9 taken from CIE Publication 12.2, *Recommendations for the lighting of roads for motorised traffic*.

Edward Arnold (Publishers) Ltd, for Fig. 2.1(b) taken from *Lamps and Lighting* by S. T. Henderson and A. M. Marsden.

Professor R. L. Gregory, for Fig. 2.2 taken from *Eye and Brain* published by Weidenfeld and Nicolson.

The Illuminating Engineering Research Institute, for Fig. 6.7 taken from *The effect of light on human judgement and behaviour* by J. E. Flynn, T. J. Spencer, O. Martyniuck and C. Hendrick, IERI Project 92, 1975.

The Illuminating Engineering Society of North America, for Figs 6.4, 6.5 and 6.9 taken from 'Interim study of procedures for investigating the

effect of light on impression and behaviour' by J. E. Flynn, T. J. Spencer, O. Martyniuck and C. Hendrick, *Journal of the Illuminating Engineering Society*, **3**, 87, 1973; for Table 6.1 taken from 'A note towards the understanding of lighting quality' by R. J. Hawkes, D. L. Loe and E. Rowlands, *Journal of the Illuminating Engineering Society*, **8**, 111, 1979; for Table 6.2 taken from 'A study of subjective responses to low energy and non-uniform lighting systems' by J. E. Flynn, *Lighting Design and Application*, February 1977; and for Table 4.4 taken from the *IES Lighting Handbook*, 5th edition, 1972.
Dr E. A. Megaw, for Fig. 4.18.
Mr A. E. Mould, for Figs 2.1(a) and 4.31.
The Royal College of Art, for Fig. 4.4 taken from a report entitled *The effects of image degradation and background noise on the legibility of text and numerals in four different typefaces* by H. Spencer, L. Reynolds and B. Coe, 1977.

Preface

The response of people to lighting has been a subject of study for many years, and rightly so. It is an important subject because it influences lighting practice and hence affects people, in terms of their comfort and task performance, and economics, in terms of building costs and energy consumption. This book provides an up-to-date survey of what is known about the interaction of people and lighting, in buildings and on roads, for natural and artificial light. It is intended to be of interest to those who are involved in the design and specification of lighting, and to those, such as architects, ergonomists and psychologists, who are concerned with people's reactions to the physical environment.

The book is divided into four, largely self-contained parts. Part 1 deals with the basics of light and vision. Part 2 covers the relationship between light and work. Part 3 discusses people's perception of lighting in terms of impression and discomfort. Part 4 examines the characteristics of documents advising on lighting practice and looks to the future. Those new to the subject are advised to read Part 1 before proceeding further. Those familiar with the quantitative aspects of lighting and the attributes of the visual system can read wherever their curiosity takes them.

It is a pleasure to thank the many people who have helped with the preparation of this book. Mr J. B. Collins, Dr V. H. C. Crisp, Mr D. J. Gazeley, Miss M. B. Halstead, Dr B. L. Hills, Mr J. A. Lynes, Dr E. A. Megaw, Mr W. A. Price, Mr G. Ratcliff and Mr P. T. Stone have all read and made valuable comments on parts of the text. Mrs J. Hughes has typed the many drafts, corrections, modifications and amendments, quickly, accurately, and without complaint. She has also dealt with the author's occasional harrassment with unfailing good humour. The Electricity Council Research Centre has supported work on people's responses to lighting for many years. Without such support and encouragement this book could not have been written.

P. R. BOYCE

Contents

Part 1:
Foundations

1

Light

1.1 INTRODUCTION

This book is concerned with people's responses to lighting. It is therefore fitting to start by considering what light is, how it may be quantified and how it can be produced. These are the subjects of this chapter.

1.2 PHOTOMETRY

1.2.1 Light and Radiation

To the physicist, light is simply a very small part of the electromagnetic spectrum, sandwiched between ultraviolet and infrared radiation. The electromagnetic spectrum ranges from cosmic rays with wavelengths of the order of 10^{-15} m to radio waves with wavelengths of the order of 10^4 m (Fig. 1.1). Light occupies the wavelength region from about 380×10^{-9} to 760×10^{-9} m, a very small section indeed. What distinguishes this part of the electromagnetic spectrum from the rest is that radiation in this region is absorbed by the photoreceptors of the human visual system and thereby initiates the process of seeing.

If the visual effect of radiation at each wavelength within the visible range were the same there would be no need for a special set of quantities to characterise light. The fundamental radiometric measures applicable to electromagnetic radiation [1] could be used. However, the visual effect of radiation varies markedly with wavelength. Therefore the quantities used to characterise light have to be derived from those used for electromagnetic radiation, by weighting them by the spectral sensitivity of the human visual system.

1

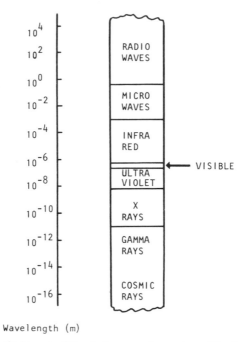

Wavelength (m)

Fig. 1.1.   A schematic diagram of the electromagnetic spectrum. The divisions between the different types of electromagnetic radiation are indicative only.

The principle used for the measurement of the human spectral sensitivity is equivalence of visual effect. The starting point for this measurement is the experimentally verified fact that the visual effect of monochromatic radiation is a function of wavelength and radiance. An observer viewing two equally presented fields with the same wavelength and the same radiance will say he sees them as equal in all respects, i.e. they have the same visual effect. If the wavelength of the fields is the same but the radiances are different, the field with the higher radiance will appear brighter. When both wavelength and radiance differ, the best that can be achieved is brightness equivalence. This condition can be reached by changing the radiance of one of the fields until the two fields appear equally bright. If L_{e1} and L_{e2} are the radiances of the fields at wavelengths λ_1 and λ_2 the brightness equivalence can be represented by the expression

$$V_1 L_{e1} = V_2 L_{e2}$$

where V_1 and V_2 are the weighting factors necessary to make the equation correct for the measured radiances. Since the only measured values are the radiances, each equivalence established produces a ratio V_1/V_2 applying to λ_1 and λ_2. By establishing equivalences of many pairs of wavelengths and using the transitive principle of mathematics, i.e. $V_1/V_3 = V_1/V_2 \times V_2/V_3$, it is possible to express the sensitivity of the visual system at each wavelength relative to its sensitivity at an arbitrarily chosen standard wavelength, i.e. $V_\lambda/V_{standard}$. The standard wavelength usually chosen is the one for which sensitivity is a maximum. Then, by giving $V_{standard}$ a value of unity and plotting the resulting V_λ against wavelength, a curve can be produced which quantifies the relative efficiency of the different wavelengths in producing the same visual effect. Such a curve is a relative spectral sensitivity curve of the human visual system. It contains the information necessary to convert the fundamental radiometric quantities into quantities suitable for describing light.

1.2.2 The CIE Standard Observers

Unfortunately there is no such thing as a single relative spectral sensitivity curve for the human visual system which is applicable to all people in all conditions. What there is is international agreement. The human visual system actually has two broad classes of photoreceptor. Their properties will be considered in more detail in Chapter 2; but for the moment it is sufficient to know that one class operates when light is plentiful in what are called photopic conditions, whilst the other operates when very little light is available in what are called scotopic conditions. These two photoreceptor types have very different spectral sensitivities and it is these spectral sensitivities which have been the subject of international agreement. The body which organises these agreements is the Commission Internationale de l'Eclairage. In 1924 the CIE, as it is usually known, adopted the CIE standard photopic observer. In 1951 the CIE adopted the CIE standard scotopic observer. The relative spectral sensitivity curves for these two creatures are shown in Fig. 1.2 [2]. These two curves, usually called the V_λ and V'_λ curves but more formally known as the relative photopic and scotopic luminous efficiency functions, form the basis of conversion from radiometric quantities to the quantities used to characterise light, namely the photometric quantities.

1.2.3 Photometric Quantities

The fundamental quantity used in photometry is luminous flux. This is a measure of the rate of flow of radiant energy modified for its effectiveness in creating the sensation of seeing. In other words luminous flux is simply radiant flux multiplied by the relevant spectral sensitivity of

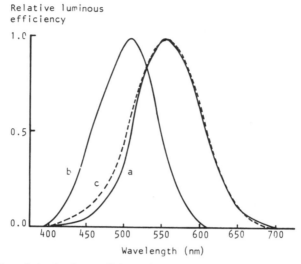

Fig. 1.2. The relative luminous efficiency functions for (a) the CIE standard photopic observer (V_λ), and (b) the CIE standard scotopic observer (V'_λ). Also shown (c) is the provisional relative luminous efficiency function for a $10°$ field in photopic conditions. (After [5].)

the visual system. This can be represented by the equation

$$\phi = K_m \int_{380}^{760} V_\lambda \phi d\lambda$$

where $\phi d\lambda$ = radiant flux in a small wavelength interval, $d\lambda$; V_λ = the relative luminous efficiency function for the appropriate standard observer; and K_m = a constant, called the maximum spectral luminous efficacy, for the appropriate standard observer.

In the SI system the radiant flux is measured in watts and luminous flux in lumens. The current values of K_m are 683 lumens watt^{-1} for the standard photopic observer, and 1699 lumens watt^{-1} for the standard scotopic observer.† It is always important to identify which of the standard

† These values derive from the decision of the Conference Internationale des Poids et Mésures that the spectral luminous efficacy (K_λ) for monochromatic radiation at a wavelength of 555 nm in air be 683 lumens watt^{-1} for both photopic and scotopic vision. For photopic vision $K_\lambda = K_m V_\lambda$, and for scotopic vision $K'_\lambda = K'_m V'_\lambda$. At 555 nm, $V_\lambda = 1.000$ and $V'_\lambda = 0.402$. Thus $K_m = (683)/(1.000) = 683$ lumens watt^{-1} and $K'_m = (683)/(0.402) = 1699$ lumens watt^{-1}.

observers is being used in any particular measurement although the photopic observer is usually assumed unless otherwise stated. This practice has led the CIE to recommend that for the scotopic standard observer all photometric terms should be preceded by the word scotopic, e.g. scotopic luminous flux.

Luminous flux is an important quantity in determining the total light output of light sources and luminaires but it is not the only useful quantity. For describing the distribution of light from a luminaire, luminous intensity is used. Luminous intensity is the luminous flux emitted/unit solid angle in a specified direction. This quantity is not as might be expected measured in lumens steradian^{-1} but in units called candelas, a procedure which owes more to history than logic. However, the Conference Internationale des Poids et Mésures (CIPM) has recently defined the candela as the luminous intensity in a given direction of a source emitting monochromatic radiation at a wavelength of 555 nm of which the radiant intensity is $1/683$ watts steradian^{-1}. Since the CIPM has also defined the spectral luminous efficacy of monochromatic radiation at 555 nm as 683 lumens watt^{-1} it is clear that a candela is the same as a lumen steradian^{-1}. This is also apparent from the current definition of the lumen, which is the luminous flux emitted within the unit solid angle (one steradian) by a point source having a uniform luminous intensity of one candela.

Both luminous flux and luminous intensity have area measures associated with them. The luminous flux falling on the unit area of a surface is called the illuminance. The luminous intensity/unit projected area of a surface in a given direction is the luminance. The approved SI' unit of illuminance is lumen metre^{-2}, usually called the lux. The approved SI unit for luminance is candela metre^{-2}. Naturally enough there is a relationship between the light incident on a surface and that reflected from it. For a perfectly diffusely reflecting surface the relationship is given by the expression

$$\text{Luminance} = \frac{\text{illuminance} \times \text{reflectance}}{\pi}$$

When the surface being considered is not a perfectly diffuse reflector, reflectance is replaced by luminance factor. Reflectance is defined as the ratio of the reflected luminous flux to the incident luminous flux and applies only to diffuse reflection. Luminance factor is defined as the ratio of the luminance of a surface viewed from a particular position and lit in a specified way to the luminance of a diffusely reflecting white surface viewed from the same direction and lit in the same way. Table 1.1 gives more precise definitions of all these photometric quantities, the units used and some of the relationships between them. The units used in Table 1.1 are those approved by the SI system.

Table 1.1:   The photometric quantities

Quantity	Definition	Units
Luminous flux	That quantity of radiant flux which expresses its capacity to produce visual sensation	lumen (lm)
Luminous intensity	The luminous flux emitted in a very narrow cone containing the given direction divided by the solid angle of the cone, i.e. luminous flux/unit solid angle	candela (cd)
Illuminance	The luminous flux/unit area at a point on a surface	lumen metre^{-2} (lm m^{-2})
Luminance	The luminous flux emitted in a given direction divided by the product of the projected area of the source element perpendicular to the direction and the solid angle containing that direction, i.e. luminous flux/unit solid angle/unit area	candela metre^{-2} (cd m^{-2})
Reflectance	The ratio of the luminous flux reflected from a surface to the luminous flux incident on it	
For a diffuse surface	Luminance $= \dfrac{\text{illuminance} \times \text{reflectance}}{\pi}$	Luminance in candelas metre^{-2} Illuminance in lumens metre^{-2}
Luminance factor	The ratio of the luminance of a reflecting surface, viewed in a given direction to that of a perfect white uniform diffusing surface identically illuminated	
For a non-diffuse surface for a specific viewing direction and lighting geometry	Luminance $= \dfrac{\text{illuminance} \times \text{luminance factor}}{\pi}$	Luminance in candelas metre^{-2} Illuminance in lumens metre^{-2}

Photometry has had a long and chequered history which has left behind a large number of different units, many named after the pioneers in the field. This recognises their contribution but gives the newcomer little idea of what the dimensions of the quantity are. Table 1.2 lists a number of these units and gives the appropriate factor to convert to the approved SI unit. Most of the conversions involve changes in the area component but at least some of the luminance units are based on an alternative definition of luminance. This definition assumes a perfectly diffusely reflecting surface so the luminance is simply the product of the illuminance and the reflectance. Since reflectance is a ratio, the dimensions of luminance are those of illuminance, i.e. lumens/unit area. This definition of luminance has been widely used by lighting engineers in the past because it is a

Table 1.2: Some common photometric units and the conversion factors necessary to change to the approved SI units

Quantity	Unit	Dimensions	Multiplying factor to convert to SI unit
Illuminance	lux	lumen metre^{-2}	1·00
(SI unit = lumen metre^{-2})	metre candle	lumen metre^{-2}	1·00
	phot	lumen centimetre^{-2}	10 000·00
	foot candle	lumen foot^{-2}	10·76
Luminance	nit	candela metre^{-2}	1·00
(SI unit = candela metre^{-2})	stilb	candela centimetre^{-2}	10 000·00
	—	candela inch^{-2}	1 550·00
	—	candela foot^{-2}	10·76
	apostilb†	lumen metre^{-2}	0·32
	blondel†	lumen metre^{-2}	0·32
	lambert†	lumen centimetre^{-2}	3 183·00
	foot-lambert†	lumen foot^{-2}	3·43

† These four items are based on an alternative definition of luminance. This definition is that if the surface can be considered as perfectly matt its luminance in any direction is the product of the illuminance on the surface and its reflectance. Thus the luminance is described in lumens/unit area. This definition is deprecated in the SI system.

common practice to assume that the surfaces of rooms are perfectly matt even if they are not. Hence, once the illuminance on a wall for example is known, it is easy to obtain its luminance. In spite of this practical advantage this definition of luminance is now deprecated and only the SI system will be used here.

Both illuminance and luminance are widely used in lighting practice to quantify the effect of the lighting on a space. Although it is necessary to know how these quantities are defined this is not much use if there is no 'feel' for the values that occur in practice. Table 1.3 gives some relevant figures which are intended to relate to everyday experience.

Before leaving the photometric quantities it is essential to consider the range of application of the photopic and scotopic series of quantities. It is generally agreed that scotopic quantities are reasonable measures for describing the visual effect of light over a luminance range from

Table 1.3:   Typical illuminance and luminance values for various situations

Situation	Illuminance on horizontal surface ($lm\,m^{-2}$)	Typical surface	Luminance ($cd\,m^{-2}$)
Clear sky in summer in northern temperate zones	150 000	Grass	2 900
Overcast sky in summer in northern temperate zones	16 000	Grass	300
Textile inspection	1 500	Light grey cloth	140
Office work	500	White paper	120
Heavy engineering	300	Steel	20
Good street lighting	10	Concrete road surface	1·0
Moonlight	0·5	Asphalt road surface	0·01

absolute threshold (about $10^{-6}\,cd\,m^{-2}$) up to $10^{-3}\,cd\,m^{-2}$. The photopic quantities are reasonable measures for luminances above about $3\,cd\,m^{-2}$.

1.2.4 Some Limitations

Although the photometric quantities just described can be calculated or measured precisely for all conditions, it is important to realise that the conditions for which they reasonably represent visual effect are limited. The reason for this is that real observers and real viewing conditions tend to produce spectral sensitivities which are different from those of the standard observers. For a start the CIE standard observers' spectral sensitivity curves are average curves. Different individuals will depart from the average, particularly the elderly and those with some forms of defective colour vision. However, for many situations, to use the average spectral sensitivity curves represented by the CIE standard observers is reasonable even though the photopic spectral sensitivity curve at least is known not to be correct. It has been shown convincingly that the sensitivity at the short wavelengths is too low [3]. Nothing has been done about this because even after correcting for this error the visual system is still very insensitive to shorter wavelengths and few light sources produce much radiation in this region. Therefore the consequences of this error for general lighting practice are small.

Another limitation of the system of photometric quantities is the gap in the ranges of application of photopic and scotopic quantities. Over a luminance range from $10^{-3}\,cd\,m^{-2}$ to about $3\,cd\,m^{-2}$ neither of the spectral sensitivity curves for the two standard observers is appropriate. In this range, called the mesopic range, the spectral sensitivity curve of the human visual system changes in shape and position between the two standard observer curves. At present there is no exact system of photometry available for the mesopic region although proposals for dealing with it by a combination of photopic and scotopic measurements have been made [4]. The existence of the mesopic range is an important limitation on the CIE photometric system because practical situations occur in which vision is operating in the mesopic range. For example, much roadway lighting produces mesopic conditions, yet roadway lighting is almost invariably measured using instruments with spectral sensitivities matched to the CIE standard photopic observer. This can give measured values of the lighting which bear little relationship to its visual effect, but until a system of mesopic photometry is available, all that can be done is to be aware of the fact.

The accuracy of the photometric quantities is also limited by the conditions in which the spectral sensitivity curves are obtained. Different methods used to determine equivalence of visual effect lead to small differences in the measured spectral sensitivities. More important are the differences that occur with different field sizes. The spectral sensitivity for the CIE standard photopic observer was derived from data obtained using 2–3° visual fields. It is known that for larger fields, up to 10°, the standard observer is not as sensitive to short wavelengths as most real observers. To accommodate this point the CIE has produced a provisional spectral sensitivity curve for a 10° field of view (see Fig. 1.2). CIE Publication 41 [5] discusses the whole problem of the limitations of photometry in considerable detail.

By now it should be apparent that the fact that the photometric quantities can be calculated and/or measured precisely is no guarantee that they will be closely related to the visual effects produced. There are too many potential sources of deviation from the standard observers' spectral sensitivity curves for that. What needs to be considered is the size of the discrepancies these deviations can produce and the conditions under which they can occur. Kinney [6] has produced some estimates on just this point. Overall she shows that for light sources which emit a wide range of wavelengths, the discrepancies caused by using the appropriate standard observer's spectral sensitivity curve, rather than a more exact curve, are small, whatever the cause of the deviation. However, for sources of monochromatic radiation most discrepancies are large. The main concern here is with people's reactions to lighting produced by light sources with the former characteristic. For such applications the CIE photometric system can be used with confidence.

1.3 COLORIMETRY

1.3.1 Colour

None of the photometric quantities so far discussed give any weight to the combination of wavelengths that form the light received at the eye. Thus it is possible for two luminous fields to have the same luminance but to be made up of totally different combinations of wavelengths. In this situation, and provided photopic conditions prevail, the two fields will look different in one respect; they will be seen as different colours. The actual colour seen will depend on the spectral distribution of the light. If the emissions are concentrated at the long wavelength end of the visible spectrum the light will appear red, if the concentration is at the centre it will appear yellow/green, if the concentration is at the short wavelength end it will appear blue. The appearance of any coloured luminous field can be described by its classification according to three separate subjective attributes. The first attribute is the hue of the colour, i.e. whether it is basically red, yellow, orange, green, blue or purple. The second is the strength of the colour or colourfulness. The third is the brightness of the colour, i.e. the extent to which the luminous field emits more or less light. Although the perception of colours in the abstract, e.g. as luminous fields, can be described in terms of hue, colourfulness and brightness, the perception of colours of objects depends on different but related attributes. For describing their colour appearance, objects are divided into luminous and non-luminous classes, depending on whether they appear to emit or absorb light. Most objects in the real world are non-luminous although light sources usually appear as luminous. A further division is made for non-luminous objects depending on the conditions in which they are seen. Non-luminous objects seen in isolation are classed as unrelated colours, whilst non-luminous objects seen in relation to other colours are classed as related colours. The two object classes of interest here are luminous colours, e.g. light sources, and non-luminous related colours, e.g. surfaces in a room. For objects of both classes hue is still a meaningful attribute of their colour, but colourfulness and brightness require some modification. For both non-luminous and luminous objects saturation is an important attribute of how colours are perceived. Saturation is the extent to which an object has more or less colour, judged in proportion to its brightness. Saturation is thus relative colourfulness. For non-luminous related objects the proportion of incident light reflected from an object provides an important clue to its identity. The perception of this proportion is called the lightness of the colour. Lightness is a form of relative brightness. From this it would appear that to describe the appearance of a luminous object the attributes hue and saturation are enough, but for a non-luminous related object, hue, saturation and lightness are necessary. This is true, but for non-luminous related objects there is an alternative attribute of relative colourfulness—perceived chroma. Perceived chroma is colourfulness judged in proportion to the average brightness of the surroundings. The extent to which saturation or perceived chroma is the more meaningful for a non-luminous object, and the ease with which lightness can be judged, depend on the circumstances in which the object is seen. Perceived chroma and lightness are most easily assessed when flat objects are viewed under uniform illumination. For three-dimensional objects in very non-uniform lighting, saturation and brightness are more obvious.

To summarise the above discussion, the basic attributes of perceived colour are hue, brightness and colourfulness. However, for the recognition of colours attached to objects the derived attributes of saturation, lightness and perceived chroma become important. It should be noted that there is nothing inevitable about these attributes. Their only justification is that they reliably relate to the way people are able to identify and separate colours. They are important because it is these attributes which systems of colour description attempt to quantify. A complete discussion of these terms is given in a valuable paper by Hunt [7].

1.3.2 Colour Classification

The three attributes, hue, lightness and perceived chroma enable an observer to classify his perception of a non-luminous, related, coloured surface. They also form the basis of the simplest practical method of identifying and classifying colours: the colour atlas. The most widely used of these, the *Munsell Book of Colour* [8], consists of approximately 1200 small chips of different colours, each one classified by three terms, called Hue, Value and Chroma. These three terms can be approximately assigned to the attributes of hue, lightness and perceived chroma respectively. Figure 1.3 shows the schematic layout of the Munsell system to form a colour space. The Hue, Value and Chroma scales are laid out in equal perceptual steps. The Hue circle is divided into five main colours and five intermediate colours with 10 steps between each pair of colours. The Value scale is divided into 10 steps and the Chroma scale has up to 16 steps. Any particular colour is given a Munsell reference in the form of Hue/Value/Chroma. Thus a strong red will be 7·5R/4/12 and a pale blue will be 5B/9/1. This standard method of notation reveals

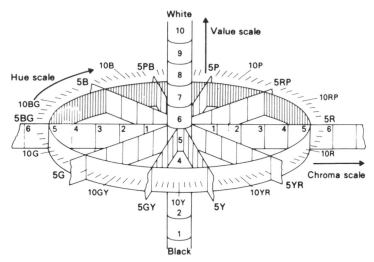

Fig. 1.3. The organisation of the Munsell colour system. The hue letters are B = blue, PB = purple/blue, P = purple, RP = red/purple, R = red, YR = yellow/red, Y = yellow, GY = green/yellow, G = green, BG = blue/green.

the purpose of the Munsell system. It is the very practical aim of enabling people to communicate about colours in a more precise way than words permit. Instead of a man in New York attempting to tell one in London that he wants 'a lightish yellowish green material' he can simply say that he wants one to match Munsell reference 5YG/8/2 under specified conditions. The Munsell system has the great advantage of making colours manifest rather than attempting to express them in terms of their spectral reflectance. A spectral reflectance curve may be more precise but it takes a highly trained mind to know what it will look like. There are other colour atlas systems available which work on the same basis although they use different methods of classification [9, 10]. In addition there exist series of standard colours for particular applications (e.g. reference [11]).

1.3.3 Colorimetric Quantities
All these atlases and similar items aim to classify colours so that the user can evaluate and communicate about them. But this is not sufficient for all the needs of those concerned with colour. There is no quantification involved in the use of a colour atlas and yet it is frequently desirable to predict the colours that will be seen using surfaces and light sources

that do not physically exist. To do this some form of colour quantification is needed. The accepted system for doing this has been produced by the CIE and is called the CIE colorimetric system [12]. The basis of this system is again the definition of a standard observer which means it is subject to many of the limitations discussed for the CIE photometric system. The standard observer's performance is again derived from a series of equivalence matches, but this time involving colour. The standard observer ignores the variations in colour perception that occur in practice and are caused by changes in viewing conditions, as well as the individual differences for normal and colour-defective observers. The CIE standard observer simply reproduces the average colour matching responses of a group of observers with normal colour vision.

The CIE colorimetric system depends on two facts. The first is that most people have very similar colour responses. The second is that any colour of light can be matched by a combination of not more than three spectral wavelengths of light lying in the red, green and blue wavelength regions. The full development of the CIE colorimetric system is complicated and is not required here (see reference [9] for details). It is sufficient to say that it is based on the assumption of a standard observer with three relative spectral sensitivity curves. These three curves, called the CIE colour matching functions $\bar{x}_\lambda$, $\bar{y}_\lambda$, and $\bar{z}_\lambda$ are completely imaginary. This does not matter because the CIE colorimetric system is not concerned with reproducing the operation of the visual system, but only its end effect. Its ultimate aim is to ensure that each colour has a separate value in the colorimetric system and that two colours which have the same value match, no matter which wavelengths are used to create them. In fact there are two sets of colour matching functions, i.e. two standard observers. The CIE 1931 standard observer is used for colours occupying visual fields up to $4°$ angular subtense. The CIE 1964 large-field standard observer is used for visual fields greater than $4°$. Figure 1.4 shows the colour matching functions of the CIE 1931 standard observer.

Once these colour matching functions have been accepted it is a simple matter to quantify the colour properties of a light source. The values X, Y and Z, called the tristimulus values of the colour, are obtained from the expressions

$$X = h \sum \phi_\lambda \bar{x}_\lambda \, \Delta\lambda$$
$$Y = h \sum \phi_\lambda \bar{y}_\lambda \, \Delta\lambda$$
$$Z = h \sum \phi_\lambda \bar{z}_\lambda \, \Delta\lambda$$

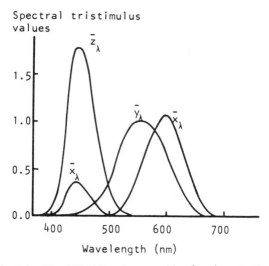

Fig. 1.4.   The CIE 1931 colour matching functions, $\bar{x}_\lambda$, $\bar{y}_\lambda$, $\bar{z}_\lambda$.

where $\phi_\lambda \Delta\lambda =$ the radiant flux in a wavelength interval $\Delta\lambda$, $\bar{x}_\lambda$, $\bar{y}_\lambda$, and $\bar{z}_\lambda =$ the spectral tristimulus values of the appropriate colour matching functions, and $h =$ constant.† These equations simply express X, Y and Z as the sum over an appropriate wavelength interval of the product of the radiant flux and the colour matching function.

Having obtained the tristimulus values of the colour, X, Y and Z, the next step is to express their individual values as proportions of their sum, i.e.

$$x = \frac{X}{X + Y + Z} \qquad y = \frac{Y}{X + Y + Z} \qquad z = \frac{Z}{X + Y + Z}$$

The values x, y and z are called the chromaticity co-ordinates of the colour. Since $x + y + z = 1$, only the x and y values are usually given. It is these x and y chromaticity co-ordinates that are conventionally used as the quantities that define the colour of a light source, but there is no reason why the method should not be applied to reflecting or transmitting surfaces provided the spectral reflectance or transmittance

† h is an arbitrary constant. If only relative values of X, Y and Z are required, an appropriate value of h is one that makes $Y = 100$. If absolute values of X, Y and Z are required it is convenient to take $h = 683$ since then the value of Y is the luminous flux in lumens. For surface colours, an appropriate value of h is one that makes $Y = 100$ for a reference white. Then the calculated Y is the percentage reflectance of the surface.

is known. For a reflecting or transmitting surface the equation for X becomes

$$X = h \sum \phi_\lambda p_\lambda \bar{x}_\lambda \Delta\lambda$$

where p_λ = spectral reflectance or transmittance, $\phi_\lambda \Delta\lambda$ = the incident radiant flux in the wavelength interval $\Delta\lambda$, and $\bar{x}_\lambda$ = the spectral tristimulus values of the colour matching function.

Similar equations apply to Y and Z. Thus chromaticity co-ordinates can be used to define the colours of either light sources, or light transmitted through filters, or surfaces which reflect light. Since the whole process is one of calculation there is no need for either surface or light source to actually exist let alone exist in a particular place. This means that the effect of all manner of variables can be calculated.

Now if a colour can be represented by a two-co-ordinate position then all colours can be represented on a plane. Figure 1.5 shows a two-dimensional diagram, with the x and y chromaticity co-ordinates as axes. This is called the CIE 1931 chromaticity diagram. Philosophically it is inherently flawed. The only thing a set of chromaticity co-ordinates tells us about a colour is that colours with the same co-ordinates will match. It tells us nothing about the appearance of the colours or about the differences between colours with different chromaticity co-ordinates. Nonetheless, it has been found that a red colour does plot in one part of the diagram, and a green in another part, and so on. Thus, although the CIE 1931 chromaticity diagram is not theoretically pure, it is useful in indicating approximately how a colour will appear, or rather, what its hue and colourfulness are, and how different it will look from another colour on these attributes.

It is possible to plot a number of interesting features on the CIE 1931 chromaticity diagram. In Fig. 1.5 there can be seen a large curve closed by a straight line. This curve is the spectrum locus. All pure colours, i.e. those which consist of a single wavelength, lie on this curve. The straight line is the line of purples. At the centre of the diagram is a point called the equal energy point. This is the point at which a colourless surface will be located. The colourfulness, and saturation, of a colour increases as it gets closer to the spectrum locus and further from the equal energy point. The hue it has determines the direction in which it moves from the equal energy point towards the spectrum locus.

The CIE chromaticity co-ordinates are used for a number of practical purposes and the CIE 1931 chromaticity diagram is a convenient way

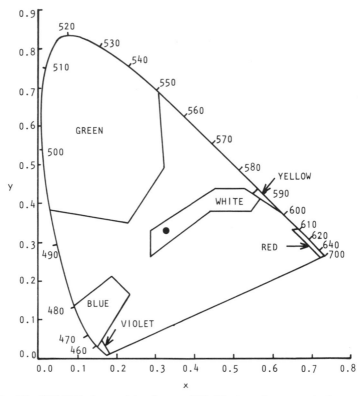

Fig. 1.5. The CIE 1931 chromaticity diagram [12]. The boundary curve is the spectrum locus, with the wavelengths (nm) marked. The filled circle is the equal energy point. The enclosed areas indicate the chromaticity co-ordinates of light signal colours that will be indentified as the specified colours. More restricted areas are recommended when a higher probability of identification is required and/or observers with abnormal colour vision are considered [13].

of displaying the information. By way of example, Fig. 1.5 shows the boundaries of chromaticity adopted by the CIE for signal colours [13] so that they will be perceived as red, blue, green, etc.

The philosophical difficulties associated with expressing colour differences by using the chromaticity co-ordinates have already been mentioned, but there is a more practical difficulty. Unfortunately equal spacings in different parts of the CIE 1931 chromaticity diagram do not correspond to equal perceptual differences. In an attempt to reduce this problem the CIE first introduced the CIE 1960 uniform chromaticity scale (UCS) diagram. This has been followed by another variation, the

CIE 1976 uniform chromaticity scale diagram [14]. Both are simply the CIE 1931 chromaticity diagram redrawn with the x and y axes subjected to a linear transformation. The equations for the CIE 1960 UCS diagram are

$$u = \frac{4x}{(-2x + 12y + 3)}$$

$$v = \frac{6y}{(-2x + 12y + 3)}$$

where x and y are the chromaticity co-ordinates. The equations for the CIE 1976 UCS diagram are

$$u' = \frac{4x}{(-2x + 12y + 3)}$$

$$v' = \frac{9y}{(-2x + 12y + 3)}$$

where x and y are the chromaticity co-ordinates. Therefore $u' = u$ and $v' = 1{\cdot}5v$. Figure 1.6 shows the CIE 1960 UCS diagram with the mean just noticeable colour differences about a number of colours marked [15]. Although the CIE 1960 UCS diagram is not perfectly perceptually

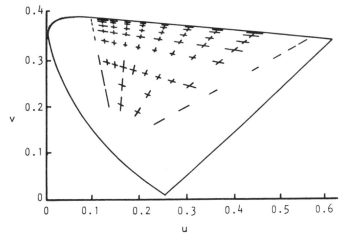

Fig. 1.6. Mean just noticeable colour differences plotted on the CIE 1960 uniform chromaticity scale diagram. The ends of the crosses indicate the chromaticity co-ordinates of the colours which were just noticeably different from the colour defined by the centre of the cross. (After [15].)

uniform, it is an improvement on the CIE 1931 chromaticity diagram. The CIE 1976 UCS diagram is a further improvement. Even so, it is still not completely perceptually uniform. Fortunately there is little point in trying to improve the uniformity of these diagrams because they have another defect.

The CIE UCS diagrams and all the other chromaticity diagrams only cover two of the three attributes on which colours are classified, namely

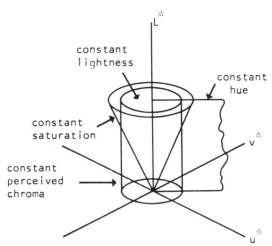

Fig. 1.7. The CIE 1976 $L^*u^*v^*$ colour space. Various surfaces in this space approximate to constant values of different attributes of object colours. (After [7].)

hue and colourfulness. For accurate estimation of the appearance and differences between colours, the third dimension, brightness, has to be considered. This is done by the use of a three-dimensional colour space. In 1978 the CIE recommended the use of two colour spaces, called the CIE 1976 $L^*u^*v^*$ and the CIE 1976 $L^*a^*b^*$ colour spaces [14]. As an example, Fig. 1.7 shows the CIE 1976 $L^*u^*v^*$ space. The vertical axis is the CIE 1976 psychometric lightness L^*. This is given by the expression

$$L^* = 116(Y/Y_n)^{0.33} - 16$$

where Y and Y_n are the Y tristimuli values of the colour of interest and a specified reference white object colour respectively. The other two axes u^* and v^* are obtained by the expression

$$u^* = 13L^*(u' - u'_n)$$
$$v^* = 13L^*(v' - v'_n)$$

where u', v' and u'_n, v'_n are the co-ordinates on the CIE 1976 UCS diagram of the colour of interest and the specified reference white object colour respectively. The CIE 1976 $L*u*v*$ colour space is the ultimate development of the CIE colorimetric system, the chromaticity diagrams described previously being only partial developments. That it is a part of the CIE colorimetric system can be seen by the fact that all the axes of the CIE 1976 $L*u*v*$ colour space are derived ultimately from the three tristimulus values of the colour and so link back to the standard observer. The CIE 1976 $L*u*v*$ colour space has two valuable properties. The first is that it is reasonably perceptually uniform, i.e. equal distances represent equal perceived colour differences. The second is that various surfaces in the space relate to the attributes of perceived object colour. In Fig. 1.7 vertical planes with the $L*$ axis as one edge approximate to surfaces of constant hue, cylinders with $L*$ as their axis follow surfaces of constant perceived chroma, cones having $L*$ as their axis and their point at the origin approach surfaces of constant saturation, and horizontal planes through the solid indicate surfaces of constant lightness. Measures based on the CIE 1976 $L*u*v*$ and CIE 1976 $L*a*b*$ colour spaces have been derived to quantify these attributes so differences between colours can be quantified in terms of hue, lightness and perceived chroma [7,14]. All this is intellectually interesting and augurs well for the future. Positions in such spaces can be used to quantify the appearance of colours of objects and the differences between them more precisely than a simple chromaticity diagram. It even appears possible to use colour space to allow for the variations in colour appearance that occur with changes in viewing conditions [16]. This colour space approach is undoubtedly the most complete and accurate method of quantifying colour but so far has been little used in lighting. It remains to be seen how widely it will be used and what problems that use will reveal. In considering the CIE colorimetric system it should always be remembered that its sole justification is that it is useful for specifying colours and for describing their appearance. If a better system was produced then it could replace the CIE colorimetric system. In fact an alternative approach has recently been proposed [17]. This proposal is based on the retinex theory of Land [18], which suggests that colour perception is better described as an evaluation of the lightnesses of surfaces in a complex scene. It will be interesting to see how and whether this approach develops.

To summarise, at the moment there exist two broad approaches to describing colours. One is by the extremely practical method of the colour atlas in which samples of different colours are classified. This has the advantage of being simple to use and understand but does not allow any theoretical examination of the effects of changes in the variables that determine a colour. The other is the more exact mathematical approach developed by the CIE into a system of colorimetry. This is essentially a calculation procedure which allows colours to be specified by their chromaticity co-ordinates on either the CIE 1931 chromaticity diagram or the CIE 1976 UCS diagram, or more completely described in one of the CIE 1976 colour spaces. The CIE colorimetric system has many uses but it should always be remembered that its mathematical precision is deceptive. As a method for describing colour appearance it is inevitably approximate to some degree.

1.3.4 Applications

Having specified how colours as such can be quantified, it is now necessary to consider the practical problems of describing the colour properties of light sources and the way they affect surfaces. The first thing to be clear about is that it is possible to distinguish between what is called the colour appearance of a light source and the colours it gives to surfaces. The colour appearance of a light source is governed by the spectral emission of the lamp and is conveniently quantified by its chromaticity co-ordinates. It is important to realise that two light sources can have different spectral emissions and still have the same colour appearance, i.e. the same chromaticity co-ordinates. The reasonableness of this will be apparent when it is remembered that having the same chromaticity co-ordinates means having the same proportions of radiant flux through the three colour matching functions. As can be seen in Fig. 1.4 the three colour matching functions are broad, which allows plenty of scope for variations in spectral emission without changing the proportions under each function. This applies to light seen directly. However, when light is seen by reflection from a surface, the spectral reflectance of the surface and the spectral emission of the lamp interact to produce the final spectrum received by the eye. If two light sources have different spectral emissions they may have different interactions with the spectral reflectance of a given surface, so the surface may have different chromaticity co-ordinates under the two light sources. Thus it is possible for two light sources to have the same colour appearance but produce different colours from the same surface.

As a final twist to this tortuous tale comes the subject of surface colour metamerism. Metameric colours are colours which match under one light source but do not match under another. Figure 1.8 shows the spectral reflectances of such a pair. Under daylight these colours match but under incandescent lamps they do not. The reason for this is that the interaction between the spectral reflectance and the spectral emission of the light source leads to the same chromaticity co-ordinates for the two colours under one light source, even though the spectral reflectances are different, but does not do so for the other light source.

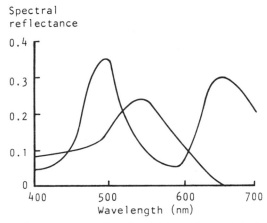

Fig. 1.8.   Spectral reflectance curves of two dyed cloths which match in daylight but do not match under light from an incandescent lamp.

Such surface colour metamerism is an important consideration in manufacturing industry, particularly where components are assembled under one light source but may be sold and/or used under another, or where it is necessary to produce the same surface colours on materials which inevitably contain different colouring matter.

From the above discussion it should be clear that in lighting a room the colour appearance of the light source and the way the surface colours are perceived need to be considered separately. This shows itself in the different measures used to describe the performance of light sources. The colour appearance of all light sources could be expressed in terms of their chromaticity co-ordinates, but in practice this is rare. For the light sources used in interiors, and the various forms of daylight, a simpler measure is used—correlated colour temperature. The basis of this measure is the fact that the spectral emission from a full radiator is a function of its temperature. Figure 1.9 shows the CIE 1960 uniform chromaticity scale diagram with the full radiator locus on it. The locus is the curved line joining the points representing full radiators at different temperatures. For light sources whose chromaticity co-ordinates lie on the full radiator locus, e.g. incandescent lamps, their colour appearance is conventionally given by the colour temperature, i.e. the temperature of the full radiator which has the same chromaticity co-ordinates as the light source in question. For light sources whose chromaticity co-ordinates lie close to the full radiator locus the colour appearance is given by the correlated

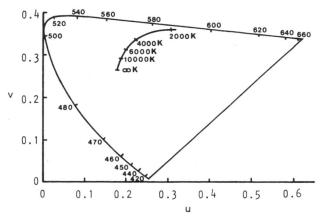

Fig. 1.9.   The CIE 1960 uniform chromaticity scale diagram and the full radiator locus.

colour temperature, i.e. the temperature of the full radiator whose chromaticity on the CIE 1960 UCS diagram lies closest to the chromaticity of the light source in question.† Correlated colour temperature is a very convenient and easily understandable measure of light source appearance. As a rough guide nominally white light sources have correlated colour temperatures from about 2700 K to 6500 K. A 2700 K correlated colour temperature is seen as 'reddish', which is usually described as having a warm appearance, whilst 6500 K is seen as 'bluish' and described as having a cool appearance.

As for the appearance of surfaces, the colours can be specified by their chromaticity co-ordinates or by their position in one of the CIE colour spaces and differences can be calculated from the colour difference

---

† Although the CIE 1960 UCS diagram has been largely superseded by the CIE 1976 UCS diagram it is still used for the definition of correlated colour temperature. This is because replacing it by the CIE 1976 UCS diagram would necessitate changes in the correlated colour temperatures of existing light sources, and there is some reluctance to do this without strong evidence that such a change is meaningful to people.

formulae [14]. This is reasonable if a specific set of colours is being considered but for general lighting, where many different but unspecified colours are used in interiors, more general advice is desirable. This is usually given in terms of the Colour Rendering Index of the light source [19]. The assumption behind this Index is that it is usually possible to specify how colours should appear. If this can be done then it is also possible to express how close to the objective a particular light source will make the colours appear, i.e. how well the colours are rendered. As a general rule some form of daylight, or a full radiator, is taken to be the reference light source and hence the way the colours appear under this source is assumed to be the aim of the lighting. The CIE Colour Rendering Index measures how accurately the colours are reproduced by the light source in question relative to the reference source. The actual calculation involves obtaining the positions of a colour in colour space under the reference light source and under the light source of interest, and expressing the difference between the two positions on a scale that gives perfect agreement between the positions a value of 100. The CIE recommend a series of test colours including some for such special subjects as human skin and vegetation. The result of the calculation for a single test colour is called the Special Colour Rendering Index. A set of eight test colours distributed around the hue circle is recommended for general use. When these eight test colours are used the Special Colour Rendering Index for each test colour is calculated and then the eight Special Colour Rendering Indices are averaged, the average being called the General Colour Rendering Index. It should be appreciated that light sources which have the same General Colour Rendering Index do not necessarily render each of the eight test colours in the same way. This fact, together with the need to choose and specify the reference source used, the limited range of test colours available and the method of correcting for chromatic adaptation, has led to some criticism of the CIE Colour Rendering Index [20] but as a general guide it is of practical value. However, where specific colours are important in a specific task it is better to use one of the colour spaces and to apply the colour difference equations to the actual colours [14].

There is one other feature of colour rendering that needs to be mentioned. The Colour Rendering Index is concerned with the accuracy of colour rendering. This is not always what is required. For some activities it is not necessary to be able to identify colours in an absolute sense but only to determine that they are different from each other. To make this easier some form of distortion of the colours may be useful. To

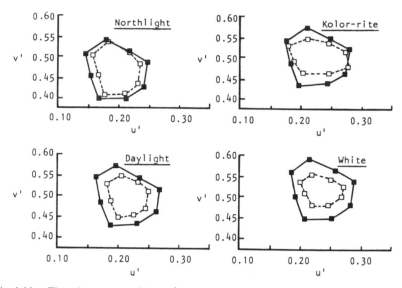

Fig. 1.10. The colour gamuts of four different types of fluorescent lamp (open squares). Also shown are the preferred positions of the test colours (filled squares). The difference in positions for each test colour indicates the extent to which the lamp will give the colour its preferred appearance.

measure the extent to which a light source will do this a colour discrimination index has been suggested [21]. The approach is to calculate the positions of the eight CIE test colours, used in calculating the General Colour Rendering Index, on the CIE 1976 UCS diagram. The positions of the eight colours usually form an elliptical enclosure called a colour gamut. A great deal can be learned from a colour gamut. From a consideration of its shape and the spacing between the positions of the individual test colours, the extent to which different parts of the hue circle can be distinguished is apparent. From its location on the diagram the appearance of the colours can be appreciated to some degree and from the area enclosed, called the gamut area, a general guide to the size of the differences between colours produced by the light source can be obtained. Figure 1.10 shows a number of colour gamuts for different tubular fluorescent light sources.

Another point of interest is that if the lighting is being provided for the sake of appearance rather than because of any need to reproduce colours accurately or to distinguish between them, then it has been established that people prefer colours to be somewhat more colourful than they would be in reality [22]. In other words grass is preferred as greener and meat is preferred as redder than occurs in reality. This knowledge has led to a suggestion that a colour preference index could be produced based on the eight CIE test colours [23]. The approach is to produce a colour gamut for the preferred appearance of the eight CIE test colours on the CIE 1976 UCS diagram. By plotting the actual positions of the test colours on the same diagram when under the light source of interest the extent to which the colours will be seen as preferred can be evaluated. Figure 1.10 shows this arrangement for four different tubular fluorescent light sources.

It must be emphasised that whilst the Colour Rendering Index is an official CIE method [19] the ideas of a colour discrimination index and a colour preference index both based on colour gamut are just that— ideas—but ideas which can be useful. After all most of the established quantities of colour description: chromaticity co-ordinates, colour space, correlated colour temperature and colour rendering index started out as ideas. We shall be hearing more of all these quantities later.

1.4 SOURCES OF LIGHT

Having discussed the various quantities used to measure the characteristics of light it is now possible to consider the properties of the more widely used sources of light. The most fundamental division between the sources of light is between the natural light of day and the artificial sources, represented almost exclusively by electric lamps. Properties of natural light are determined by nature and are therefore largely uncontrollable. Properties of artificial light can be controlled by man and so are adjustable.

1.4.1 Natural Light
The characteristic of natural light as a light source that most clearly distinguishes it from artificial light is its variability. Natural light varies in magnitude, spectral emission and distribution with different meteorological conditions, at different times of the day and year, at different places on the Earth's surface. The sun emits a wide spectrum of electromagnetic radiation including the visible region. When this visible radiation reaches the Earth's atmosphere it is scattered and absorbed. It is this scattering of light that gives the sky its blue appearance as compared with the blackness of space. Natural light can be divided into two components, sunlight and skylight. Sunlight is the direct light received from

the sun after losses caused by scattering in the atmosphere. Therefore sunlight comes from a point source of light and produces strong, sharp-edged shadows. Skylight is light from the sun which has been scattered in the atmosphere. Skylight can be considered to come from a diffuse light source and hence produces only weak shadows. It should be noted that the term daylight is usually taken to mean skylight and any available sunlight.

The illuminances produced on the Earth's surface by natural light vary considerably with atmospheric conditions, this variation being on top of the regular and predictable changes related to the time of day and the season caused by the relative motion of the sun and the earth. To indicate the size of these variations it is sufficient to say that on a sunny day in summer the illuminance on the ground may exceed 100 000 lx, but on a heavily overcast day in winter the illuminance may never exceed 1000 lx.

The spectral composition of natural light varies with the nature of the atmosphere through which it passes and the path length. The correlated colour temperature can vary from about 4000 K for overcast sky to 40 000 K for clear blue sky, but the most common value is around 6000 K. This is reflected in the five standard phases of daylight determined by Judd, MacAdam and Wyszecki [24] to represent the typical spectral distributions of irradiance produced on the Earth's surface by the sun. The correlated colour temperatures of these phases are 4800 K, 5500 K, 6500 K, 7500 K and 10 000 K. The relative spectral irradiances are shown in Fig. 1.11 from which it can be seen that as the correlated colour temperature increases, i.e. as clear sky conditions prevail, the blue end of the spectrum becomes dominant.

It would seem that the mention of spectral composition raises the question of the accuracy with which colours are rendered by natural light. However this question is not really relevant because it is common practice to take an appropriate phase of daylight as a reference source. Thus the colour rendering of natural light will conventionally be almost perfect.

Similarly, luminous efficacy is not really of concern since there is no electrical energy supplied to produce the light. However, this should not be taken to mean that the natural light is without cost when used as a light source in buildings. The provision of windows is usually more expensive than the plain brick wall or cladding they replace. In addition, in admitting natural light windows also admit infrared radiation, which is likely to be an unwanted heat gain in summer but a useful contribution

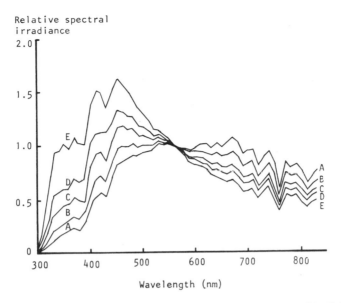

Fig. 1.11.   Relative spectral irradiance distributions of the standard phases of daylight. The distributions are normalised to a value of unity at 560 nm. The standard phases of daylight have correlated colour temperatures of A = 4800 K, B = 5500 K, C = 6500 K, D = 7500 K, E = 10 000 K. (After [24].)

to heating in the winter. Thus natural lighting has both good and bad features for building users. But cost and technology are not everything. It is well known that people prefer some natural light in buildings wherever this is possible (see Chapter 7).

The luminance distribution of natural light in the sky has been standardised by the CIE for a completely overcast sky in 1955 [25] and for a clear sky in 1967 [26]. The luminance distribution for a completely overcast sky is given by the expression

$$L_\alpha = \frac{L_z}{3}(1 + 2\sin\alpha)$$

where L_α = luminance of the sky at an altitude of α degrees above the horizon, and L_z = luminance of the sky at the zenith. The maximum luminance occurs at the zenith, the luminance distribution being symmetrical about the zenith and independent of the actual altitude of the sun.

The luminance of an element of clear sky (L) is given by the expression

$$L = \frac{L_z(1 - e^{-0.32 \cos \alpha})(0.91 + 10e^{-3\psi} + 0.45\cos^2 \psi)}{0.274(0.91 + 10e^{-3z_s} + 0.45\cos^2 z_s)}$$

where I_z = the luminance of the sky at the zenith, α = altitude of the element above the horizon, ψ = the angle between the element and the sun, and z_s = the angle of the sun from the zenith. This distribution is such that except around the sun itself the sky is brightest near the horizon.

In practice the luminance distribution of the sky varies between these two standards but some assumption is necessary to facilitate the design of daylighting in buildings. In temperate zones it is usual to assume a CIE standard overcast sky with the luminance distribution given above and producing an illuminance on the ground of 5000 lx. This can be used to calculate the illuminances in interiors from different window forms by a number of methods [27, 28, 29, 30]. There also exist standard sunpath diagrams which can be used to predict the position of a patch of sunlight in a building for any specific window size and orientation [27, 28, 31]. Natural light is a feature of many buildings. It is highly regarded by people, at least in cold and temperate climates. In all climates it has a considerable influence on the design and layout of buildings.

1.4.2. Light Sources—Artificial
The first form of artificial light used by man was firelight, created by the combustion of wood. Developments in basic technology led to the creation of the candle and the oil lamp, and then the gas lamp, all of which depend on combustion. Combustion is still with us. Even in what may appear as a piece of high technology, the photoflash bulb, the light is produced by a controlled combustion of aluminium wire. However combustion is not involved in any of the more common light sources used in modern lighting practice. These light sources can conveniently be divided into two broad classes, incandescent lamps and discharge lamps. Both rely on electricity as a source of energy but they have different basic mechanisms.

1.4.2.1 Incandescent Lamps
The most common form of incandescent lamp is the household bulb. This works by heating a thin wire (or filament) of tungsten to incandescence

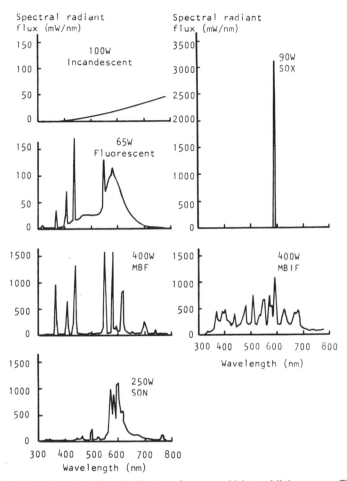

Fig. 1.12. Spectral radiant flux distributions for some widely used light sources. The light sources shown are a 100 W incandescent lamp, a 65 W White low pressure mercury (fluorescent) discharge lamp, a 90 W low pressure sodium (SOX) discharge lamp, a 400 W high pressure mercury (MBF) discharge lamp, a 400 W high pressure mercury metal halide (MBIF) discharge lamp and a 250 W high pressure sodium (SON) discharge lamp. Note the different scales on the ordinates.

whilst it is in an inert gas atmosphere. The main characteristic of the incandescent lamp is that the spectral emission forms a continuum (Fig. 1.12). The actual spectral emission of an incandescent lamp is governed by the temperature of the filament because it behaves almost as a full radiator. As the temperature increases the peak of its emission moves from the red end of the spectrum to the blue. The spectral emission curve in Fig. 1.12 is for a filament temperature of 2850 K, typical of commercially available products. Incandescent lamps are relatively inefficient converters of electrical power into light because much of their emission is outside the visible spectrum in the infrared region. Typical luminous efficacy figures are of about 12 lm W^{-1}. However they do have the advantage of being very cheap and do not require any equipment to control the electricity supply. The life of an incandescent lamp is a matter of balancing life against light output. Longer life can be achieved if less efficient conversion from electricity to light is accepted. Present practice is to produce domestic incandescent lamps with lives of either 1000 or 2000 h. Incandescent lamps are used where capital cost and convenience are important, as in domestic lighting, or where their small physical size and the consequent ease with which the light distribution can be controlled are of value, e.g. spot lights.

A more recent development in incandescent lamps has been the introduction of a halogen into the gas filling. Such lamps are called tungsten halogen lamps. This lamp can be used at a higher filament temperature than the conventional incandescent lamp because, although the tungsten is evaporated off the filament, and deposited on the glass envelope at a faster rate at the higher temperature, the halogen then chemically reacts with the tungsten to form a tungsten halogen which diffuses back towards the filament, where it separates into tungsten and halogen thereby depositing tungsten back onto the filament. Because of the higher operating temperatures the luminous efficacy of this lamp is increased to, typically, 20 lm W^{-1}. Their small physical size makes such tungsten halogen lamps popular for use in projectors and floodlights of various types.

Both tungsten and tungsten halogen lamps have good colour rendering properties by definition, because their spectral emission is very similar to the full radiator chosen as a reference source.

Table 1.4 summarises the important characteristics of the incandescent lamps commonly used for interior lighting.

1.4.2.2 Discharge Lamps

All discharge lamps operate by ionisation of a gas. The process is one in which electrons are first created in the gas. For virtually all the lamp types used in interior and roadway lighting these electrons are obtained by thermionic emission from a hot cathode. The electrons are accelerated through the gas by the potential difference between the electrodes. The accelerated electrons collide with the atoms of gas producing two effects of interest. The first is the ionisation of the atom into electrons and a partially changed particle called an ion. This increases the electron concentration and hence maintains the discharge. The second possible outcome is that the atom absorbs most of the energy of the electrons and thereby raises the energy state of one of its captive electrons to a higher level. These energy levels are discrete, and when the captive

Table 1.4: Summary of the properties of artificial light sources

Source	Luminous† efficacy ($lm\ W^{-1}$)	Correlated colour temperature (K)	General Colour Rendering Index (CRI) group‡	Typical application
Incandescent				
Tungsten	8–18	2 700	1	homes
Tungsten halogen	17–25	2 900	1	floodlighting
Discharge				
Low pressure mercury (fluorescent lamp)	20–70†	2 700–6 500	1–3	offices
Low pressure sodium	90–150†	not applicable	not applicable	roadways
High pressure mercury	35–70†	4 000	2–3	factories and offices
High pressure sodium	55–110†	2 100	3	factories

† For discharge lamps the luminous efficacies are lumens/circuit watt.
‡ Group 1, CRI ≥ 85; group 2, 70 ≤ CRI < 85; group 3, CRI < 70.

electrons shortly afterwards decay back to their resting energy level, energy is radiated at a wavelength determined by the energy level structure. For some gases the energy level structure is such that the energy emitted is in or close to the visible region. It is this process of absorbtion of electron energy and subsequent collapse of electrons from high energy levels that produces light from gas discharges. The characteristic which most readily distinguishes discharge light sources from incandescent is the fact that they tend to have discontinuous spectral emissions (Fig. 1.12), as might be expected from the discrete nature of the energy level structure. The actual spectral emission obtained is a function of the gas used in the lamp and its pressure. The effect of the gas used is felt through its energy level structure. The effect of gas pressure is felt by the re-absorbtion of light emitted in the discharge and the broadening of spectral lines because of mutual interference between the closely packed atoms. This has the effect of introducing a continuous element back into the spectral emission which acts as a background to the spectral lines emitted by the original process.

Two gases are commonly used in commercially available discharge lamps, mercury and sodium, both at low and high pressure. The low pressure mercury discharge lamp is better known as the fluorescent lamp (or strip light as it is sometimes called). The discharge in this type of lamp produces radiation mainly in the ultraviolet region and hence would be of little use if it was not for another feature of discharge lamps—a phosphor coating on the inside of the glass envelope. The function of this phosphor is to absorb the ultraviolet radiation from the discharge and to emit radiation in the visible region, usually in a continuous band. This is a form of the physical phenomenon of luminescence called fluorescence. The phosphor coating on a fluorescent lamp increases the efficiency of the lamp in terms of the lumens emitted/watt of electrical power applied. Also, by varying the composition of the phosphor, different spectral emissions can be produced and hence lamps with different colour appearances and colour rendering properties can be produced. An extensive range of fluorescent lamps is available commercially. They are very widely used in industry and commerce, where their greater cost is more than offset by their higher luminous efficiency (typically 20–70 lm W^{-1} depending on type), and their longer life (typically 7500 h).

The other form of low pressure discharge lamp uses sodium gas and no phosphor. The spectral emission is concentrated in what are called the sodium D lines which are near the peak wavelength sensitivity of the visual system (Fig. 1.12). This light source is widely used for roadway lighting. As might be expected from a light source with a spectral emission concentrated in one region of the visible spectrum, the colour rendering is very poor, but it can have a very high luminous efficacy, about 140 lm W^{-1} being typical. Life is of the order of 8000 h.

The high pressure mercury discharge lamps often have spectral emissions characterised by a combination of strong spectral lines, and a continuous background produced by line broadening and phosphor excitation (Fig. 1.12). These lamps are available in a variety of forms with different phosphors which produce different spectral emissions, although some also have another modification present. This modification is the addition of some metals into the gas discharge. These improve the overall colour properties by introducing spectral lines where the spectrum from the discharge and phosphor is lacking. This variety are known as metal halide mercury discharge lamps (Fig. 1.12). High pressure mercury discharge lamps are available in a wide range of wattages and colour properties. They are of small physical size, compared with the fluorescent lamp, have roughly the same life and range of luminous efficacies as the fluorescent lamp and are used in places as disparate as roads and large offices.

The high pressure sodium discharge lamp emits a more continuous spectrum (Fig. 1.12) than its low pressure relative, and consequently has better colour properties although it is by no means as good as the mercury lamps, both high and low pressure. Against this disadvantage it does have the advantage of a high luminous efficacy, typically about $100 \, \text{lm} \, \text{W}^{-1}$. It is being increasingly used in industry and for roadway lighting in city centres, where its better colour properties over low pressure sodium are considered worth the drop in luminous efficacy. The life of these lamps is of the order of $7000 \, \text{h}$.

All discharge lamps need an electrical control circuit because the discharge itself is unstable. In addition, some need a starting circuit to initiate the discharge. These facilities consume power without producing light and, to allow for this, luminous efficacies for discharge lamps are usually quoted as lumens/circuit watt. Table 1.4 summarises the properties of these types of discharge light sources. The properties considered are the range of luminous efficacies, the correlated colour temperatures, the CIE General Colour Rendering Index group, and typical areas of application. Lamp development is a continuing story. It is likely that this table will soon become out of date. Therefore it should not be used as a basis for choosing lamps for a particular application. For this the lamp manufacturers should be consulted. However for the purpose of introducing the broad classifications of lamp types available, it is more than adequate.

1.5 CONCLUSIONS

There are two main conclusions to be drawn from this discussion of light. The first is that there are a large number of different sources of light available, each with different properties. The second is that the quantities used to measure these properties are both precise and inaccurate. The precision arises because the quantities can be calculated or measured exactly and if they are regarded as simple physical measures then they can be considered to be accurate. But they are not simple physical measures. The whole reason for having photometric and colorimetric quantities is to quantify the visual effect of light. Because of the complexity of the human visual system and the differences between individuals, any one standardised measure of visual effect is inevitably an approximation. Nonetheless the photometric and colorimetric quantities discussed are the best approximations so far devised. For general lighting conditions they can be used with considerable confidence but it is always necessary to remember their limitations.

1.6 SUMMARY

This chapter is concerned with what light is, how it can be measured and how it can be produced. Light is the part of the electromagnetic spectrum between 380 and 760 nm. What differentiates this wavelength region from the rest is that the human visual system responds to it. The actual human spectral response has been standardised in an internationally agreed form represented by the CIE standard observers. The form of this spectral sensitivity changes from daytime (photopic) conditions to night-time (scotopic) conditions. Using the appropriate spectral sensitivity curve the four basic photometric quantities can be derived— luminous flux, luminous intensity, illuminance and luminance.

These measures are all concerned with the overall effect of light and not with its colour properties. To deal with colour the two approaches of the colour atlas and colour measurement are considered. This leads to a description of the CIE colorimetric system, including the chromaticity diagrams and colour spaces. By using some of these measures the appearance and differences between object colours can be quantified, and the colour appearance and colour rendering properties of light sources characterised by correlated colour temperature and Colour Rendering Index.

Having considered how light can be measured the physical principles and properties of natural and artificial light sources are considered. Natural light is principally characterised by its variability in both quantity and spectral distribution. Artificial light sources are more stable but differ considerably in their efficiency at converting electricity to light, their life, and their colour properties.

REFERENCES

1. British Standards Institution. BS 4727, *Glossary of electrotechnical power, tele-communication, electronic, lighting and colour terms*, part 4, Terms particular to lighting and colour, group 01, radiation and photometry, 1971.

2. Commission Internationale de l'Eclairage. *Principles of light measurement*, CIE Publication 18, 1970.

3. Judd, D. B. Report of US Secretariat Committee on Colorimetry and Artificial Daylight, *Proc. CIE 12th session, Stockholm*, 1951.

4. Palmer, D. A. Standard observer for large field photometry at any level, *J. Opt. Soc. Amer.*, **58**, 1296, 1968.

5. Commission Internationale de l'Eclairage. *Light as a true visual quantity: principles of measurement*, CIE Publication 41, 1978.

6. Kinney, J. A. S. Degree of applicability and consequence of inappropriate uses of units of light, *Appl. Optics*, **6**, 1473, 1967.

7. Hunt, R. W. G. Colour terminology, *Colour Research and Application*, **3**, 79, 1978.

8. Munsell Colour Company. *Munsell Book of Colour*, Baltimore, USA, 1973.

9. Wyszecki, G. and Stiles, W. S. *Colour Science*, John Wiley, London, 1967.

10. National Bureau of Standards. *Colour: universal language and dictionary of names*, Special Bulletin 440, 1976.

11. British Standards Institution. BS 4800, *Paint colours for building purposes*, 1972.

12. Commission Internationale de l'Eclairage. *Colorimetry*, CIE Publication 15, 1971.

13. Commission Internationale de l'Eclairage. *Colors of light signals*, CIE Publication 2.2, 1975.

14. Commission Internationale de l'Eclairage. *Official recommendations on uniform color spaces, color difference equations and psychometric color terms*, CIE Publication 15, Supplement 2, 1978.

15. Pointer, M. R. Colour discrimination as a function of observer adaptation, *J. Opt. Soc. Amer.*, **64**, 750, 1974.

16. Hunt, R. W. G. The specification of colour appearance. I. Concepts and terms. II. Effect of changes in viewing conditions, *Colour Research and Application*, **2**, 55 and 109, 1977.

17. Corth, R. The basis for a new system of colorimetry, *J.I.E.S.*, **8**, 155, 1979.

18. Land, E. H. The retinex theory of colour vision, *Scientific American*, **237**, 6, 108, 1977.

19. Commission Internationale de l'Eclairage. *Methods of measuring and specifying colour rendering properties of light sources*, CIE Publication 13.2, 1974.

20. Lynes, J. A. *The interpretation of colour rendering data*, Plymouth Polytechnic Report No. 1/74, Plymouth, UK, 1974.

21. Thornton, W. A. Colour discrimination index, *J. Opt. Soc. Amer.*, **62**, 191, 1972.

22. Judd, D. B. A flattery index for artificial illuminants, *Illum. Engng*, **62**, 593, 1967.

23. Thornton, W. A. A validation of the colour-preference index, *J.I.E.S.*, **4**, 48, 1974.

24. Judd, D. B., MacAdam, D. L. and Wyszecki, G. W. Spectral distribution of typical daylight as a function of correlated colour temperature, *J. Opt. Soc. Amer.*, **54**, 1031, 1964.

25. Commission Internationale de l'Eclairage. *Proc. CIE 13th session, Zurich*, Committee 3.2, 1955.

26. Kittler, R. Standardisation of outdoor conditions for the calculation of daylight factor with clear skies, in *Proc. CIE Intersessional Conference on Sunlight in Buildings*, ed. R. G. Hopkinson, Bouwcentrum International, Rotterdam, 1967.

27. Lynes, J. A. *Principles of Natural Lighting*, Applied Science Publishers, London, 1968.

28. Hopkinson, R. G., Petherbridge, P. and Longmore, J. *Daylighting*, Heinemann, London, 1966.

29. British Standards Institution. BS CP3, Chapter 1, Part 1, *Daylighting*, 1964.

30. Commission Internationale de l'Eclairage. *Daylighting: international recommendations for the calculation of natural daylight*, CIE Publication 16, 1970.

31. British Standards Institution. Draft for development, *Basic data for the design of buildings: Sunlight*, 1980.

2

Vision

2.1 INTRODUCTION

Light is necessary for the visual system to operate. The preceding chapter described how light is produced and measured. This chapter describes the construction and operational characteristics of the visual system. It is important to realise the implications of the term 'visual system'. It is much more than just the eye, for physiologically the visual system involves both eye and brain, and psychologically, in involves both immediate stimulation and past experience. It is the mechanism of our foremost sense. It can operate over a wide range of conditions with a high degree of precision. Without it we are severely handicapped, with it we are capable of fine discrimination between and within many different components of our world. It has been the subject of study for hundreds of years and yet it remains largely a mystery.

The anatomy of the eye was known to the ancient Greeks, although the first detailed diagram appears in an Arab text of the 9th Century [1]. In spite of this early knowledge an accurate understanding of the optics of the eye was not achieved until the 17th Century. The 20th Century's main contribution to understanding the visual system has been to reveal the complexity of the operation of the photoreceptor surface of the eye. Currently, research in vision is concentrated on defining the way in which signals reaching the brain from the eye are coded at various stages in the system, and how this coding affects perception. As might be expected this is an extremely complicated task. Space, and the present state of knowledge, do not allow the full story to be given here. Rather the approach is first to describe the structure of the visual system, what connects with what, and what the functions of the various components are, then to examine the operational characteristics of the visual system in terms of its limits of performance and the characteristics of the perceptions that occur, and finally to consider the more common defects of the visual system. No attempt will be made to describe how the performance capabilities or the perceptual properties arise. Those interested in such information should read some of the excellent books on vision that are available [2,3,4,5].

2.2 THE STRUCTURE OF THE VISUAL SYSTEM

The eye is the obvious starting point for an examination of the visual system. It can be considered as a bag of jelly held in shape by the pressure of the fluids it contains. When someone looks at an object the eye is first turned so that the image of the object strikes a particular part of the photoreceptor surface, known as the fovea. This movement is made by the combined effects of the six extra-ocular muscles attached in pairs to each eyeball. Figure 2.1 shows the location of these extra-ocular

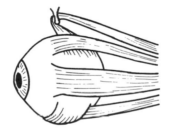

Fig. 2.1(a).   The eye and the extra-ocular muscles used to move it.

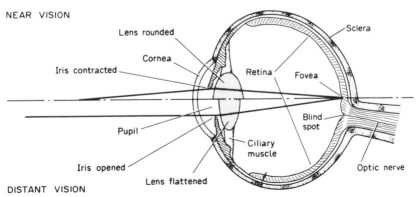

Fig. 2.1(b).   A vertical section through the eye when adjusted for near and distant vision.

muscles. It also shows a section through the eye. Rays of light enter the eye through the cornea and are brought to focus by the lens onto the photoreceptor layer, the retina. In more detail, the rays of light first cross the cornea, which is the transparent portion of the sclera, the fibrous material containing the ocular fluids. In fact the curvature of the cornea is somewhat greater than the sclera. The exact form of the curvature is important because refraction at the air/cornea surface contributes a major part to the overall power of the eye as an optical system. After the cornea the light passes through a chamber of transparent fluid, rather quaintly called the aqueous humour, and an aperture in a coloured diaphragm. The diaphragm is called the iris and the aperture is the pupil. The size of the pupil is determined by several different interacting factors. It is primarily controlled by feedback from the photoreceptor surface, the retina. If the amount of light reaching the retina is too great for the state of adaptation prevailing then the pupil size is reduced. If the amount of light is too small then the pupil size is increased. But feedback from the retina is not the only factor that influences pupil size, emotion also affects it [6] as does the distance at which the eye is focused. Pupil diameter can vary from about 2 mm to 8 mm. Nominally this should change the luminous flux on the retina over a range of 16:1 but because of the directional sensitivity of one of the two photoreceptor types [7], in certain conditions the maximum change in pupil size only produces an effective change of 10:1 in luminous flux at the retina. In either case, a comparison with the operating range of the visual system, which is about 10^{10}:1 reveals that the pupil variation is a minor aspect of the adaptation of the visual system to the prevailing conditions.

After passing through the pupil, light reaches the lens. The lens is fixed in position but varies its focal length by changing its shape. As mentioned above it is not the only optical component that controls the total power of the eye's optical system, the cornea is also involved, but it is the flexible part that allows the eye to bring objects at different distances to focus at a constant distance, i.e. on the retina. Young people with normal vision can focus an object at any distance from infinity to about 15 cm from the front of the eye. This process of achieving focus at different object distances is called accommodation. The change in the shape of the lens which is necessary for accommodation to occur is produced by the ciliary muscle. For objects close to the eye the lens is fattened, for objects far away it is flattened. If there is no object to focus on, a situation which occurs in a dense fog or on a very dark

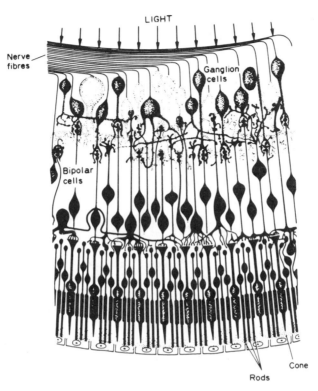

Fig. 2.2. A schematic illustration of the retinal structure.

night, the lens tends to focus at a distance of about 1 m from the eye. [8].

The space between the lens and the retina is filled with another transparent material, the jelly-like vitreous body. After crossing this the light reaches the retina, a location where it undergoes a photochemical transformation. The structure of the eye is often likened to a camera and it is easy to see why this should be so. The eye produces an image on a photochemically sensitive surface by means of a lens, and the amount of light admitted is controlled by an aperture. However, as soon as the retina is considered the analogy breaks down. The retina is a very complicated structure, as can be seen from the section shown in Fig. 2.2. Basically it can be considered as having three layers: a layer of photoreceptors, which are conventionally divided into two functional types, named rods and cones from their structure; and a layer of bipolar cells which provide interconnections amongst themselves and from the photo-receptors to the third layer, the ganglion cells. Each ganglion cell links to one nerve fibre which is then carried away from the eye up the optic nerve to the brain. It will be apparent from this brief description that the bipolar layer of the retina allows for the possibility of some analysis of the information contained in the retinal image. That such analysis does occur is suggested by recordings of electrophysiological signals from single ganglion cells in the cat and monkey [9]. In these creatures ganglion cells have been found which produce a response whenever a small spot of light falls within a given area. In other words each ganglion cell has what is called a receptive field. Further, such receptive fields have particular characteristics, i.e. they respond differently to different patterns and directions of movement of the light across the field. It is usually assumed that the retinal organisation of man is similar to that of cats and monkeys. The degree of signal modification that occurs in the retina, as well as the histology, supports the suggestion that the retina can legitimately be considered as an extension of the brain.

It should be realised that the response to light is not the same all over the retina. This is partly because the rods and cones have different spectral sensitivities and different absolute sensitivities to light and are not evenly distributed across the retina. Figure 2.3 shows the distribution of rods and cones across the horizontal meridian of the eye. The cones are concentrated on the visual axis in an area called the fovea. Outside the fovea there is first a rapid increase in the number of rods and then a steady decline until, by the far periphery, there are almost the same number of rods and cones. The distribution is symmetrical about the visual axis except for the blind spot where the nerve fibres from the ganglion cells leave the retina so that there are no photoreceptors (Fig. 2.1). Another feature indicating non-uniformity of response across the retina is that the receptive fields tend to be small in the fovea and increase with deviation from it. This can be explained by the interconnections between the layers in the retina. At all locations in the retina there is both convergence and divergence of connections. A single ganglion cell connects with many photoreceptor cells and each photoreceptor connects with several ganglions. The extent of this divergence and convergence varies across the retina but is at a minimum in the fovea.

The structure of the retina has another interesting feature. From Fig. 2.2 it can be seen that the rods and cones are not facing the incident light but rather lie at the bottom of a jumble of bipolar cells, ganglion cells and optic nerve fibres. These elements of the retina are transparent but

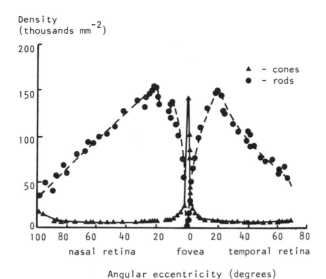

Fig. 2.3.   Density of rods and cones across the retina on a horizontal meridian. (After Osterburg, G., Topography of the layer of rods and cones in the human retina, *Acta Ophthal.*, **13**, Suppl. 6, 1935.)

the blood in the blood vessels which traverse the retina is not and this therefore produces images at the photoreceptor level. However, the visual system operates to eliminate these images. The eye is in continual motion even when the observer fixates an object, i.e. stares directly at the object without conscious deviation. This means that the retinal image is in continual movement. If the movement is cancelled out optically, i.e. a stabilised retinal image is created [10], vision rapidly fails. Since the blood vessels on top of the photoreceptors move with the retina any images they produce will be stabilised and hence will not be seen.

By now it should be apparent that the retina is much more than a film. Its structure differs from place to place, it undertakes a degree of signal modification and it requires continual movement to operate. Functionally it is considered as part of the brain, with all that organism's mysteries.

The modified signal from the retina travels up the optic nerve along the route shown in Fig. 2.4. The first point of interest is the optic chiasma. It is here that the fact that man has two forward facing eyes becomes important. It is at the optic chiasma that the signals from the two eyes are combined. In fact signals from the left half of each retina come together and proceed to the part of the visual cortex in the left hemisphere

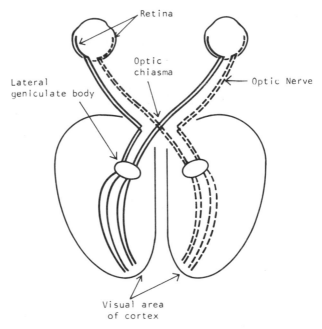

Fig. 2.4.   A schematic diagram of the binocular nerve pathways.

of the brain; signals from the right half of each retina also combine but these go to the part of the visual cortex in the right hemisphere. The retinal images in the two eyes are not identical, not even when the two eyes are focused on the same object, because they view from slightly different positions. The differences in the two images, together with other clues such as the masking of one object by another and differences of texture, enable perception of depth to occur. It is interesting to note that man is one of a group of creatures which have forward facing eyes and hence have such depth perception. This group includes the predators, e.g. hawks. Another group does not have depth perception because their two eyes are located on either side of their head, but they do have a very wide field of view. This group includes many common prey, e.g. pigeons.

Returning now to the visual system, after the optic chiasma the signals from the two eyes proceed through the lateral geniculate body and thence to the visual cortex. Electrophysiological studies have revealed a complicated pattern of signal modification in the cortex but this is really only scratching the surface. Ultimately the brain takes the signals from the eye and interprets them on the basis of past experience. How this is done is still a mystery. It is in this area that psychology and physiology meet. The current approach to providing a common base for both disciplines is the structuralist approach. Structuralism says that knowledge of the world enters the mind not as raw data but in a structured form. Currently physiologists are working on how the retinal image is coded into structures. The psychologists are attempting to derive the underlying structures from the visual perceptions.

Our knowledge of the visual system ranges from a good understanding in the eye to almost complete ignorance at the visual cortex. Electrophysiological studies have made considerable progress in recent years in tracing the paths and modifications of the signals arising in the retina and the nature of the stimuli which produce them. Even so the way in which perception is achieved is still unknown. Fortunately, for the purposes of this book it is sufficient to know the performance characteristics of the visual system and the form of the perceptions it produces. Knowledge of both these end results is fairly well established.

2.3 PERFORMANCE CAPABILITIES OF THE VISUAL SYSTEM

The uneven distribution of rods and cones across the retina has already been described (Fig. 2.3) but there are other differences between rods and cones. Figure 2.5 shows the spectral sensitivity of rods, foveal cones and peripheral cones. It can be seen that the spectral responses of the two cone types are similar and both are different from that of the rods. The rods are much more sensitive than the cones and differ in the wavelengths to which they are sensitive. The peak sensitivity for cones is about 555 nm but that for rods is about 505 nm. Since both rods and cones respond to incident light in the same way, i.e. by using the quanta to initiate a photochemical reaction, differences in spectral sensitivity suggest that rods and cones contain different photopigments. This view is supported by another distinct difference between rods and cones. It can be broadly said that cones operate during the daytime, at which level of radiant flux nearly all the photopigment in the rods will be 'bleached', so that the rods emit few if any signals [11]. However, at night-time there is insufficient radiant flux to cause the cones to operate, so vision is mediated by the rods alone. The relevant differences between daytime and night-time vision and hence between rods and cones is well known—in daytime we can see colours but at night no colours are visible.

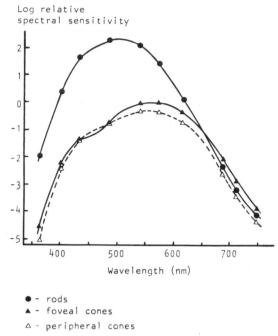

Fig. 2.5.   Log relative spectral sensitivity of rods, foveal cones and peripheral cones plotted
against wavelength. (After Wald, G., *Science*, **101**, 653, 1945.)

There can be little doubt that rods and cones contain different photopigments. Indeed the photopigments of human rods is by now well established [11]. However, the exact nature of the photopigment of human
cones is still a matter of argument, presumably because the cones do
not contain a single photopigment, as do the rods, but rather three
different photopigments [12]; although only one is thought to occur
in each cone. This makes the problem of identification much more difficult.

Fundamentally, man can be considered to have two separate retinas.
A night retina consisting of rods only and a day retina which uses only
cones. For field luminances below about $10^{-6}\,\mathrm{cd\,m^{-2}}$ there is insufficient
light for the visual system to operate at all. Between 10^{-6} and $10^{-3}\,\mathrm{cd\,m^{-2}}$
only the rods are operating, a condition conventionally referred to as
scotopic. For fields of luminance above about $3\,\mathrm{cd\,m^{-2}}$ the rods are
saturated and only cones are important, this condition being called
photopic. As is evident in Fig. 2.5 the spectral response of the eye changes
on moving from scotopic to photopic conditions, a change called the

Purkinje shift. For visual field luminances between 10^{-3} and $3 \, \text{cd m}^{-2}$ an intermediate state exists, called mesopic, in which both cones and rods operate. Spectral sensitivity is intermediate between the scotopic and photopic extremes. This classification of conditions may seem rather academic but it does relate to some major changes in visual capabilities. In scotopic conditions colours are not visible. As the mesopic condition is reached, colour vision starts to appear and strengthens until, when photopic conditions are reached, complete colour vision is available. In scotopic conditions only the rods are operating so the fovea is blind, in other words the retinal area used for fine discrimination of detail is not functioning. The best discrimination that can be achieved is in the near periphery where the size of receptive fields is greater than in the fovea. This means that discrimination of detail is reduced in scotopic conditions. Again there is no abrupt transition. In the mesopic state, discrimination of detail in the fovea improves above that for the scotopic condition, but is not as good as for photopic conditions. It is important when considering the interaction of people and lighting to identify the balance between the retinas that exists under the conditions being studied.

It will be apparent from the above discussion that the visual system can operate over a very large range of luminances. But it cannot operate over this range simultaneously. Anyone who has walked from daylight into a cinema is aware of this. Several minutes elapse before discrimination of detail is possible. This process of adjustment of the visual system to the prevailing conditions is called adaptation. When the eye is adapted to a given luminance (adaptation luminance) much higher luminances appear as glaringly white and those much less as black shadows. An indication of the limits within which differences in luminance and hence details can be discriminated for any particular adaptation luminance is given in Fig. 2.6. The areas marked are very approximate but they do indicate the general pattern.

The actual process of adaptation has three components (a) a rapid phase, presumably involving neural mechanisms, (b) a medium time phase of adjustment of the pupil for size and (c) a slow phase governed by the rates of the photochemical processes involved in reaching an equilibrium. The overall rate of adaptation to a steady state stimulus is governed by the slow photochemical phase, the actual time taken depending on the starting and final adaptation luminances. This is because the adaptation processes for rods and cones have different time constants, of the order of seven to eight minutes for rods, and two minutes for cones. Therefore, when both the starting and final adaptation luminances

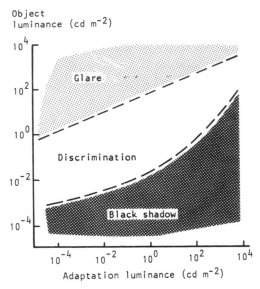

Fig. 2.6. A schematic illustration of the range of object luminances within which discrimination is possible for different adaptation luminances. The boundaries are approximate. (After Hopkinson, R. G. and Collins, J. B., The Ergonomics of Lighting, Macdonald, London, 1970.)

are in the photopic range, adaptation is rapid, typically a few minutes, since only the cones are involved. When the starting luminance is in the photopic range and the final adaptation luminance is in the scotopic range, a much longer two stage process occurs, the first stage involving the cones, the second the rods. Complete adaptation to darkness from a high photopic luminance can take up to an hour. When both starting and final adaptation luminances are in the scotopic range then only the rods are involved. Again adaptation is fairly rapid, typical values being of the order of several minutes. Observers who are adapted to the photopic state are usually said to be 'light adapted' while those adapted to the scotopic state are said to be 'dark adapted'.

The above has been concerned with adaptation to quantity of light. Adaptation to colour also occurs, although whether this involves a similar process of stabilisation of photopigment balance in the cones or is done at a higher level is still an open question. The direction of the adaptation is to make any colour appear less saturated, i.e. to move its appearance towards white.

Once the visual system is adapted to the prevailing conditions and assuming these are not too extreme, it is possible to discriminate detail. This enables the visual system to fulfill its function, namely to identify the objects that fill the world around us in all their varieties of texture, pattern, colour and form. The next stage in examining the performance of the visual system is therefore to describe the characteristic limits for such activities. A convenient measure of such capabilities is threshold performance. What this is can best be illustrated by an example. Consider a uniform field of luminance with a central spot of luminance super-imposed. Now if the difference in luminance between the central spot and the uniform field (ΔL) is high enough the spot can be detected every time it is presented. However, if ΔL is zero then there is no difference and the spot can never be detected. Therefore, if the spot is presented with a wide range of ΔL values it will be possible to obtain the values of ΔL for which the probability of detection is zero and 100% and for all points in between. The value of the ΔL for which the probability of detection is 50% is conventionally taken as the threshold. It is such threshold values that will be considered as the limits of performance.

The early studies of the capabilities of the visual system used visual acuity as a measure. Visual acuity is simply the angle subtended at the eye by the detail that can be discriminated on 50% of the occasions it is presented. For such measurements the contrast of the target is made very high. Visual acuity needs to be considered at a number of different levels. The simplest is the detection of presence. Hecht and Mintz [13] showed that a line only 0·5 sec arc wide could be detected on a uniform field. But this is hardly a common situation. Most often we are concerned not with telling that something is there but rather what it is. To do this, discrimination of detail is necessary. The test object most widely used to measure discrimination of detail has been the grating, i.e. alternate bars of black and white of equal width. The visual acuity measure with this target is the size of the bars when the observer can see the striations in the grating on 50% of the occasions it is presented. Schlaer [14] showed that with this target, visual acuity was 0·5 min arc, considerably greater than was required for the simple detection of presence. Also, in an experiment involving the recognition of four different high contrast shapes constructed from blocks of equal size, Brown [15] showed that the size of the block at the threshold for recognition of the shapes was 1 min arc. This supports the view that conventional visual acuity measurements, such as grating acuity, are meaningful for the discrimination of detail. However, it must be realised that the recognition of an object depends on many factors in addition to the size of detail. The number

of alternatives, the complexity of form and the background against which
it is seen, the observer's experience and his expectations are all involved.
Overington [16] provides a good summary of what little is known about
thresholds for recognition.

It is worth noting that the acuity measurements are better than would
seem likely when one considers the operation of the eye. The retinal

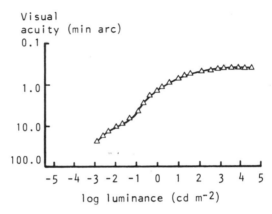

Fig. 2.7.   The variation in visual acuity with luminance. (After [14].)

image suffers from diffraction, spherical aberration and chromatic
aberration; it is sometimes out of focus and it is in perpetual motion. The
fact that visual acuity is so good suggests that the visual system operates
in some way to correct for the optical failings of the eye. This view
is supported by the precision of vernier acuity. Berry [17] showed that
for two lines, a misalignment of about 1·0 sec arc is identifiable. This
misalignment is much less than the diameter of a single cone photoreceptor.

It should also be realised that these values of visual acuity were obtained
at high adaptation luminances and with the target fixated. Under different
conditions inferior acuity levels will be found. Figure 2.7 shows the variation in visual acuity with adaptation luminance [14]. The improvement
in visual acuity with increased adaptation luminance is obvious.

Whilst visual acuity is an interesting quantity and is widely used for
sight testing, there are other important characteristics. One of these is
contrast. Contrast is conventionally defined by the expression

$$\text{contrast} = C = \left| \frac{L_t - L_b}{L_b} \right|$$

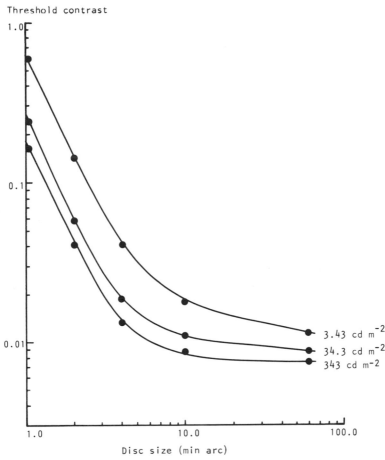

Fig. 2.8.   Threshold contrast plotted against disc size for detection of a disc presented for
1·0 s, at different adaptation luminances. (After [18].)

where L_t = luminance   of   the   target,   and L_b = luminance   of   the
background.

It is usually assumed that the observer is adapted to the luminance
of the background. There have been many different measurements of
threshold contrast for different conditions. Typical is an extensive study
by Blackwell [18] who measured the threshold contrast of discs of different
sizes at different luminances exposed for different times. Figure 2.8 shows
threshold contrasts for different disc sizes for different adaptation
luminances for a presentation time of 1·0 s. Figure 2.9 shows threshold

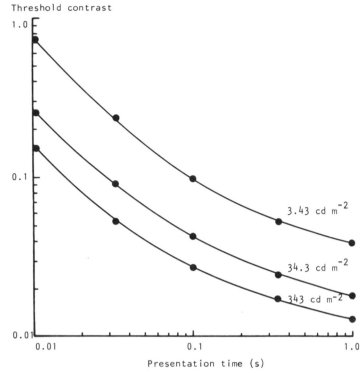

Fig. 2.9.   Threshold contrast plotted against presentation time for detection of a 4 min arc
disc at different adaptation luminances. (After [18].)

contrasts against presentation time for different adaptation luminances
for a disc of size 4 min arc. From Fig. 2.8 it can be seen that threshold
contrast decreases with increasing disc size and increasing luminance.
The curves can be divided into two sections: for discs less than about
4 min arc diameter the threshold contrast is inversely proportional to
the area of the disc (Ricco's Law). For discs of greater size this law
does not apply. The contrast threshold is less and soon becomes independent of disc size. Figure 2.9 shows that threshold contrast decreases
with increasing presentation time and higher adaptation luminances.
Eventually threshold contrast becomes independent of presentation time.

A common feature of all the relationships shown in Figs 2.8 and
2.9 is a law of diminishing returns. Equal step changes in size, luminance
or presentation time become steadily less and less effective in changing
threshold contrast. At very low luminances, very small sizes or very

short times, a small change in any of the variables will be very effective but at high levels of the variables the same changes will have little effect. These results have been derived from a simple disc target. Fortunately Guth and McNelis [19] have shown that the same relationship between threshold contrast and luminance holds for a range of different targets, such as printed words, gratings, etc. It seems reasonable to suggest that the pattern of results shown in Figs 2.8 and 2.9 is representative of a wide range of targets.

The results discussed above have been found for optimal conditions of viewing, a sharp, well-defined target occurring in a known position on a uniform luminance field. Real viewing is seldom like this. One of the most important differences is that targets occur off the visual axis, i.e. away from the fovea. Figure 2.10 shows the target luminance required to detect the presence of circular targets of different sizes occurring for 500 ms on a field of luminance $3 \cdot 2 \, cd \, m^{-2}$ at different deviations from the visual axis [20]. It can be seen that the target luminance is least and hence sensitivity is greatest in the fovea, but the important point is that sensitivity decreases as eccentricity increases. Figure 2.10 also shows the target luminances required for 83% correct recognition of whether a circle or a square target is presented during the 500 ms, where the essential difference in size between the targets is 0·15 times the diameter of the circle. The circle or square target is again presented at various deviations from the visual axis on the same $3 \cdot 2 \, cd \, m^{-2}$ field [20]. As before, the most sensitive position is in the fovea and the sensitivity decreases with increasing eccentricity, particularly for the small targets.

Another important difference between ideal and real conditions is the degree of blur present in the stimulus. Blur is an important influence on the threshold contrast of small targets. Ogle [21] has shown that for discs less than 40 sec arc in diameter, threshold contrast is rapidly reduced by blur, but that for objects of greater than 20 min arc diameter there is no effect on the detection of presence. Whether the same would be true for recognition threshold remains an open question.

Recently a more general way of considering the detection of targets has been used. This is in terms of the spatial modulation transfer function of the visual system. This is a rather grand name for what is essentially a simple piece of information, namely the frequency response of the visual system to spatial variations in luminance. Figure 2.11 displays some spatial frequency responses for one observer [22]. It shows the contrast sensitivity, which is the reciprocal of threshold contrast, plotted

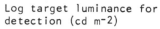

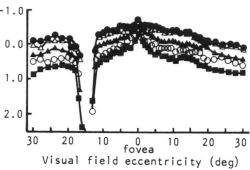

Log target luminance for
recognition (cd m⁻2)

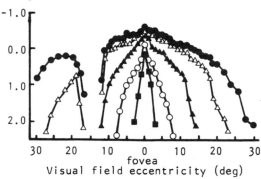

Fig. 2.10.   Log target luminances for detection of circular targets on a field of luminance $3\cdot2\,\text{cd m}^{-2}$ and the recognition of a circular or a square target on the same field, for targets of different sizes presented at different eccentricities from the visual axis. Target sizes are: ■, 10 min arc; ○, 17 min arc; ▲, 26 min arc; △, 42 min arc; ●, 66 min arc. The gap around 15° in the temporal field corresponds to the blind spot. (After [20].)

against spatial frequency in cycles per degree for two stimuli, a sinewave and a square wave grating, for two states of adaptation, one at $0\cdot05\,\text{cd m}^{-2}$ the other at $500\,\text{cd m}^{-2}$. For these stimuli, contrast is defined as

$$\text{contrast} = \frac{L_{\text{max}} - L_{\text{min}}}{\bar{L}}$$

where L_{max} and L_{min} are the maximum and minimum luminances of

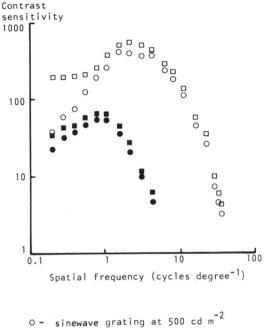

O - sinewave grating at 500 cd m^{-2}
● - sinewave grating at 0.05 cd m^{-2}
□ - squarewave grating at 500 cd m^{-2}
■ - squarewave grating at 0.05 cd m^{-2}

Fig. 2.11.   Contrast sensitivity for sinewave and square wave gratings at mean luminances of $0.05\,\mathrm{cd\,m^{-2}}$ and $500\,\mathrm{cd\,m^{-2}}$ [22].

the grating, and $\bar{L}$ is the average luminance. The spatial modulation transfer functions are given by the spatial frequency responses to the sinewave grating.

The value of all this apparently esoteric information is that any variation in luminance across a space can be represented as a wave form, any wave form can be represented as a series of sinewaves of different amplitudes and frequencies, and the sensitivity of the visual system to different frequencies is given by the appropriate spatial modulation transfer function. By multiplying the input amplitudes of different frequencies by the visual system sensitivity, the output amplitudes at the frequencies can be determined. From this resulting set of amplitudes and frequencies the appearance of the luminance distribution can be predicted by Fourier synthesis. This whole process is shown in Fig. 2.12 for the Mach band

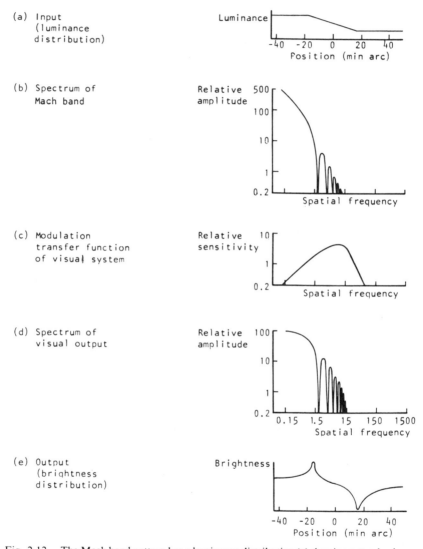

Fig. 2.12.   The Mach band pattern has a luminance distribution (a) showing a steady change from a high luminance to a low luminance. This pattern can be represented by a spectrum of relative amplitudes at different spatial frequencies (b). This spectrum is modified by the modulation transfer function of the visual system (c). The resulting output spectrum (d) corresponds to the pattern (e) which predicts the appearance of a bright and dark band on the pattern. These are the Mach bands. (After [3].)

effect. This analysis also explains the similarity at high spatial frequencies and the differences at low spatial frequencies between the sine and square waves in Fig. 2.11. For the $500 \, cd \, m^{-2}$ adaptation at spatial frequencies above 3 cycles degree^{-1}, the frequency terms that distinguish a square wave from a sinewave (i.e. the third and higher harmonics), are below the theshold contrast of the visual system and so are not seen. Therefore, square waves with a fundamental spatial frequency above 3 cycles degree^{-1} are seen as sinewaves, at least as regards threshold contrast. However, for low frequencies the third harmonic and other terms fall into the frequency region of maximum sensitivity whilst the sinewave which has only the fundamental frequency is entering the region of low sensitivity. Therefore the square wave sensitivity stays at a high level whilst the sinewave declines at low spatial frequencies.

The response for the sinewave stimulus shown in Fig. 2.11 also explains a number of well-known phenomena. It can be seen that for the $500 \, cd \, m^{-2}$ adaptation the peak sensitivity occurs at about 3 cycles degree^{-1}. On either side of this, contrast sensitivity declines. Thus very low spatial frequencies, such as occur across a wall lit from a side window, or very high spatial frequencies, such as occur in small details, are difficult to see.

Although the results in Fig. 2.11 and a number of other effects can be explained by hypothesising a Fourier frequency analysis, that is not proof that the visual system does operate in this way. However, there is some evidence that such frequency selective mechanisms do exist in the visual system [23]. Unfortunately for simplicity, there is also evidence that in suprathreshold conditions the simple spatial modulation transfer function approach breaks down, possibly because of the non-linearity of the visual system over the large range of luminances occurring [3]. This whole question of the spatial frequency approach to the operation of the visual system is currently the subject of much interest.

So far all the performance limits considered have been related to the characteristics of the target itself, its size, contrast, etc., when it is seen on a uniform background. However, the size, luminance and uniformity of the background itself can also be important. Foxell and Stevens [24] found that visual acuity steadily increased as the uniform background field was increased in size up to about $6°$. For bigger fields there was little change in acuity. They also studied the change in acuity for a slightly more realistic situation where the target was seen against a very small background, called the surround, and the luminance of the rest of the background could be varied independently. They found that the

Fig. 2.13.   A demonstration of simultaneous contrast. The two parts of the square annulus have the same luminance but appear to be of different brightness because of the influence of the background.

optimum visual acuity occurred when the luminance of the background was equal to the luminance of the immediate target surround, tending to slightly less at high surround luminances. If the luminance of the background was more or much less than the luminance of the immediate surround then the visual acuity was much reduced. This is one of the reasons for the practice of lighting background areas slightly less brightly than task areas.

Having considered the effect of the size and luminance of the background field it is also useful to consider its uniformity. As regards lighting, the problem here is that of high luminances in the field of view. These have two effects upon the visual system. The first is caused by the scattering of light in the eye. The effect of the scattering of light is to place a veil of light across the retinal image of anything that is close to the source of the scattered light. Such a veil will reduce the contrast of the retinal image. The second effect occurs in the retina and is assumed to be due to the neural interconnections. An example is shown in Fig. 2.13, the actual luminance of the two parts of the square annulus is the same but the half surrounded by black looks much lighter than that surrounded by white. The effect is enhanced if a dividing edge is laid along the black/white boundary although this will have a very small effect on the retinal image. This phenomenon is called simultaneous contrast, or, more generally, induction. It simply means that the response of one part of the retina is influenced by the response of another.

For the practice of lighting, non-uniformity in the visual field is of little importance until extreme conditions occur. When such extreme conditions are present, e.g. when meeting an oncoming car with headlights on at night, what is called disability glare occurs. The extent of disability glare is expressed in terms of equivalent veiling luminance, which is simply the luminance of a uniform veil which if placed over the target would raise its luminance difference threshold by as much as the actual glaring situation does. Fry [25] has shown that the equivalent veiling luminance can be predicted from the expression

$$L_v = \sum_n \frac{9 \cdot 2 E_n}{\theta_n(1 \cdot 5 + \theta_n)}$$

where L_v = equivalent veiling luminance (cd m^{-2}), E_n = the illuminance produced by the nth glare source at the observer's eye in the plane of the pupil (lux), θ_n = the angle of separation of the nth glare source from the line of sight (degrees).

Fry claims that this formula demonstrates that the scattering of light is the major factor influencing equivalent veiling luminance. However, the above formula is known to break down for high luminances within about 1·5° of the line of sight, the increase in luminance difference threshold being much greater than predicted. For this region the expression

$$L_v = \sum_n \frac{9 \cdot 2 E_n}{\theta_n^{3 \cdot 44}}$$

has been proposed where the terms have the same meaning as above [26]. It seems reasonable to suppose that within about 1·5° of the line of sight, induction is combining with scattered light to alter the contrast of the target still further. From the two equations above it should be clear that disability glare rapidly diminishes with deviation from the line of sight.

Another factor that has not been considered so far is the possible movement of the object being sought. The performance of the visual system for moving targets can be considered to have two different aspects: the detection and identification of the movement itself and the discrimination of detail within a moving object. The visual system is very good for detecting movement. Even in the absence of a frame of reference, velocities as low as 7 min arc sec^{-1} can be detected in the fovea [27]. With a frame of reference, a velocity of about 1 min arc sec^{-1} can be

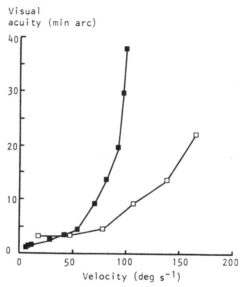

Fig. 2.14.   Visual acuity for different target velocities (open squares after [29], filled squares after [30]).

detected at 10° eccentricity [28]. For targets that appear further out into the periphery the threshold velocity is greater. However the increase in threshold velocity with eccentricity is less than the increase in threshold contrast for simple detection of presence. This implies that the periphery is relatively much better at motion detection than discrimination of detail but the fovea is better at both.

Discrimination of detail in a target that is obviously moving depends on whether the eye can lock on to the target. If it can, then the performance limits are almost the same as for a stationary target; but if it cannot, then the performance limits rapidly worsen. The velocity at which locking on becomes difficult is about 50° s^{-1}. Figure 2.14 shows the visual acuity measured for different target velocities for two experiments. The curves are different because the Miller and Ludvigh data [29] are for constant exposure time whilst the Rose data [30] are for constant field of view which means a reduced time of exposure for higher velocities. It should be noted that velocity is not the only factor that influences the ability to lock on to movement. The predictability of the movement is also important. Trying to follow the erratic movements of a bat in flight

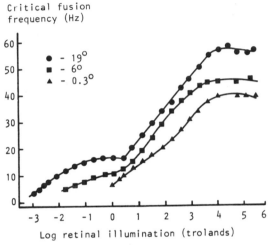

Fig. 2.15. Critical fusion frequency plotted against log retinal illumination for three different test field sizes [31].

is much more difficult than following a bird. As a general rule the more predictable the movement pattern the easier it is to lock on and hence to see the object.

Even a moving target can be regarded as steady state in one respect, its characteristics do not change with time. But sometimes the visual system has to deal with temporally varying targets. The case of a single flash has already been discussed as the effect of presentation time (Fig. 2.9). However, if a series of flashes occur, then as the frequency increases the light appears to flicker. The strength of this sensation first increases with frequency and then decreases until it disappears altogether and the light appears to be steady. The frequency of oscillation at which the flicker disappears is called the critical fusion frequency (CFF). Figure 2.15 shows the variation of CFF with retinal illumination for different sizes of flickering field [31]. Retinal illumination is simply the

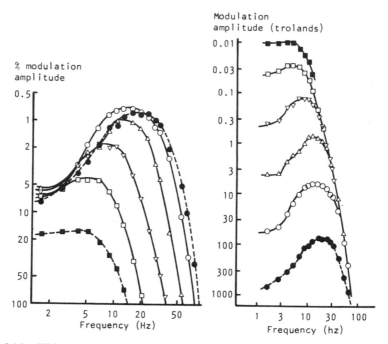

Fig. 2.16.   Flicker detection thresholds for a large field at different retinal illuminations and frequencies. The detection thresholds are expressed as percentage modulation (left diagram) and absolute modulation (right diagram). The retinal illuminations are: ●, 9300 trolands; ○, 850 trolands; △, 77 trolands; ▽, 7·1 trolands; □, 0·65 trolands; ■, 0·06 trolands. (After [33].)

product of the luminance of the field and the pupil area in square milli-metres, and is expressed in units called trolands.† Figure 2.15 shows that for large fields CFF increases with retinal illumination in both the scotopic and photopic conditions but is virtually independent of retinal illumination in the mesopic region. It is also apparent that CFF increases with increasing field area. For the optimum conditions for flicker detection, i.e. large, uniform, 100% modulation, flickering fields at high adaptation luminances, the CFF is about 75 Hz [32].

Another way of considering flicker is by measuring the modulation required for threshold detection at different frequencies. Figure 2.16 shows some results of this type for a 60° diameter, uniformly illuminated disc whose luminance could be varied sinusoidally [33]. The results are plotted

† For example, a field luminance of 1 cd m⁻² seen through a pupil of 19·7 mm² gives a retinal illumination of 19·7 trolands.

in two ways. The left-hand diagram in Fig. 2.16 has the percentage modulation plotted against frequency for different retinal illuminations. The results show that increasing the retinal illumination increases relative sensitivity and changes the frequency of peak sensitivity from about 5 Hz to 20 Hz. The other important feature is that virtually all the curves come to a common locus at low frequencies. This implies that the low frequency response of the visual system is consistently related to the percentage of modulation, not to the absolute level. For the experimental conditions examined, about 6 % modulation is necessary for low frequency modulation to be detected. The same results are shown in the right-hand diagram of Fig. 2.16 but this time with the absolute modulation amplitude plotted against frequency. The shift in frequency for peak sensitivity with increasing retinal illumination is again apparent but the interesting feature now is that the high frequency end of the response forms a common envelope. This means that the high frequency response of the visual system is consistently related to the absolute modulation of the oscillation not the percentage modulation. Thus a fixed high frequency modulation amplitude that can just be seen to flicker at a given retinal illumination can still be seen to flicker even when steady light is added to it.

The curves shown in Fig. 2.16 can serve as temporal modulation transfer functions in an analogous way to the transfer function used for spatial modulation. To predict the appearance of a particular temporal waveform one can describe the temporal waveform by a Fourier series and multiply each component of the Fourier series by the appropriate temporal modulation transfer function value. Synthesis of the resulting Fourier series shows how the incident waveform will appear. As mentioned earlier the validity of this process depends on an assumption of linearity of response to different levels of stimulation. It also assumes constancy of phase shift. The common envelope shown for high frequencies in the right-hand diagram of Fig. 2.16 suggests that in its response to high frequencies the visual system is linear but at low frequencies it certainly is not [3]. This means that a Fourier analysis approach can only be used for predicting the effects of waveforms that involve high frequencies. Fortunately the frequencies at which flicker occurs in lighting systems are usually high and close to the threshold. As proof of its validity the Fourier analysis approach can be used to explain Bloch's law, i.e. that all light flashes lower than some critical duration cause equal effects when the total incident flux is the same [3]. It can also explain why the CFFs of both square wave and sinewave oscillations are the same except at very low

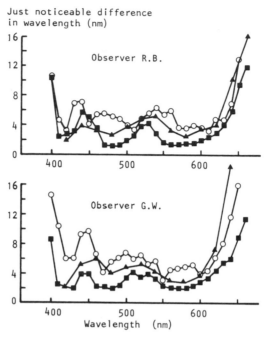

Fig. 2.17. Just noticeable differences in wavelength for various test field arrangements producing different retinal illuminations. ○, Data for two 1·5 min arc discs, separated by 40 min arc, producing 2000 trolands; ▲, data for two 12 min arc discs separated by 24 min arc, producing 500 trolands; ■, data for a 1 degree bi-partite field producing 100 trolands. (After [37].)

frequencies [3]. It should be apparent that the Fourier analysis approach has a lot to offer as a method of predicting how the different forms of temporal waveform that can occur in lighting practice will appear.

The final aspect of the performance capabilities of the visual system that will be considered is that of colour. All the performance limits discussed so far have been derived using achromatic targets lit by nominally white lights. But in photopic conditions man has a highly developed colour sense. There are three separate aspects of the way colour affects the performance capabilities of the visual system. The first is the effect of using coloured lights rather than white light. Strongly coloured light might be expected to improve visual acuity because it reduces chromatic aberration in the eye. However, once the white light and coloured light have been equalised for luminance, any effects of coloured light on acuity are very small [34] except for very small targets [35] for which coloured

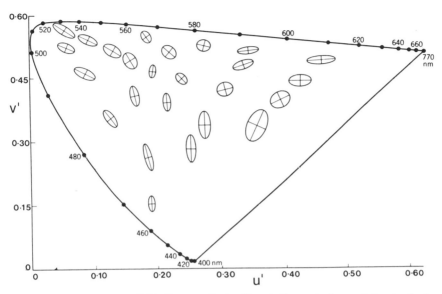

Fig. 2.18. The CIE 1976 UCS diagram. Each ellipse indicates the locus of the standard deviation of matches to the colour represented by the centre point of the ellipse. For ease of presentation the ellipses shown are ten times their true size. (After [38].)

light gives better acuity than white light. This conclusion applies to targets which are achromatic or in which both target and background have the same colour. However, when the target and background are coloured differently, then colour can make an important contribution to distinguishing the object from its background by providing colour contrast. When the luminance contrast between an object and its background is small, then colour contrast is often the only difference that is discernible between the object and its background. Eastman [36] suggests a value of luminance contrast of about 0·4 below which colour contrast becomes important. But colour can also have a meaning of its own. In electrical wiring for example, the colours of the conductors identify what the particular wire is for. When comparing colours side-by-side the visual system is capable of fine discriminations. Figure 2.17 shows the difference in wavelengths that are just noticeable when two monochromatic colours are seen side-by-side [37]. Discrimination is poor at the extremes of the visible spectrum but is better than 5 nm for most wavelengths. A more general way to consider this precision with which colours can be discriminated is given in Fig. 2.18. This figure shows a number of ellipses, each ellipse being centred on a specific chromaticity point. These ellipses, which are ten times enlarged, are loci of the standard deviation of matches to the central chromaticity [38].

To summarise, this section has attempted to map out the performance limits of the visual system. These limits have been studied by many people for many years and the foregoing only gives an outline of what is known. More detailed information is available on almost any of the variables in the literature [4, 5, 16]. Nonetheless, it should be apparent that the visual system covers an enormous range of luminances and can adjust itself to make fine discriminations of detail and structure for stationary and moving, for plain or coloured, for steady or flickering, targets. It is hardly surprising that it is the prime means by which we gather information from the world about us.

2.4 PERCEPTUAL CHARACTERISTICS OF THE VISUAL SYSTEM

Studies of the visual system tend to fall into one of three classes: (a) the visual system as a detector of signals in a 'noisy' environment, (b) the visual system as an organiser of information contained in the retinal image, (c) the visual system as an interpreter of information gleaned from the visual world. The performance limits discussed in the previous section fall into the first class. Also involved in this class are many of the electrophysiological studies of the visual system. However, electrophysiology overlaps into the second class depending on the stage of the visual system that is being examined. Examples of this second class are studies which examine the validity of the Fourier analysis approach to handling signals from particular stimuli [39]. The dominant characteristic of both these classes is that they examine the operation of the visual system in conditions which are extremely simple; uniform luminance fields containing simple two-dimensional patterns often without colour. The third class of studies considers the operation of the visual system in the real world with all its variety in both kind and degree of stimulation. At the moment there is little clear relationship between the first two classes and the third. To understand this failing it is useful to consider a musical analogy. A piece of music played by an orchestra can be analysed in terms of the physics of the sounds produced, and even in terms of the psychophysical response of the ear, but this would not give any degree of understanding of the perception of the music, the appreciation of melody and tonality, etc. Music and vision can be examined at many different levels and our knowledge of different levels is by no means at the same stage of development. The visual system considered as a detector has been extensively studied. The visual system as an organiser of information is now the subject of much research activity but no clear conclusions have emerged as yet. The study of the visual system as an interpreter of information is even less advanced. Nonetheless some understanding of its basic principles has been achieved. When considering how we perceive the real world the overwhelming impression is one of stability in the face of constant variation. As the eyes move in the head and the head itself moves about, the retinal images of the objects in the real world move across the retina and change their shape and size according to the laws of optics. Further, throughout the day the spectral emission and distribution of daylight changes as the sun moves across the sky and meteorological conditions vary. In spite of all these variations in the retinal image our perception of objects is rarely changed, a chair is still a chair no matter how it is seen. This invariance of perception goes under the name of perceptual constancy.

The most relevant of these constancies as far as lighting is concerned is lightness constancy. Suppose an observer is looking at two uniform surfaces of different reflectance, one grey and the other white. By giving the grey surface a high illuminance and the white surface a low illuminance it should be possible to get both surfaces to the same luminance. Since the stimulus received by the retinal photoreceptors is related to the luminance of the surface and the luminances of the two surfaces are the same, it might be thought that they will look the same. They do, but only under very special circumstances. These circumstances are that the surfaces are viewed through a blackened tube in such a way that the whole field of view is covered by the surface. In these circumstances a brightly lit grey surface cannot be distinguished from a dimly lit white surface. However, if we take away the tube and view the two surfaces under more usual conditions, e.g. in a normal room lit by either daylight or electric light, the difference in reflectance becomes obvious; the grey surface is seen as grey and the white as white, even though they have the same luminance. But how can this be, since it is luminance of the surface that is received at the retina? The answer is that the surfaces do not occur in isolation, (except when viewed through the blackened tube). They exist within a frame of reference which contains other objects and surfaces, well defined textures, contours, contrasts and colours. It is the interpretation of all this information that ensures that a piece of coal near a window is still seen as darker than a piece of white paper in shade even though the former has the higher luminance. Whether this interpretation is an inevitable consequence of a visual system physiologically designed to pick out invariant relationships between different aspects of the visual scene (ratios of luminance in the case of lightness constancy), or involves a knowledge of the objects to some extent, is still a matter of controversy [2, 3, 40]. Whatever is the explanation, lightness constancy occurs and is influenced by the whole visual field. It is an important factor to consider for the practice of lighting.

Much work on the subjective impression given by a uniform surface of known luminance under a given state of adaptation has been done [41] under the name 'apparent brightness'. The essential aim of this work has been to produce a method of designing lighting so that it will give a room a predicted appearance. The idea is first to decide on the apparent brightness desired for the surfaces. Then, after estimating a suitable state of adaptation, the equations relating apparent brightness to luminance are used to obtain the luminances of the surfaces. Once this is done the illuminance pattern necessary to create these luminances can be derived [42]. The problem with this approach is that the existence of lightness constancy suggests that under conventional lighting conditions people perceive reflectance and illumination separately. This in turn suggests that apparent brightness, which is related only to their combined effect, luminance, is of limited use in predicting the appearance of an interior [43]. However lightness constancy is not an all or nothing phenomenon. It holds over a range of conditions and then starts gradually to break down. This is shown in Fig. 2.19. The lines show the apparent Munsell Value of surfaces seen against a background of reflectance of 0·2 for a wide range of illuminances [43, 44]. Munsell Value is a measure of the lightness of a surface, and equal steps in the scale represent equal perceptual differences (see Chapter 1). For perfect lightness constancy each surface should give a horizontal line since the reflectance does not change as the illuminance increases. From Fig. 2.19 it can be seen that the lines are not horizontal but have a small slope. This means that lightness constancy is not absolute but the change in apparent Munsell Value is still much less than the change in illuminance. It should also be remembered that the range of illuminances likely to occur in many interiors is small, and in this situation the change in apparent Munsell Value is small, i.e. lightness constancy is approximately true. This implies that apparent brightness is likely to be of only limited use for conventional interior lighting design. However, where extreme ranges of illuminance occur, then lightness constancy will increasingly break down and apparent brightness become of greater value.

It should be realised that the experimental situation in which the results in

Fig. 2.19 were derived was very simple. In real interiors there could well be other cues, such as texture, which would anchor lightness constancy over a wider range of illuminances. Lynes [45] has examined the conditions for which lightness constancy is likely to break down. He concludes that diffuse uniform lighting such as is provided in most buildings ensures lightness constancy; but where strong contrasts occur across the room and the

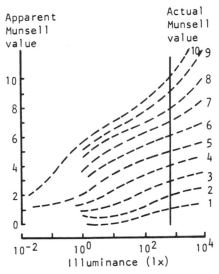

Fig. 2.19. Apparent Munsell Value at different illuminances for surfaces seen against a background of reflectance 0·2. The vertical line at an illuminance of 786 lx indicates the reference condition. At this illuminance the apparent Munsell Values of the surfaces have been normalised to their actual Munsell Values. (After [43].)

sources of light are not visible, then lightness constancy will break down and objects lit to high illuminances will become apparently self luminous. In fact the breakdown of lightness constancy is the basis of much display lighting.

In spite of this detailed discussion of lightness constancy, it should not be imagined that this is the only form of perceptual constancy. Colour, size and shape constancy all occur. Physically, the colour of a surface depends on the spectral emission of the light source illuminating it and the spectral reflectance of the surface. However, under normal viewing conditions, quite large changes in spectral emission of the light source do not cause much change in the perceived colours of surfaces, i.e. colour constancy occurs, provided both surfaces and background are similarly illuminated. This is usually true for light sources that have continuous spectral compositions. It is less true for discharge light sources which tend to have discontinuous spectral compositions.

Size constancy refers to the phenomenon that as an object gets further away its retinal image diminishes in size but the object size does not appear to change. There is plenty of evidence that size constancy breaks down if all cues to distance are eliminated or made misleading. The cues to distance include gradients in texture leading from the observer to the object, overlapping of objects, motion parallax, and binocular information.

The other perceptual constancy of importance is shape constancy. This constancy is illustrated by the observation that under normal viewing conditions circular plates always look like circular plates no matter at which angle they are viewed, even though the retinal image changes from a circle to an ellipse as the viewing angle changes. Such shape constancy depends on information being available from other parts of the visual field, particularly that which demonstrates the plane in which the object is lying.

Throughout this discussion of constancies a recurring theme has been the importance of the frame of reference, i.e. of the environmental information that is coincidental with the object being examined. This suggests that perception in normal conditions is very much a matter of mutual support between all the objects and surfaces in the visual world. The richer the information available from the visual world the more likely it is that constancy will be maintained. Constancy is likely to break down whenever it is asked to operate over too big a range or whenever there is insufficient or misleading information from the surrounding visual field. This latter mechanism of failure of constancy provides a convenient link to the earlier section on the performance capabilities of the visual system. Conditions under which high levels of system performance (e.g. good acuity, high contrast sensitivity) can be achieved are likely to be those most suitable for obtaining accurate information about the visual world, and hence for maintaining constancy.

2.5 VISUAL DISABILITIES

Just as people differ in their height, weight and strength so they differ in their visual capabilities. The distribution of visual capabilities is usually divided into normal and defective vision, the dividing line between the two parts and the function to which it is applied varying somewhat depending on the activities for which vision is required. This section is concerned with the more common visual disabilities, what form they take, how they affect sight and the extent to which they usually arise. Visual capabilities can be restricted from birth or reduced during development, or diminished by injury, by infection, or by the process of ageing. The changes that occur by injury or by infection will not be considered here.

Table 2.1:   Types and properties of defective colour vision

Type	Monochromatic wavelength discrimination	Spectral sensitivity function and wavelength of peak sensitivity	Percentage frequency of occurrence	
			Males	Females
Monochromat				
Rod monochromat	No discrimination	Scotopic (505 nm)	0·003	0·002
Dichromat				
Protanope	Absent above 520 nm	Insensitive to deep red light. Peak spectral sensitivity shifted (540 nm)	1·0	0·02
Deuteranope	Absent above 530 nm	Almost photopic (560 nm)	1·1	0·01
Tritanope	Absent from about 445 to 480 nm	Almost photopic (555 nm)	0·002	0·001
Trichromat				
Protanomalous	Reduced from red to yellowish green but varies considerably in different cases	Insensitive to deep red light, peak sensitivity wavelength shifted (540 nm)	1·0	0·02
Deuteranomalous		Almost photopic (560 nm)	4·9	0·38
Tritanomalous	Reduced for green to blue	—	—	—

It is convenient to divide the visual disabilities to be considered into those that can be treated clinically and those that cannot. The most common form of untreatable visual disability amongst those who have some form of vision is defective colour vision. A classification of the various types that exist, their frequency of occurrence and some relevant properties are given in Table 2.1. The most extreme form of defective colour vision occurs in the rod monochromat. Such a person can make no colour discriminations at all and has only a scotopic spectral sensitivity function. Further he has very poor acuity in the fovea. All these factors suggest that such people have only the rod system and their visual capabilities are limited by the absence of cones. Rod monochromats occur very rarely. Cone monochromats, i.e. people with apparently only one cone photopigment also occur. These people, too, can make no colour discriminations but are able to produce good acuity in the fovea. Such people are so rare that little else is known about the properties of their colour vision, but Weale [46] describes observations on a few individuals.

The next form of defective colour vision is the dichromat. It is thought that dichromats have only two of the cone photopigments present. Dichromats can be divided into three classes: protanopes, who lack sensitivity at the red end of the visible spectrum and who cannot discriminate between wavelengths above about 520 nm; deuteranopes, who have almost normal photopic spectral sensitivity but who cannot discriminate between wavelengths above about 530 nm; and tritanopes, who have a spectral sensitivity curve similar to the photopic curve but with a slight insensitivity to blue light, and who cannot discriminate between wavelengths of 445 and 480 nm.

An examination of colour matches [47] has led to the conclusion that dichromats, such as protanopes and deuteranopes give colour matches which are located far beyond the usual limits of the normal distribution for such matches. However, closer to the ends of the distribution are those with the most common form of deficient colour vision—the anomalous trichromats. These people apparently have all three cone photopigments present in some measure but suffer from some degree of weakness somewhere in the visual system. Therefore they can make colour discriminations in all parts of the visual spectrum, something that dichromats cannot do, but they still confuse some colours and are less sensitive to some wavelengths than people with normal colour vision. There are three classes of anomalous trichromat: protanomalous, deuteranomalous and tritanomalous trichromats, the last occurring very rarely. The trichromats have reduced colour discrimination in the same spectral regions as the analogous classes for the dichromats but to a lesser extent. There is considerable variation between individuals for these types of defects. Deuteranomalous trichromats have by far the most common form of defective colour vision.

Finally, the occurrence of night blindness should be mentioned. This appears to be caused by an absence of rods, photopic vision is normal but fails completely when luminance falls to scotopic levels.

All these visual disabilities occur in moderate numbers, are usually inherited and are untreatable. The other main form of untreatable visual disability arises from retinal degeneration. Retinal degeneration is an inevitable consequence of ageing but fortunately its effects usually only become apparent in very old age. The effect such degeneration has on sight depends on its location. One form, retinitis pigmentosa, causes a loss of peripheral vision. Another, senile macula degeneration, produces loss of central vision, since the macula is located over the fovea.

As for the treatable visual disabilities, the most common form affects the ability to focus an image on the retina. These disabilities go under the name of refractive errors. All can be corrected by using the appropriate spectacles. The three main types of refractive error are: (a) the limitation in the range over which focusing can be achieved that comes with middle age (presbyopia); (b) variations in length of the eyeball which, when it does not match the power of the eye's optical system, ensures either long sight (hypermetropia) or short sight (myopia); and (c) variations in curvature of the cornea in different meridians (astigmatism).

The near point of the eye, i.e. the shortest distance from the eye for which objects can be focused, recedes with age. By about the age of 45 years the near point is sufficiently far away for some print to be difficult to read. It is for this reason that most people in middle age require convex spectacles. As the near point continues to recede, more powerful spectacles are called for. Eventually, when the range over which focusing can be achieved by the eye is very small, different spectacles will be required for seeing objects at different distances.

When there is a mismatch between the length of the eye and the power of the optical system such that the retinal image of an object close to the eye is formed behind the retina, hypermetropia is occurring. Young people can often overcome this by increasing the power of the eye's optical system, i.e. by fattening the lens. However, this becomes more difficult with age as the lens stiffens and then a simple convex lens spectacle will be effective. When there is a mismatch between the length of the eye and the power of the optical system such that the retinal image of a distant object is formed in front of the retina, myopia is occurring. Concave spectacles are the answer in this case.

Most people have some degree of astigmatism but normally this is insufficient to be noticed. However, in extreme cases, vision is blurred in some parts of the visual field, irrespective of the overall focus. Astigmatism can be alleviated by wearing spectacles with appropriately different curvatures in different meridians.

Another treatable eye problem is a misalignment of the visual axes of the two eyes outside the normal range of convergence. In extreme cases this is called a squint. Such misalignments can arise either from a control or a motor failure in the system that controls the relative position of the two eyes. In extreme cases double vision occurs unless the person centrally suppresses the vision from one eye. If this suppression is maintained for several years then that eye is likely to remain permanently poor sighted. Most squints develop during childhood. The treatment for squint usually consists of forcing the child to use the eye which is being suppressed, with spectacles to correct for any refractive error present. If this fails, then an operation is carried out to relocate and shorten the appropriate eye muscles. Misalignment can also lead to eyestrain [48] and in some cases can cause problems in depth perception because it leads the observer to use monocular vision. This does not mean that depth perception is completely absent, there being plenty of monocular cues to depth available, but it does mean the richness of the depth cues is reduced.

Two other visual disabilities which can be treated are cataract and glaucoma. Both are common amongst the partially sighted. The main symptom of cataract is the gradual failure of sight, so that reading becomes more difficult yet bright lights are dazzling. This is caused by the growth of an opacity in the lens. Cataract can be caused by a number of different problems, e.g. injury, endocrine imbalance or infrared radiation, but in all cases the only treatment is surgical removal and replacement by some other form of lens. Cataracts are another of the hazards of old age.

Glaucoma is caused by increased intra-ocular pressure which eventually destroys some of the retinal structure. The main symptom of glaucoma is development of a blind spot which gradually extends until only the fovea remains unaffected, and eventually even this fails. Glaucoma can occur suddenly but more frequently it is a feature of old age. The treatment of glaucoma is aimed at reducing the intra-ocular pressure. This can be done by drugs or when necessary by surgery. This will stop the development of glaucoma but will not restore vision to the damaged area.

This brief review of some common forms of visual disability has ranged from the relatively minor, which diminish one aspect of vision without discomfort, to the extremely serious, which ultimately result in blindness. Treatment can also range widely from the simple fitting of spectacles to the extremes of surgery. Virtually everyone who lives an average life span can expect to need glasses at some stage. Happily not everyone who lives for the same time need anticipate the surgeon's knife. All the visual disabilities discussed affect some aspect of everyday life. Colour vision, depth perception, acuity and the size of the visual field are all affected by these disabilities. People who have them will not be suitable for certain specific activities. For example, railway engine drivers need good colour vision and crane drivers need good depth perception. Simple vision screening of the relevant functions is usually sufficient to identify people who do not have the necessary visual capabilities.

2.6 SUMMARY

The visual system is extremely complicated. Various aspects of it have been the subject of numerous studies over many years and yet much of it is still a mystery. This chapter outlines the structure of the visual system, its performance capabilities, the characteristics of the perceptions it produces and the more common visual defects that occur. No attempt is made to explain the operation of the visual system in detail.

The visual system involves both eye and brain. Images of the external world are formed on the retina by a simple optical system. The retinal image initiates signals from the photoreceptors which are modified by the neural organisation and then transmitted to the brain where they are interpreted on the basis of past experience. The visual system can adapt to a wide range of luminances, and once adapted is capable of fine discrimination of many different aspects of the visual world. Factors influencing the threshold values of visual acuity, contrast sensitivity, movement flicker and wavelength discrimination are considered. A common feature of all these performance limits is that they depend on the retinal location of the stimulus, the state of adaptation of the visual system and the presentation time.

An alternative way of considering the visual system is as an interpreter of information gathered from the external world. The overall impression is one of stability in the face of wide variations in physical conditions. In spite of changes in the retinal image, objects in the real world maintain their apparent size, shape, colour and lightness. This phenomenon is known as perceptual constancy. Such constancies depend on the visual system having information available from the setting in which the object appears. Conditions which enable fine discrimination to be made are the most suitable for maintaining constancy.

Finally the more common failings of the visual system are discussed. These can range from the very serious leading to eventual blindness, e.g. retinal degeneration, through those such as misalignment of the two eyes which can cause discomfort, to the minor and treatable conditions such as presbyopia. Many of these disabilities are a product of ageing.

REFERENCES

1. Meyerhof, M. *The Book of the Ten Treatises on the Eye*, ascribed to Hunain Ibn Ishaq, Government Press, Cairo, 1928.
2. Von Fieandt, K. and Moustgaard, I. K. *The Perceptual World*, Academic Press, London, 1977.
3. Cornsweet, T. N. *Visual Perception*, Academic Press, London, 1970.
4. Carterette, E. C. and Friedman, M. P. (eds). *Handbook of Perception, Vol. 5: Seeing*, Academic Press, London, 1975.
5. Davson, H. (ed.). *The Eye, Vols 1–4*, Academic Press, London, 1962.
6. Duke-Elder, W. S. *Textbook of Ophthalmology, Vol. 1*, C. V. Mosby & Co., St. Louis, 1944.
7. Pirenne, M. H. Directional sensitivity of the rods and cones, in *The Eye, Vol. 2*, ed. H. Davson, Academic Press, London, 1962.
8. Leibowitz, H. W. and Owens, D. A. Night myopia and the intermediate dark focus of accommodation, *J. Opt. Soc. Amer.*, **65**, 1121, 1975.
9. Robson, J. G. Receptive fields: neural representation of the spatial and intensive attributes of the visual image, in *Handbook of Perception, Vol. 5: Seeing*, eds. E. C. Carterette and M. P. Friedman, Academic Press, London, 1975.
10. Ditchburn, R. W. and Ginsborg, B. L. Vision with a stabilised retinal image, *Nature*, **170**, 36, 1952.
11. Jameson, D. and Hurvich, L. M. *Handbook of Sensory Physiology, Vol. 7/4, Visual Psychophysics*, Springer-Verlag, Berlin, 1972.
12. Vos, J. J. and Walraven, P. L. An analytical description of the line element in the zone fluctuation model of colour vision—1. Basic concepts, *Vision Research*, **12**, 1327, 1972.
13. Hecht, S. and Mintz, E. V. The visibility of single lines at various illuminations and the retinal basis of visual resolution, *J. Gen. Physiol.*, **22**, 593, 1939.
14. Schlaer, S. The relation between visual acuity and illumination, *J. Gen. Physiol.*. **21**, 165, 1937.
15. Brown, M. B. The effect of complex backgrounds on acquisition performance, *AGARD Conference Proceedings*, No. 100 B5-1, London, 1972.
16. Overington, I. *Vision and Acquisition*, Pentech Press, London, 1978.
17. Berry, R. N. Quantitative relations among vernier, real depth and stereoscopic depth acuity, *J. Exp. Physiol.*, **38**, 708, 1948.
18. Blackwell, H. R. Specification of interior illumination levels, *Illum. Engng*, **54**, 317, 1959.
19. Guth, S. K. and McNelis, J. F. Threshold contrast as a function of target complexity, *Am. J. Optom.*, **46**, 98, 1969.
20. Johnson, C. A., Keltner, J. L. and Balestrery, F. Effects of target size and eccentricity on visual detection and resolution, *Vision Research*, **18**, 1217, 1978.
21. Ogle, K. N. Foveal contrast thresholds with blurring of the retinal image and increasing size of test stimulus, *J. Opt. Soc. Amer.*, **51**, 862, 1961.
22. Campbell, F. W. and Robson, J. G. Applications of Fourier analysis to the visibility of gratings, *J. Physiol.*, **197**, 551, 1968.
23. Blakemore, C. and Campbell, F. W. On the existence of neurons in the human visual system selectively sensitive to the orientation and size of retinal images, *J. Physiol.*, **203**, 237, 1969.
24. Foxell, C. A. P. and Stevens, W. R. Measurements of visual acuity, *Brit. J. Ophthal.*, **39**, 513, 1955.
25. Fry, G. A. A re-evaluation of the scattering theory of glare, *Illum. Engng*, **49**, 98, 1954.
26. Hills, B. L. Visibility under night driving conditions: Derivation of $(\Delta L, A)$ characteristics and factors in their application, *Ltg Res. and Technol.*, **8**, 11, 1976.
27. Boyce, P. R. The visual perception of movement in the absence of an external frame of reference, *Optica Acta*, **12**, 47, 1965.

28. Salaman, M. *Some experiments on peripheral vision*, MRC Special Report 136, HMSO, London, 1929.
29. Miller, J. W. and Ludvigh, E. The effect of relative motion on visual acuity, *Survey Ophthal.*, **7**, 83, 1962.
30. Rose, A. *Proc. Armed Forces NRC Vision Committee*, Washington, DC, 1952.
31. Hecht, S. and Smith, E. L. Intermittent stimulation by light. VI. Area and the relation between critical frequency and seeing, *J. Gen. Physiol.*, **19**, 979, 1936.
32. Collins, J. B. and Hopkinson, R. G. Intermittent light stimulation and flicker sensation, *Ergonomics*, **1**, 61, 1957.
33. Kelly, D. H. Visual responses to time-dependent stimuli. 1. Amplitude sensitivity measurements, *J. Opt. Soc. Amer.*, **51**, 422, 1961.
34. Baker, K. E. Some variables influencing vision acuity. I. Illumination and response time, II. Wavelength of illumination, *J. Opt. Soc. Amer.*, **39**, 567, 1949.
35. Nikane, Y. and Katsuzo, I. Study on standard visual acuity curves for better seeing in lighting design, *J. Light and Visual Env.*, **2**, 38, 1978.
36. Eastmann, A. A. Colour contrast versus luminance contrast, *Illum. Engng*, **63**, 613, 1968.
37. Bedford, R. E. and Wyszecki, G. Wavelength discrimination for point sources, *J. Opt. Soc. Amer.*, **48**, 129, 1958.
38. MacAdam, D. L. Visual sensitivities to colour differences in daylight, *J. Opt. Soc. Amer.*, **32**, 247, 1942.
39. Maffei, L. and Fiorentini, A. The visual cortex as a spatial frequency analyser, *Vision Research*, **13**, 1255, 1973.
40. Epstein, W. *Stability and Constancy in Visual Perception*, Wiley Interscience, New York, 1977.
41. Hopkinson, R. G., Waldram, J. M. and Stevens, W. R. Brightness and contrast in illuminating engineering, *Trans. Illum. Engng Soc. (London)*, **6**, 37, 1941.
42. Waldram, J. A. Designed appearance lighting, in *Developments in Lighting*, ed. J. A. Lynes, Applied Science Publishers, London, 1978.
43. Jay, P. A. Lighting and visual perception, *Ltg Res. and Technol.*, **3**, 133, 1971.
44. Saunders, J. E. *Measurements of apparent brightness with special reference to their application in lighting design*, Ph.D. Thesis, University of London, London, 1969.
45. Lynes, J. A. Lightness, colour and constancy in lighting design, *Ltg Res. and Technol.*, **3**, 24, 1972.
46. Weale, R. A. Cone-monochromatism, *J. Physiol. (London)*, **121**, 548, 1953.
47. Pickford, R. W. *Individual Differences in Colour Vision*, Routledge and Kegan Paul, London, 1951.
48. Bedwell, C. H. The eye, vision and visual discomfort, *Ltg Res. and Technol.*, **4**, 151, 1972.

Part 2:
Light and Work

3

Methods, Models and Results

3.1 INTRODUCTION

One of the principal functions of lighting is to enable work to be done quickly, accurately and easily. This in itself is sufficient to explain an interest in the relationship between various aspects of lighting and work performance. But an altruistic desire to provide good lighting is not the only source of motivation. Money is also involved. If good lighting enables work to be performed more quickly, accurately and easily, then good lighting is an aid to productivity. If good lighting can be shown to imply more lighting, then more lighting equipment and more energy to operate that equipment can be sold. Further, it is a general rule that the cost of lighting is such that only a very small change in output is necessary to justify an investment in lighting. In these circumstances it is hardly surprising that the relationship between light and work has been the topic of investigations by many people over many years. The purpose of this chapter is to describe the methods of investigation adopted and to outline the overall picture of the relationship between light and work that has emerged.

3.2 THE METHODS USED

Before going into details of methods it is worth while contemplating some of the variables that lie behind those two small but expressive words, light and work. Light involves all aspects of lighting. Amongst those that might be expected to be important for work performance are the illuminance on the task, the colour of the light, its directional properties and its spatial distribution about the task. Work includes an even bigger range of variables. For a start most tasks can be considered to have a visual and a non-visual component, e.g. in proof reading the task is strongly visual but in shovelling coal the visual component is so small that the task could be, and often is, done almost in the dark. Within any visual component the important features include the size, contrast, and colour of the details to be discriminated, the complexity and uniformity of the background against which they are seen, the extent to which form and texture are relevant and the direction and duration of any movement. The wide variety of variables that need to be considered explain why studies of the effects of lighting on work have been undertaken for many years without any comprehensive and conclusive result. At best an outline of the relationships between light and work has been established.

The studies that have been made can be conveniently classified into two broad groups; those undertaken by 'the practical men' and those done by another collection of people—'the model makers'. The difference between these two groups is essentially one of application and immediacy. The approach adopted by the practical men has been to take a specific task, either in the field or simulated in a laboratory, and to examine the effects of different lighting variables on the performance of that task. Usually the lighting variables have been those that can most easily be changed in practice using conventional lighting equipment, e.g. illuminance, colour of light and spatial distribution of light. The results obtained apply strictly to the specific task investigated. This is useful in itself but the practical men also hope that by doing enough studies of this type it will be possible to obtain an overall pattern of results which will allow the conclusions to be generalised to other tasks. In other words a process of generalisation by analogy is anticipated.

The approach adopted by the model makers has been to get down to the detail of the operation of the visual system when it performs a very simple task. Their aim has been to construct a model which will allow the actual performance obtained to be predicted and then, by a series of refinements, to extend the applicability of the model to more complex visual work. In this way it is hoped to build a model of the relationship between light and work which can be used to predict the level of work performance for any combination of task and lighting conditions. Thus the model maker's approach also attempts to produce a result that can be generalised, but this time by synthesis using the systematic and proven development of a model.

Both these approaches have their advantages and limitations but they are not mutually exclusive. If the model makers cannot predict the results of the practical men then their models are of little use. Without the generality provided by the model the results of the practical men have a limited application. Eventually a model may be fully developed and the ideal state will be reached where the effect of any lighting variable on any task can be predicted. Unfortunately this state is still some way off. At the moment there are available a number of different studies of real or simulated tasks done under different lighting conditions, and two models of the relationship between light and work. The details of these studies and of the models and the implications they have for lighting practice will now be considered.

3.3 FIELD STUDIES OF LIGHT AND WORK

Some of the most widely quoted results obtained in the field and involving light and work are the initial studies in what have become known as the Hawthorne experiments [1]. These experiments eventually became concerned with the influence of working times and method of payment on work rate but they did not start like that. The origins of the Hawthorne experiments were in a lighting study. The Western Electric Company at Hawthorne, Chicago, were involved in electrical engineering, principally the manufacture of telephone apparatus. The company conducted an experiment on the influence of lighting on the level and maintenance of output. The experiment used two groups of workers, one a test group, the other a control group. The control group worked throughout the experiment in the same lighting conditions as were used by the test group before the experiment. The test group experienced a wide range of lighting conditions, some better, some worse than the control group. The main finding was that regardless of what was done to the lighting (increased, decreased, made better or made worse), the work output of the test group increased steadily. Even more surprising was the fact that the output of the control group steadily increased as well, though they had experienced no change in lighting conditions. These results are occasionally generalised to suggest lighting has no effect on work performance and that motivation and relationships between workers, and between worker and supervisor, are the only important factors. Whilst there can be no disagreement with the statement that motivation and relationships are important, the assertion that light does not affect work is nonsense. There is a continuum of performance associated with lighting conditions ranging from no light to plenty of light; as the amount of light increases it becomes possible to see more and more detail. This will allow work to be done more quickly and accurately so performance should increase until all the detail essential for the work is easily available. There is no doubt that lighting affects work, the problem is to identify the range of lighting conditions that allows improvement in work output to occur. The operative word here is allow. Lighting cannot produce work output. What it can do is to make details easier to see and colours easier to discriminate without producing discomfort and distraction. The worker can then use this increased ease of seeing to produce output if he is so motivated and/or is not limited by some other aspect. In the case of the Hawthorne experiment the cause of the

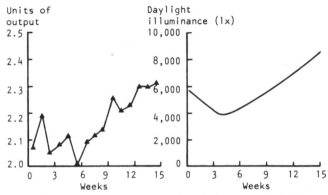

Fig. 3.1.   Output of silk weavers over a 15-week period compared with the amount of daylight typically available during those weeks. (After [2].)

increase in output is generally ascribed to a change in motivation brought about by rivalry between the two groups. This explains why the performance of both groups improved. What it does not explain is why changing the lighting had little effect. A plausible explanation for this lack of effect is that the importance of lighting to the work done in the Hawthorne experiment was less than expected. There is no measurement of the level of visual difficulty of the task, it may have been easy. Further the workers were well practised at their work so the visual component may have been small. Both these effects would tend to diminish the influence of the lighting. The relevance of these points can be seen by considering the work of an audio typist. Once started such typists only rarely look at the keys and could probably work very well in near darkness even though typing is apparently a strongly visual task.

At about the same time as the initial Hawthorne experiment was being done, there was reported a series of investigations involving the relationship between illuminance and such industrial work as silk weaving [2], linen weaving [3] and typesetting by hand [4]. Figure 3.1 shows the silk weavers' output over 15 winter weeks against the amount of daylight outdoors during those weeks in an average year. There is a clear relationship between the output and the amount of daylight available. This rather gross result was confirmed by more detailed examination of the output of linen weavers. Figure 3.2 shows the mean output for 20 weavers for the months of October, November and December, and the daylight inside the weaving shed, plotted against time of day. It can be seen that there is a decline in

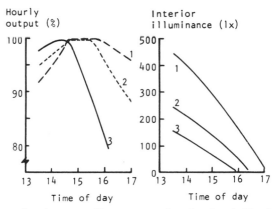

Fig. 3.2. Percentage hourly output of linen weavers plotted against the illuminance provided by daylight inside the weaving shed at different times in the afternoon during: 1, October; 2, November; 3, December. The best hourly output in each period is taken as the reference level. (After [3].)

output with decreasing illuminance and that the decline is greater for lower illuminances. Both these studies confirm what is common experience: that lighting conditions can become inadequate for a task to be seen clearly, so performance declines. The point where this happens will depend on the nature of the task and the motivation of the worker.

Similar studies can be found throughout the literature on visual performance. Some of them are well-conducted experiments [4], others are little more than assertions produced for commercial literature. Among the well-conducted studies is one by Stenzel [5] in which he measured the output from a leather factory over a four-year period in the middle of which he introduced a change in lighting installation. The work involved was punching out fault-free outer leathers, for handbags, purses, etc., from skins using iron shapes and mallets. From 1957 to 1959 the lighting was provided by daylight, supplemented by local fluorescent lighting giving an illuminance of 350 lx. From 1959 until 1961, when the investigation stopped, the daylight was virtually eliminated and a uniform 1000 lx was provided by general fluorescent lighting. The average monthly performance for 12 people who were present throughout the four years is shown in Fig. 3.3. There is a statistically significant improvement in performance with the higher illuminance lighting giving the better performance.

Although this study was well controlled, it does have two weaknesses common in field studies. First, only two lighting conditions are used and

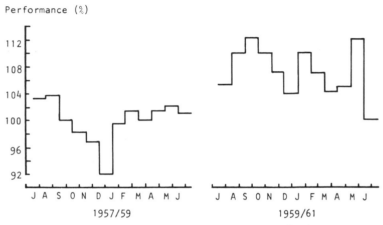

Fig. 3.3. Mean monthly performance for cutting leather shapes for the years 1957/59 and 1959/61. Performance is normalised at 100% for the average performance over the years 1957/59. (After [5].)

these are widely different. Thus these field results do little more than demonstrate that changes in lighting can influence the performance of the task studied. Certainly, there is insufficient detail to enable the lighting conditions required for optimum performance to be identified.

Second, this study takes a before and after form, the distinguishing event being the installation of a new lighting system. This is a weakness because a new installation usually changes several variables at once. In this case, in addition to providing a different illuminance on the task the new installation eliminated daylight, changed the distribution of light and modified the colour of the light. In these circumstances ascribing the changes in output to illuminance only may be misleading. But changes in lighting conditions are not the only changes that can occur in field studies. Often changes in lighting are associated with new working arrangements, new equipment and new people. In these circumstances studies which ascribe any changes in output to the lighting only, are also potentially misleading. The main problem with field trials is simply the degree of experimental control required [6]. Ideally the experimenter needs to be able to control the characteristics of the lighting installation, how it is used, the type of work done in detail, the methods of payment and the people involved. It is only rarely that this degree of control is available. It can be concluded that worthwhile field trials are difficult. They are expensive, time consuming and troublesome to interpret but ultimately they are a necessity. Any model of the relationship between lighting and work will have to pass the test of predicting the results of field trials.

3.4 SIMULATED TASKS IN LABORATORIES

The lack of control in the field has forced many experimenters to return to the laboratory where tests simulating real life tasks can be done under strictly controlled conditions. This has the advantage of allowing the experimenter to eliminate extraneous factors but the disadvantage that the situation is less real. The whole context of the task will be changed from everyday work to that of a laboratory. This is likely to have a marked effect on the worker's motivation. People who come to laboratories expect to be tested and are usually extremely well motivated to succeed. The same cannot be said for people going about their everyday work. However, as long as the high level of motivation associated with people taking part in laboratory experiments is realised, the laboratory is a very convenient place to examine the effect of lighting.

Typical of studies using simulated tasks is one by Stenzel and Sommer [7]. They had a group of people either sorting screws into different classes or crocheting stoles for two-hour periods under different illuminances. The same lighting system was used throughout the test, with the illuminance on the task as the variable. The differences in illuminance were achieved by switching different lamps. The order of experiencing the conditions was varied systematically. The results showed the expected trend for both tasks of an increase in output as the illuminance on the task increased from 100 to 300 to 700 to 1700 lx. The error scores gave the expected decrease with increased illuminance for the crocheting but not for the screw sorting task. Here the errors decreased as the illuminance increased from 100 to 300 to 700 lx but increased as the illuminance went up to 1700 lx, although this could be related to the very different performance pattern of one of the subjects.

A rather more usual task was used by Smith and Rea [8]. They had eight

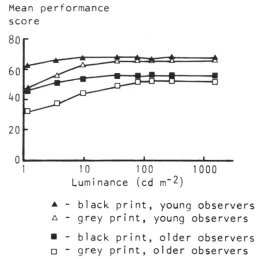

Fig. 3.4.   Mean performance score for young (20–24 years) and older (60–69 years) observers checking columns of figures, printed in black and grey ink on white paper. The performance score was calculated by the equation

$$\text{performance score} = \frac{(20 - \text{number of errors})}{(\text{time taken in seconds} + 5)} \times 100$$

(After [8].)

young and eight older subjects checking a list of 20 numbers for agreement with a comparison list. This was done at eight different task luminances, produced by changing the task illuminance, for lists printed in black or grey ink on white paper. The mean performance scores achieved at each task luminance are shown in Fig. 3.4. The results reveal an increase in performance with increasing task luminance for both tasks although the luminance at which improvement in performance ceases depends on the task. Specifically, the performance saturates at a lower task luminance for the high contrast, black on white task. It should also be noted that the performance of the older subjects is worse than that of the younger subjects. The applicability of such results is obvious. Checking lists of numbers is a common activity in many offices. From such results it is possible to get an idea of the illuminance necessary for such a task. Bennett *et al.* [9] have produced information on the effects of illuminance for a number of practical tasks.

Lion *et al.* [10] used two simulated inspection tasks to examine the effects of having either incandescent lighting or fluorescent lighting. The work involved inspecting a steady stream of plastic discs or buttons as they came along a conveyor belt. The inspector was asked to do two tasks. One was to remove from the stream all those discs which had a broken rather than a complete loop marked on them. The other was to remove from the stream all the buttons with off-centre holes. The lighting conditions were arranged to produce the same illuminance on the conveyor belt, namely 320 lx, with the luminaires in the same positions in the room. Significantly better levels of inspection were found when sorting the discs with broken loops under the fluorescent lighting than under the incandescent lighting. No significant effects of the lighting were found for the button hole sorting. The explanation given for the significant difference in performance for the broken loop inspection was that the shadowing produced by the subject on the task was greater for the incandescent lighting than for the fluorescent. An alternative explanation is that as the plastic discs on which the loops appeared were somewhat specular, reflections of the light sources in the disc would be worse for the incandescent lighting, which uses point sources, than for the fluorescent lighting, which uses extended sources. It seems unlikely that the spectral emissions of the light sources were important although they were very different. The loop was engraved as white on a black surface so the contrast of the detail to be examined was high and without colour. The explanation given for the failure to find any effect of lighting on the button hole task was that this was a task unfamiliar to the subjects. Equally important could be the degree of discrimination required. It may well be that identifying four button holes off-centre is visually easier than searching a loop for a small gap. This study was carefully done with all the right experimental controls and yet it produced a set of rather confusing results. It illustrates well the need for exact specification of the lighting conditions and the inherent limitations on generalising the results of such simulated tasks. Regarding this latter point it will be appreciated that the fact that the two sorting tasks gave different results means that the conclusion cannot be applied to sorting tasks in general. The effects of lighting will all depend on the details being sought and the background against which they occur. Thus the results obtained apply to sorting discs with the particular pattern of loops on them, and if the explanation involving specular reflections is correct then they only apply to the particular lighting installations. A different arrangement of luminaires might well have produced no veiling reflections towards the person inspecting.

It can be concluded that the results obtained on simulated tasks in laboratories tend to be very specific. They cannot always legitimately be generalised to other tasks of a similar type or maybe even to other lighting layouts. This summarises the dilemma of the practical men. The results obtained either from field studies or laboratory studies doing simulated work are only useful for the particular task being studied. Their main contribution outside this task is to add to the stockpile of knowledge of the effects of lighting against which any model can be tested.

3.5 THE ANALYTICAL APPROACH

The first attempt to produce a model of the effects of light on work was made by Beutell [11]. He was principally interested in specifying the illuminances necessary for different tasks, and in 1934 he suggested a very simple but very effective method for doing this. The basis of his method was to define a standard task. The effect of lighting on this standard task could be thoroughly investigated and the illuminance for any desired level of performance identified. Then the illuminance required for any other task could be obtained by introducing a series of multiplying factors that allowed for the differences between the task of interest and the standard task. The multiplying factors were related to the size and contrast of the critical details of the task, any relative movement between the observer and the task, and the degree of emphasis to be given to the task in its setting. Although this concept has been very influential in the specification of lighting standards, the point of interest here is the implication that the visual difficulty of many tasks can be quantified by the size and contrast of their critical detail. It is this aspect of analysing task difficulty which gives the approach its name.

Beutell's suggestion was exploited by Weston [12] who developed it into one of the most widely used methods of investigating the effects of lighting on work. What Weston did was to devise a very simple task which was largely visual and in which the critical detail was easy to identify and measure. This task is usually known as the Landolt ring chart. As can be seen from Fig. 3.5 it consists of a series of C shapes, called Landolt rings, with gaps orientated in one of the eight cardinal directions of the compass. The observer is asked to read through such a chart and mark, in some way, those Landolt rings which have a gap orientated in a specific direction. The time taken to do this and the number of errors made under different lighting conditions can be easily measured. The critical detail of the Landolt ring is the gap. The critical size is therefore the size of the gap, the critical contrast is the contrast of the gap with the Landolt ring. A practical advantage of this task is that it can be reproduced in large numbers by printing and the

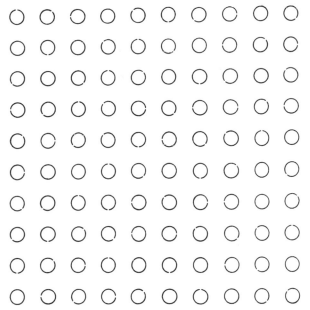

Fig. 3.5.   A typical Landolt ring chart. Different charts will have rings of different sizes and contrasts. It should be noted that the rings in such charts do not always follow the definition used in ophthalmology that the gap size and the thickness of the ring should both be one-fifth of the outer diameter of the ring.

critical size and contrast can be easily changed. Even colour contrast can be introduced if required. It is little wonder that such a flexible experimental instrument has been widely used to examine the effect of lighting conditions over a wide range of task difficulty.

Of course, the first set of important investigations using the Landolt ring chart were those of Weston himself [12]. He examined the effects of the illuminance on the task for a number of different sizes and contrasts. The actual charts used had 256 Landolt rings on them, each with a gap in one of the eight main compass directions. The subjects were asked to read through the chart and cancel all the rings which had a gap in a specified diagonal direction. The total time taken and the numbers of errors made were recorded. In an effort to remove any manual component contained in the time taken, the subjects were also asked to repeat the task with all the rings which had to be cancelled filled in with red ink, it being argued that such marking would make the rings very easy to see and the time recorded would then be almost entirely that required for the manual operation of cancelling. This manual time was subtracted from the time taken when the

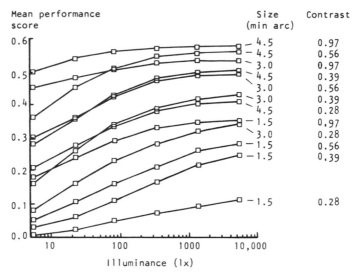

Fig. 3.6.　Mean performance scores for Landolt ring charts with rings of different critical size and contrast plotted against illuminance. The performance score for each individual is given by the expression

$$\left[\left(\frac{\text{Total time taken}}{\text{Number of rings correctly cancelled}}\right)\right.$$

$$\left.\times\left(\frac{\text{Total number of rings to be cancelled}}{\text{Number of rings correctly cancelled}}\right) - \text{manual time per ring}\right]^{-1}$$

(After [12].)

rings were not marked in red, modified by the accuracy of cancelling, to produce a score of overall visual performance.

The results obtained by Weston by this method are summarised in Fig. 3.6. There are a number of conclusions that can be drawn from these results. First it is apparent that for any task the effect of increasing illuminance follows a law of diminishing returns, i.e. that equal step increases in illuminance lead to smaller and smaller changes in visual performance until saturation occurs. Second it can be seen that the point where this saturation occurs is different for different sizes and contrasts. The smaller the size and the lower the contrast then the higher is the illuminance at which saturation occurs. Third is the observation that larger improvements in visual performance can be achieved by changing the task, i.e. increasing size and contrast, than by increasing the illuminance, at least over any illuminance range of practical interest. Fourth is the point that it is not possible to make a visually difficult task, i.e. small size and contrast, reach the same level of performance as a visually easy task simply by increasing the illuminance. Although the results discussed are those of Weston [12] these conclusions have been confirmed by different people in different countries, in some cases using similar but different tasks [13, 14, 15].

Various forms of Landolt ring chart have been used to study different lighting conditions. Rowlands *et al.* [16] used Landolt ring charts to examine the effects of fluorescent lamps with different spectral emissions on the performance of achromatic and coloured tasks. They found that the different spectral emissions had no effect on the performance of the achromatic visual tasks, a result confirmed by others [8, 17]. The results for the Landolt rings which had colour contrast, e.g. red rings on a green background, were very variable. No clear relationship with the different spectral emissions of the light sources was found, although there was the expected improvement in performance with increasing illuminance.

Stone and Groves [18] used Landolt ring charts as a task for subjects to do in an investigation of the effect on task performance of having an area of high luminance in the visual field. In fact no significant effects were found. Even when the luminance was such that the glare produced was subjectively rated as 'just intolerable' the task performance was maintained, which incidentally suggests the strength of the motivation associated with taking part in an experiment.

Reitmaier [19] reported a study in which a Landolt ring task was used to investigate the effects of veiling reflections on task performance. What is meant by veiling reflections can be seen from Fig. 3.7. When a specular surface is viewed from such an angle that a high luminance surface, such as a luminaire, is reflected from it into the eyes of the observer, veiling reflections occur. Reitmaier's results showed that increasing the veiling reflections caused a reduction in task performance. This is what would be expected because one of the effects of veiling reflections superimposed on a task is to decrease its contrast, and reducing the contrast will usually mean a reduction in performance (Fig. 3.6).

So far all the variables discussed have resulted from different lighting conditions. But the Landolt ring chart has also been used to examine the effects of differences in tasks other than those of size and contrast. An example of this is a study [20] of the effect of task illuminance on performance on a conventional Landolt ring chart, and on one made up of what are called complex Landolt rings (Fig. 3.8). The size and contrast of the critical detail are the same in both tasks but the complexity of the task,

psycho   variables

a   b

psycho   

c   d

Fig. 3.7. The effect of specular reflections on the visibility of words for two different lighting arrangements and two different tasks. When the task is specular and the lighting arrangement is such that specular reflections occur towards the observer then strong veiling reflections are present and visibility is reduced. (a) Matt ink on matt paper lit from above and behind the observer. (b) Gloss ink on semi-matt paper lit from above and behind the observer. (c) Matt ink on matt paper lit from above and in front of the observer. (d) Gloss ink on semi-matt paper lit from above and in front of the observer.

in terms of the number of alternative locations of the critical gap, is much greater for the complex Landolt ring than the simple Landolt ring. An alternative way to consider complexity is as information transfer. The greater the number of alternative locations of the gap, the greater is the information gained each time a gap position is identified. It is well established that reaction times increase with greater information transfer [21] so the complex ring task should take longer to do. The question that arises is how does a change of illuminance affect the performance of the simple and complex Landolt ring tasks. The results showed, as expected,

Fig. 3.8. Simple and complex Landolt rings, as used in the Landolt ring charts.

that the complex ring task took significantly longer than the simple ring task, but the change in performance with illuminance was similar for both tasks. In fact, the time taken to do the complex ring task at each illuminance was a constant multiple of the time taken for the simple Landolt ring task at the same illuminance. This result suggests that the visual difficulty for the task was adequately described by the size and contrast of the gap being sought. The complex rings were not more visually difficult, because the gap size and contrast were identical to the simple ring, but they did contain more information because there were more possible locations of the gap. The effect of illuminance over the range examined only changes the visual difficulty and not the complete task difficulty. It is important to keep this limitation in mind when assessing the effectiveness of any proposed change in lighting.

Although the above results have supported the view that size and contrast of critical detail are important, it should be realised that other factors are also relevant. From the performance capabilities of the visual system that were discussed previously (Chapter 2) it seems reasonable to suggest that any off-axis work, any blurring of target edges, any colour contrast when luminance contrast is low, and any task movement may all affect the visual difficulty of a task. Unfortunately most of these factors still await investigation. Size and contrast are the two variables that have been extensively examined. From his results Weston [12] derived a relationship between size and contrast which indicated their relative importance. He showed that the performance on the Landolt ring charts was correctly ranked according to the index (apparent size $^{1 \cdot 5}$ × contrast). Apparent size is the angle subtended at the eye by the critical detail in minutes of arc.

There can be little doubt that the analytical approach has been very successful—and with good reason. It enables a rapid survey to be made of the influence of any particular lighting variable over a wide range of visual difficulty. Further, it provides a direct if somewhat tenuous link to the real world where real tasks can be classified by their size and contrast of critical detail. The results obtained tend to support the results found in the field trials or at least provide some basis for understanding. In all these aspects the analytical approach is valuable, but it is not the only model available. The analytical approach has the disadvantage that it essentially starts from a description of the visual difficulty of the task of interest. But this is not a measure which describes the main purpose of lighting. Lighting is provided to enable people to see well. Therefore any measure of its effectiveness should aim at quantifying how well an observer can see. This has been the basis of the other model—the visibility approach.

3.6 THE VISIBILITY APPROACH

3.6.1 Introduction

The visibility approach is related to the activities involved in visual work. These can be summarised as searching the visual field to establish the presence of objects and then discriminating details within objects so as to identify them. Often these activities are carried on against either a self-imposed or external time limit which implies that search information is collected off the visual axis. The historic development of the visibility approach [22, 23] has led to the two aspects of visual work, i.e. discrimination of detail for on-axis and off-axis targets, being considered separately. The first activity to be examined is discrimination of detail for on-axis targets.

3.6.2 Visibility Level

The essence of the problem faced by the visibility approach is to identify a measure which can be used to quantify how easy it is to discriminate details. If this can be done then the effectiveness of different lighting conditions can be evaluated in terms of this measure. The visibility approach uses luminance contrast as the fundamental visual characteristic of any target. Intuitively there is some justification for this because, if there is no luminance contrast between a target and its background and there is no colour contrast, the target cannot be seen no matter how large it is. The actual measure used to describe how easy a target is to see is called its Visibility Level. This is defined as a ratio of contrasts as follows:

$$\text{Visibility Level} = \frac{\text{equivalent contrast}}{\text{threshold contrast}}$$

In the visibility approach, threshold contrast is defined as the luminance contrast of a target when it can only just be seen. This is different from the conventional 50% performance definition of a threshold condition but, since its function in the visibility approach is simply to act as a convenient baseline, this difference does not matter. The actual target used as a visibility reference task in the visibility approach is a 4 min arc luminous disc presented against a uniform luminance field. The disc is surrounded by markers to aid fixation, so the threshold contrast is measured for on-axis viewing conditions. Further the disc is seen only for 0·2 s glimpses which are meant to simulate the duration of an eye fixation. Mean values of threshold contrasts for this target for different adaptation luminances (the luminance of the uniform background) determined by a group of 20–30-year-old

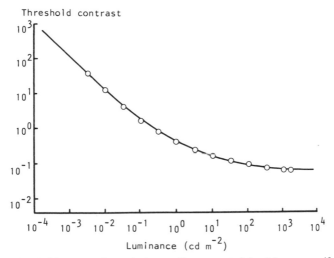

Fig. 3.9.   Threshold contrast for a 4 min arc disc presented for 0·2 s on a uniform field, plotted against luminance [24].

observers are shown in Fig. 3.9 [24]. As might be expected, the threshold contrast decreases as the adaptation luminance increases, i.e. the observers become more sensitive. The threshold contrast of the 4 min arc disc at the particular adaptation luminance associated with the target of interest is used as the denominator in the calculation of the Visibility Level of the target. The problem now is that to be consistent the numerator in the calculation of Visibility Level, i.e. equivalent contrast, also has to describe the target of interest in terms of the 4 min arc disc.

Equivalent contrast is properly measured with the aid of an instrument called a visibility meter [22]. Figure 3.10 is a schematic diagram of its

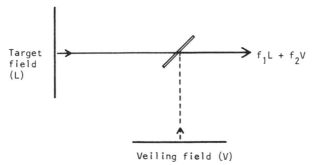

Fig. 3.10.   The operating principle of the visibility meter [22].

operating principle. The principle is that the observer views the target of interest through the visibility meter. Superimposed on the view of the target is a field of uniform luminance called the veiling field. By increasing the luminance from this veiling field and simultaneously decreasing the luminance from the target field, the target can be made to disappear, as if a layer of mist had covered it. In more detail, the procedure is first to set the maximum luminance of the veiling field (V) equal to the adaptation luminance of the target (L). Next the visibility meter is set so that none of the light from the veiling field reaches the observer. In other words, the visibility meter is set so that the fraction of the total luminance reaching the observer from the target field (f_1) is equal to unity, and the fraction of the total luminance reaching the observer from the veiling field (f_2) is zero. The observer then alters the mechanism for combining the two fields and thereby decreases f_1 and increases f_2. This mechanism, which can come in various forms, has the optical characteristic that $f_1 + f_2 = 1$ for all possible combinations of the two fields. This means the total luminance to which the observer is adapted is constant, and, because of the original setting, is equal to the adaptation luminance of the target, which in turn means the state of adaptation of the observer is appropriate to the target and is unchanged in making the measurement. The combination of the two fields is altered until the target can only just be seen. It is this condition which is being sought with a visibility meter.

But how can a visibility meter be used to measure the equivalent contrast of any target of interest? The idea is first to examine the target of interest through the visibility meter and to adjust the combination of the two fields until the target can only just be seen. The veiling luminance ($f_2 V$) when the target can only just be seen is recorded. Having established the condition when the target of interest can only just be seen, the next stage is to view the 4 min arc disc used to determine the threshold contrast curve (Fig. 3.9), through the visibility meter. The background luminance of the 4 min arc disc is adjusted so that it is the same as the adaptation luminance of the target of interest, and the veiling luminance is set to the value it had when the target of interest could only just be seen ($f_2 V$). By changing the luminance of the 4 min arc disc its contrast with the background can be adjusted. By doing this, the luminance contrast of the 4 min arc disc that can only just be seen when the veiling luminance is such that the target of interest can only just be seen, is established. This contrast is the equivalent contrast of the target of interest. To repeat, it is the luminance contrast of the visibility reference task that can only just be seen when the veiling luminance is such that the target of interest can only just be seen.

The attraction of equivalent contrast should be clear. It allows the visual difficulty of targets of different sizes, shapes, degrees of blur and requiring different levels of information, e.g. detection, identification or recognition, to be quantified on a common scale in terms of contrast. Further, the way in which it is measured ensures that the results obtained represent the integration of all the various components of the task, together with their differing degrees of importance, as they appear to a human observer. This is a considerable advance over the useful but somewhat artificial method of describing the visual difficulty of targets by their physical characteristics such as size, contrast, blur, etc., each measured separately.

We now have both the components of the Visibility Level measure. Visibility Level is simply the ratio of the equivalent contrast of the target of interest to the threshold contrast of the visibility reference task at the same adaptation luminance. Since equivalent contrast is actually measured with the same visibility reference task the ratio simply related two contrasts of the same target.

What Visibility Level measures is how far the equivalent contrast is above the baseline represented by the threshold contrast of the visibility reference task. It seems intuitively reasonable to suppose that the higher the Visibility Level the easier it is to see the target represented by the equivalent contrast. This can be supported by common experience. Because equivalent contrast is based on making a match between two targets which have the same adaptation luminance it should not vary greatly with adaptation luminance. However threshold contrast does decrease with increasing adaptation luminance (Fig. 3.9). The net result is that, for any specific target of interest, Visibility Level will increase with greater adaptation luminance. It is common experience, up to a certain point, that higher luminances do make things easier to see.

3.6.3 Departure from Reference Conditions

It is important to realise that threshold contrast and equivalent contrast and, hence, Visibility Level are obtained under rather special conditions. Specifically they are measured under what are called reference lighting conditions by a reference group of people. Reference lighting is defined [23] as completely diffuse and unpolarised illumination at a colour temperature of 2856 K and a field of uniform luminance. There is nothing special about these conditions, they simply form a convenient reference condition. The reference population is defined [23] as observers of 20–30 years of age, with no visual, motor or cognitive defects and with any necessary ocular correction. Again there is nothing special about these people, they simply form a convenient reference population. In fact the reference conditions are derived from the conditions used in obtaining the basic threshold contrast data [23]. However, practical lighting installations frequently do not produce completely diffuse illumination and other groups of people do not have the same vision as the reference population. Therefore it is necessary to have some convenient measures of the extent to which lighting conditions and other populations differ from the reference conditions. The necessary corrections to the lighting conditions are given by the contrast rendering factor (CRF) and the disability glare factor (DGF). These factors can be measured with a visibility meter in the following way [22]. The target of interest is first viewed under the actual lighting of interest and, as described earlier, the target visibility is reduced until the target can only just be seen. The proportion of the total luminance represented by the luminance from the target field (f_1) when this state is reached is expressed as a reciprocal, and called the relative visibility of the target. Obviously the smaller this proportion when the target can only just be seen, the greater the ease with which the target was seen initially. Fittingly, the smaller the proportion when the target can just be seen the greater is the relative visibility.

Having just measured the relative visibility of the target of interest under the lighting of interest, the relative visibility of the target is then measured under reference lighting conditions at the same adaptation luminance. Then CRF and DGF can be calculated from the ratio of the two relative visibilities of the target in the form: relative visibility under actual lighting conditions over relative visibility under reference lighting conditions. Contrast rendering factor quantifies the effect of veiling reflections on the task. Disability glare factor quantifies the effect of non-uniformity of luminance in the visual field surrounding the task. CRF is the more important of the two. The conditions under which poor CRF values occur in practice have been extensively studied. In addition to proving that task specularity and the relative positions of luminaire, task and observer are the two factors which primarily influence CRF values, these results have also shown that in many cases it is reasonable to replace the tiring and time-consuming visibility meter measurements by direct physical measurements of the luminance contrast of the target under the lighting of interest and the reference lighting conditions, or even by calculated contrasts [25, 26].

Contrast rendering factor is of value in its own right because it quantifies the effect of different lighting installations in producing veiling reflections. However it can also be combined with Visibility Level to form effective

Visibility Level† which quantifies the combined effects of veiling reflections and luminance on how easy any particular target of interest is to see. The effective Visibility Level is given as effective Visibility Level = Visibility Level × CRF. In theory, if the lighting produces disability glare, the effect of this can be evaluated by measuring DGF and combining it with the effect of luminance for any specific target in terms of effective Visibility Level in the same way as CRF, i.e. effective Visibility Level = Visibility Level × DGF. Where both veiling reflections and disability glare occur both effects can be allowed for in effective Visibility Level = Visibility Level × CRF × DGF. However, measurements of DGF with a visibility meter are not currently possible because existing visibility meters do not have the necessary design features. What values of DGF are available have been obtained by calculation [22], and it is these values which indicate that DGF is relatively insignificant in interiors when compared with CRF. It should also be noted that in the definitive description of the visibility approach [23] there is another measure, the transient adaptation factor (TAF), which allows for changes in adaptation when the eyes stray from fixation on the target across a non-uniform visual field. This is of little practical interest for interiors but may be relevant for roadway lighting. At the moment there is no agreed way of measuring or calculating TAF.

The other aspect of departure from reference conditions occurs when tasks are done by people other than the reference population. In principle such departures can be compensated for by modifying the threshold contrast term in the calculation of Visibility Level. The idea is that different people, and particularly older people, will have contrast thresholds different from the average of the reference population shown in Fig. 3.9, so they will have different Visibility Levels. Specifically, older people will have higher threshold contrasts, and hence lower Visibility Levels, than the reference population for the same adaptation luminance. The full implications of the differences between other populations and the reference population are still being investigated [23]. It remains to be seen if this simple approach is an adequate method of dealing with differences between people.

To summarise, effective Visibility Level is a measure of how effective real lighting is in making a particular target visible to specified people. The

† In the latest CIE report [23] the term 'effective Visibility Level' has been dropped and only Visibility Level used. The justification for this is that for virtually all practical applications CRF will have to be considered so its inclusion is simply a generalisation of the definition of Visibility Level. This may be logical but it is hardly convenient for explanation. Therefore effective Visibility Level will be used here.

intrinsic visual difficulty of the task under reference lighting is given by equivalent contrast, the effect of the luminance the observer is adapted to is given by the threshold contrast, both equivalent contrast and threshold contrast are combined in Visibility Level. The effect of the actual lighting in producing veiling reflections and disability glare rather than completely diffuse lighting is given by the CRF and DGF, and, in principle at least, deviations from the reference population can be allowed for by the use of multiplying factors. Thus effective Visibility Level includes the influences of all three components involved in visual performance, i.e. task, lighting and people, in its determination.

3.6.4 Applications

The utility of effective Visibility Level can be seen from Fig. 3.11. The curve shows the relationship between mean performance score and effective Visibility Level for a series of different Landolt ring charts. In fact this curve was derived from the results of Weston's Landolt ring chart experiment [12] (Fig. 3.6). The different sizes and contrasts of the rings used, produce different equivalent contrasts and hence have different Visibility Levels, even when they have the same illuminance. The lighting used approximated to reference lighting so the CRF and DGF values have been assumed to be unity.

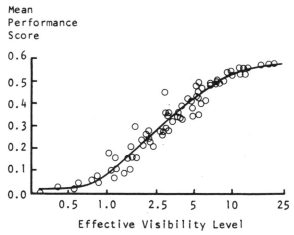

Fig. 3.11.   Mean performance scores on Landolt ring charts at different effective Visibility Levels. This figure was derived from the data shown in Fig. 3.6. (After Blackwell, H. R. *Analysis of twenty sets of visual performance data in terms of a quantitative method for prediction of visual performance potential as a function of reference luminance,* Report to the Illuminating Engineering Research Institute, New York, 1980.)

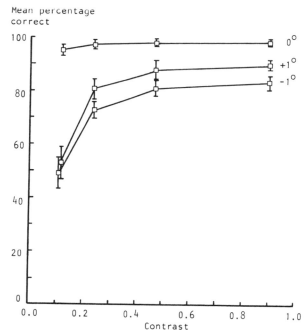

Fig. 3.12. Mean percentage correct identifications of three-letter words presented for 100 ms in the fovea (0°) and at ±1° eccentricity, at different levels of luminance contrast. Also shown are the standard errors of the means. (After [29].)

It can be seen that Visibility Level brings the results for the Landolt ring charts with different sizes and contrasts of critical detail presented at different illuminance onto a common line. This is suggestive of its value as a way of quantifying the effects of particular lighting conditions on a specific task with different visual characteristics. Similar examinations of the use of Visibility Level show it can do the same job of combining results for other people's work [22, 27, 28]. Initially this led to the conclusion that there was a universal relationship between performance of the visual component of a task and effective Visibility Level [22]. Examination of later data revealed that this was not true. The missing factor was the extent of search and scan of the visual field and the necessity for information to be gathered off-axis. Widely different relationships between visual performance and effective Visibility Level were found for tasks requiring different degrees of off-axis information. But the measurements involved in effective Visibility Level are all made with the task directly fixated, i.e. on the line of sight. How important off-axis viewing is can be seen from Fig. 3.12. This shows the percentage correct scores on a task involving the reading of three-letter words of different contrast presented for 100 ms [29]. The words were presented on-axis and ± 1 ° off-axis. It can be seen that even for the highest contrast there is a distinct reduction in performance for the off-axis presentation, and for lower contrasts the difference increases dramatically. To gain an understanding of the effect of off-axis viewing, the performance

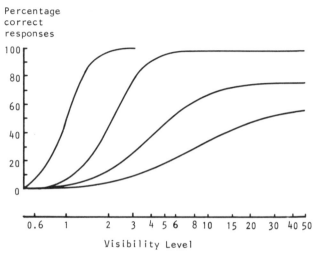

Fig. 3.13.   Percentage correct responses on the visual performance reference task plotted against Visibility Level for different degrees of task difficulty. The conditions for which each curve was obtained increase in difficulty from left to right [23, 30].

of what is called the visual performance reference task was studied. The target here consists of an array of five 4 min arc simple Landolt rings, one in a central position and the other four at the four cardinal points of the compass, equidistant from the centre. The observer is asked simply to view this display and report which, if any, of the four peripheral rings has a gap in the same orientation as the centre ring. The Visibility Level of the display can be changed by altering the luminance contrast of the rings, thereby altering the equivalent contrast, or by changing the adaptation luminance and hence changing the threshold contrast. The difficulty of the task can also be altered by changing the time for which the display can be seen, the separation between the central ring and the peripheral rings and the type of information required. Thus this task allows for the possibility of both changes in task visibility and different degrees of off-axis information collection. Data collected using this task from the reference population are shown in Fig. 3.13 in the form of percentage correct scores plotted against Visibility Level, for different viewing conditions [30]. In fact the left-most curve in Fig. 3.13 shows the performance for on-axis viewing of a single Landolt ring. This result is obtained with the visual performance reference task when there is sufficient time available for each of the peripheral Landolt rings to be fixated. The three other curves represent increasingly more difficult task conditions: either shorter presentation times or greater separations, etc. It is interesting to note that as the task requirements get more difficult, i.e. the off-axis element increases, the Visibility Level at which performance saturates increases and the maximum performance level reached decreases. This illustrates the general pattern of results obtained, and confirms the need for some measure of the extent of off-axis viewing required if Visibility Level is to be reliably related to task performance. Unfortunately no such measure is available although possibilities are being investigated [23]. This severely limits the application of the visibility approach to specific tasks but has not delayed the construction of a model of how lighting parameters affect visual performance [23].

3.6.5 The Model

The model assumes four components, three of which relate to Visibility Level. The first is what is called the sensory process. This is the part that is involved in obtaining critical information from task detail. It covers the influence of task requirements, the visual difficulty of the task, the amount of off-axis information required, the need for search and scan over the visual field, etc. The second component involves the accuracy of fixation. It has been found [31] that as Visibility Level reduces to low values the accuracy of fixation is reduced, which in turns means the extent of off-axis information gathering is likely to increase. The third component also involves the ocular motor system. At very low Visibility Levels the accuracy with which rapid eye movements can be made from one fixation point to another is reduced [31]. Thus at very low Visibility Levels the efficiency of any search and scan operations will be reduced. Figure 3.14 shows these three components separately in terms of the relative performance of each plotted against Visibility Level for a very difficult condition of the visual performance reference task. It can be seen that as the Visibility Level increases the performance of the second and third processes saturate so that above a Visibility Level of approximately five, for this task, only the sensory process is limiting performance. To obtain the relative visual performance (RVP) of the task the relative performance (P) of each of these processes is first given an appropriate weighting (W) and then they are summed, i.e. RVP = $W_1P_1 + W_2P_2 + W_3P_3$, where $W_1 + W_2 + W_3 = 1$. To obtain relative task performance a fourth process needs to be considered. This fourth process is simply the extent to which the task involves non-visual components. The performance of the three visual and one non-visual processes can then be given their appropriate weights and summed to give the relative task performance. Using this model of the effect of lighting

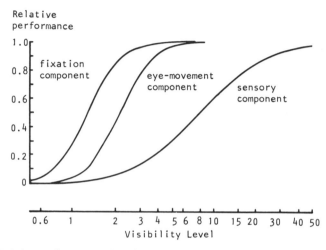

Fig. 3.14. Relative performance plotted against Visibility Level for the three visual components in the model of the relationship between visual performance and lighting conditions [23].

parameters on visual performance it is possible to calculate relative visual performance or relative task performance on a specific task for given lighting conditions, or to specify the lighting conditions necessary to achieve a required level of performance on a specified task [23]. The only problem is that to do so, at present, requires some assumptions as to appropriate values for a number of important variables.

This model is the most ambitious attempt ever made to account for the effects of differences between tasks, differences between people and differences between lighting conditions as they affect the performance of any specific task. There are undoubtedly many problems with the model, but its predictions do conform with everyday experience. For example, it demonstrates that relative task performance depends much more on the intrinsic difficulty of the task, as characterised by equivalent contrast, and on the task requirements for off-axis viewing than it does on the task illuminance. Even so it has limitations. For example, it seems probable that there should be at least a fifth process in the model to deal with the effect of low Visibility Levels on accommodation. There are also questions surrounding the equality of effect of Visibility Levels produced by different combinations of contrast and luminance [32]. In addition the precision of the model is in some doubt. Many of the measurements involved are psycho-physical in nature, which inevitably produce wide variations between people, and the definition of threshold is hardly precise. This

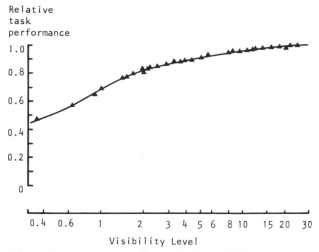

Fig. 3.15.   Relative performance on a numerical verification task [8] plotted against Visibility Level. The points are the experimental results. The line is the relationship between relative task performance and Visibility Level derived from the model [23].

suggests that the model is likely to be difficult to use accurately. However, the most serious defect at the moment, laying aside the point that some parts have not been validated, is the fact that the model is not predictive. It can be made to fit independently obtained sets of experimental results, and has been [23]; but this is done by varying the weighting of the processes, and choosing a particular set of results from the visual performance reference task as representative, after the performance results of the specific task have been obtained. Figure 3.15 shows the fit of the model applied to some results of Smith and Rea [8] involving checking lists of numbers (see Fig. 3.4). The points are the experimental points, the line is the output of the model. If the line had been a true prediction rather than a *post hoc* manipulation there would have been cause for celebration.

Obviously there is still much to do to perfect the model. Nevertheless it is a step forward. It represents the most complete and systematic approach to measuring the combined effects of task, lighting and people variables on task performance. It has introduced into the study of the relationship between lighting and work a number of new quantities which are of considerable value; for example, equivalent contrast as a measure of intrinsic visual difficulty, contrast rendering factor as a measure of the extent to which lighting produces veiling reflections, and effective Visibility Level as a measure of the effect of specific lighting conditions on how difficult a specific target is to see. None of these measures is perfect but they do represent a great contribution to the subject of light and work. It is fitting therefore to acknowledge here the efforts of Professor H. R. Blackwell who, for many years, has been the driving force behind the developments in this model.

3.7 CONCLUSIONS

The question to be considered now concerns the present state of knowledge of the relationship between light and work. The answer is that some of the important variables have been identified, or rather the variables that directly affect the visual difficulty of the task have been identified, but few of their effects have been quantified in detail except for a few specific tasks.

Of the lighting variables that have been examined, by far the most thoroughly investigated has been illuminance. The effects of illuminance on a wide range of different tasks have been examined by field experiment, laboratory experiments using simulated tasks, and by the analytical approach. The most comprehensive set of data has been obtained by the analytical approach. The others have produced little of general importance because of their limited application. However, these other studies do tend to confirm the results of the analytical studies. These latter results indicate that the effect of increasing illuminance on the performance of tasks follows a law of diminishing returns, i.e. equal increments in illuminance produce less and less improvement in task performance until performance finally saturates. The illuminance at which performance saturates depends on the visual difficulty of the task, which can be conveniently classified by the size and contrast of the critical detail. The smaller the size and contrast, the higher is the illuminance at which task performance saturates. In fact, larger changes in task performance can be obtained by increasing size and contrast than by increasing illuminance over any practically possible range. The results also show that it is not possible to make a visually difficult task be performed as well as a visually easy task simply by increasing the illuminance.

As for the visibility approach this can best be described as 'having great expectations'. At the moment it is relatively untested, but based as it is on measuring the extent to which tasks are easy to see, it seems a useful way of examining the relationship between light and work. Nor is its adoption an 'all or nothing' offer. It is perfectly possible to treat various measures as of value in their own right without accepting the full complexity of the present model. Indeed it would be a mistake to do so. The model has already undergone one major modification with the admission that there is no unique relationship between visual performance and Visibility Level. It is unlikely that this will be the last, especially as an alternative model, using off-axis detection probability as a measure of ease of seeing, has recently been outlined [33]. Nonetheless the present model does represent a fundamental way of considering the relationship between light and work, and, as such, deserves to be thoroughly probed and tested.

At the moment anyone looking for answers to questions on how lighting can affect performance on specific tasks, and who is unwilling to undertake an experiment himself, can adopt one of two courses. First, he can examine the results of the analytical approach. By classifying the task of interest in terms of the size and contrast of its critical detail an idea of the illuminance necessary to achieve a given level of performance can be obtained. By looking at the effect of changes in contrast it may also be possible to identify the influence of veiling reflections on task performance. But this still leaves one big unknown in the form of the importance of the visual component to the overall task. Establishing this calls for a detailed task analysis. How much time is spent on visual work compared with non-visual; where in the process does the visual component occur? Is any increase in time taken or errors made important? All these questions need to be considered.

Alternatively, he can do some calculations with the visibility model [23], using the data contained in the report to select appropriate values of some of the variables. Neither of these courses will give exact and accurate answers but they should indicate the order of any effect and the important variables. This seems a rather lame conclusion but it is a true reflection of the facts, except in a few cases. The effects of lighting on some common activities, e.g. driving, have been more extensively studied. The knowledge produced will be discussed in detail in the next chapter.

3.8   SUMMARY

The relationship between light and work has been the subject of prolonged study. Measurements of specific tasks under different lighting conditions, both in the field and in laboratories, have confirmed the importance of lighting and demonstrated the influence of particular variables. However, the value of these results is limited because the field studies have often been poorly controlled whilst the studies in laboratories have limited applicability.

An alternative method of examining the effect of light on work is the analytical approach. This involves adopting a standard task and examining the performance of this task at different levels of difficulty under a wide range of lighting conditions. The task chosen is the Landolt ring chart, the size and contrast of the gaps in the rings being taken to describe the visual difficulty of the task. The idea is that the effect of lighting conditions on another task can be predicted by examining their effects on the Landolt ring chart when the rings have the same size and contrast as the critical detail of the other task. Using the Landolt ring chart the effects of several different lighting variables have been examined, but by far the most popular has been illuminance. It has been established that the relationship between task performance and illuminance follows a law of diminishing returns, that the illuminance at which performance saturates depends on the visual difficulty of the task, and that larger increases in performance can be achieved by increasing size and contrast than by increasing the illuminance.

These conclusions have been confirmed by the most ambitious model of the relationship between light and work so far developed—the visibility approach. This approach is based on quantifying how difficult tasks are to see. Ideally the method of quantification involves using a visibility meter to measure the intrinsic visual difficulty of the task and hence how visible it is under given conditions. The measure of visibility, Visibility Level, has been shown to be related to the performance of a number of real tasks. However, the model is by no means complete. It still involves *post hoc* rationalisation rather than prediction, but it has already introduced a number of useful measures into the vocabulary of visual performance. It deserves careful testing to identify its weaknesses and further development to overcome them.

REFERENCES

1. Urwick, L. and Brech, E. F. L. *The Making of Scientific Management; Vol. 3, the Hawthorne Investigations*, Pitmans, London, 1965.
2. Elton, P. M. *A study of output in silk weaving during winter months*, Industrial Fatigue Research Board Report No. 9, HMSO, London, 1920.
3. Weston, H. C. *A study of efficiency in fine linen weaving*, Industrial Fatigue Research Board Report No. 20, HMSO, London, 1922.
4. Weston, H. C. and Taylor, A. K. *The relation between illumination and efficiency in fine work (typesetting by hand)*, Final Report of the Industrial Fatigue Research Board and the Illumination Research Committee (DSIR) HMSO, London, 1926.
5. Stenzel, A. G. Experience with 1000 lx in a leather factory, *Lichttechnik*, **14**, 16, 1962.
6. Hartnett, O. M. and Murrell, K. F. H. Some problems of field research, *Applied Ergonomics*, **4**, 219, 1973.
7. Stenzel, A. G. and Sommer, J. The effect of illumination on tasks which are largely independent of vision, *Lichttechnik*, **21**, 143, 1969.
8. Smith, S. W. and Rea, M. S. Relationships between office task performance and ratings of feelings and task evaluation under different light sources and levels, *Proc. CIE 19th session, Kyoto*, 1979.
9. Bennett, C. A., Chitlangia, A. and Pangrekar, A. Illumination levels and performance of practical visual tasks, *Human Factors Society, 21st Annual Meeting, San Francisco*, 1977.
10. Lion, J. S., Richardson, E. and Browne, R. C. A study of the performance of industrial inspection under two levels of lighting, *Ergonomics*, **11**, 23, 1968.
11. Beutell, A. W. An analytical basis for a lighting code, *Illum. Engr (London)*, **27**, 5, 1934.
12. Weston, H. C. *The relation between illuminance and visual performance*, Industrial Health Research Board Report No. 87, HMSO, London, 1945.
13. Khek, J. and Krivohlavy, J. Evaluation of the criterion to measure the suitability of visual conditions, *Proc. CIE 16th session, Washington*, 1967.
14. Muck, E. and Bodmann, H. W. Die bedeutung des beleuchtungsniveaus bei praktische sehtatigkeit, *Lichttechnik*, **13**, 502, 1961.
15. Boyce, P. R. Age, illuminance, visual performance and preference, *Ltg Res. and Technol.*, **5**, 125, 1973.
16. Rowlands, E., Waters, I., Loe, D. L. and Hopkinson, R. G. *Visual performance in illumination of differing spectral quality*, UCERG Report, University College, London, 1973.
17. Milova, A. The influence of light of different spectral composition on visual performance, *Proc. CIE 17th session, Barcelona*, 1971.
18. Stone, P. T. and Groves, S. P. D. Discomfort glare and visual performance. *Trans. Illum. Engng Soc. (London)*, **33**, 9, 1968.
19. Reitmaier, J. Some effects of veiling reflections in papers, *Ltg Res. and Technol.*, **11**, 204, 1979.
20. Boyce, P. R. Illuminance, difficulty, complexity and visual performance, *Ltg Res. and Technol.*, **6**, 222, 1974.
21. Hick, W. E. On the rate of gain of information, *Quart. J. Expl. Psychol.*, **4**, 11, 1952.
22. Commission Internationale de l'Eclairage. *A unified framework for evaluating the visual performance aspects of lighting*, CIE Publication 19, 1972.

23. Commission Internationale de l'Eclairage. *An analytical model for describing the influence of lighting parameters upon visual performance*, Draft CIE Publication 19/2, 1980.
24. Blackwell, O. M. and Blackwell, H. R. Visual performance data for 156 normal observers of various ages, *J.I.E.S.*, **1**, 3, 1971.
25. Boyce, P. R. Variability of contrast rendering factor, *Ltg Res. and Technol.*. **10**, 94, 1978.
26. Blackwell, H. R., Dilaura, D. L. and Helms, R. N. Application procedures for evaluation of veiling reflections in terms of ESI: IV. final validation of measurements and predetermination methods for luminaire installations, *J.I.E.S.* **2**, 299, 1973.
27. Bodmann, H. W. Illumination levels and visual performance, *International Lighting Review*, **13**, 41, 1962.
28. Boynton, R. M. and Boss, D. E. The effect of background luminance and contrast upon visual search performance, *Illum. Engng*, **66**, 173, 1971.
29. Timmers, H. *An effect of contrast on legibility of printed text*, Institute of Perception Annual Report, IPO, Eindhoven, Netherlands, 1978.
30. Blackwell, H. R. and Blackwell, O. M. Population data for 140 normal 20–30 year olds for use in assessing some effects of lighting upon visual performance, *J.I.E.S.*, **9**, 158, 1980.
31. Blackwell, H. R. and Sinton, D. J. Accuracy of steady ocular fixation and speed and accuracy of saccadic eye movements as a function of task visibility level, *J.I.E.S.*, **10**, 1980.
32. Ross, D. K. Task lighting—yet another view, *Lighting Design and Application*. **8**(5), 37, 1978.
33. Inditsky, B. and Bodmann, H. W. Quantitative model of visual search, *Proc. CIE 19th session, Kyoto*, 1979.

4

Specifics

4.1 INTRODUCTION

There is no magic formula for predicting the influence of lighting on the performance of a wide range of tasks. The differences between tasks are too great and the necessary measures too ill-defined for one to exist. However, the effects of lighting conditions on the performance of a few common activities have been studied in sufficient depth to make a detailed examination of the results worth while. The activities to be considered are reading, making colour judgements, visual inspection and driving. The purpose of each examination is not only to describe how lighting affects each activity but also to illustrate the variety of factors that influence the relationship between light and work.

4.2 READING

4.2.1 Introduction
Much of the world's knowledge is locked up in the written word. Reading is the skill which enables people to release this information. As such it is considered important, a fact made evident by the emphasis given to literacy in all countries. This section is concerned with only one aspect of reading. It is concerned with the physical features that influence the ease with which a passage can be read and understood, i.e. the type of print, its layout and form, the media in which it is set and the prevailing lighting conditions. Further, the consideration of these factors is restricted to their effect on reading by people with good eyesight and normal Western adult reading capabilities. This limitation is imposed because most of the relevant studies have used such people.

111

HUMAN FACTORS IN LIGHTING

4.2.2 The Nature of Reading

Reading is essentially a process of generating coherent messages from patterns of marks. In order to produce any sort of meaning at all, the marks have to be visible, identifiable and arranged in an expected order. All three attributes are necessary for easy reading. Separately they are not sufficient. Figure 4.1 contains three passages of print. The first passage shows the

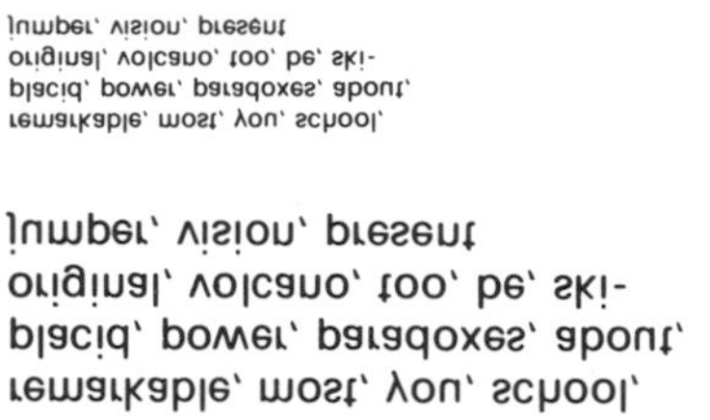

Inwper viziou' presear
oriāiual' voicauo' roo' pe' ski-
pĮaciq' bower' paradoxes' about'
remarkapĮe' most' you' school'

Inwper viziou' bresear
oriāiual' voicauo' roo' pe' ski-
pĮaciq' bower' paradoxes' about'
remarkapĮe' most' you' school'

passage visible but marks not identifiable and expected—passage unreadable

mrebkarlea, stom, oyu, hocosl, calpid, repow, dopraxses, toabu, grinoail, novalco, oto, eb, ksi-pejmur, nisvio, tnesrep

passage visible and marks identifiable but not arranged as expected—passage unreadable

remarkable, most, you, school, placid, power, paradoxes, about, original, volcano, too, be, ski-jumper, vision, present

passage visible, marks identifiable and arranged as expected. Passage is readable but meaning is unknown.

Fig. 4.1.   Visibility, identification and expectation as they affect ease of reading.

variation in visibility that can be achieved by increasing the size of the print but also demonstrates that if the marks are not identifiable the message remains unreadable. The second passage has good visibility of marks of known form, they are all in the conventional Roman script, but it is still unreadable. The reason this time is that the identifiable, visible marks are not arranged in an expected order. Only the third passage is readable. In this passage the marks are visible, identifiable and arranged in an expected order. Although it is readable in the sense that it can be verbally communicated to someone else, the message may still not make sense because the groups of marks, i.e. the words, do not fit the conventions of grammar—but this is a matter of linguistics, a study with much wider implications than those of concern here. The point to note is that reading involves much more than just making marks visible. However, visibility is

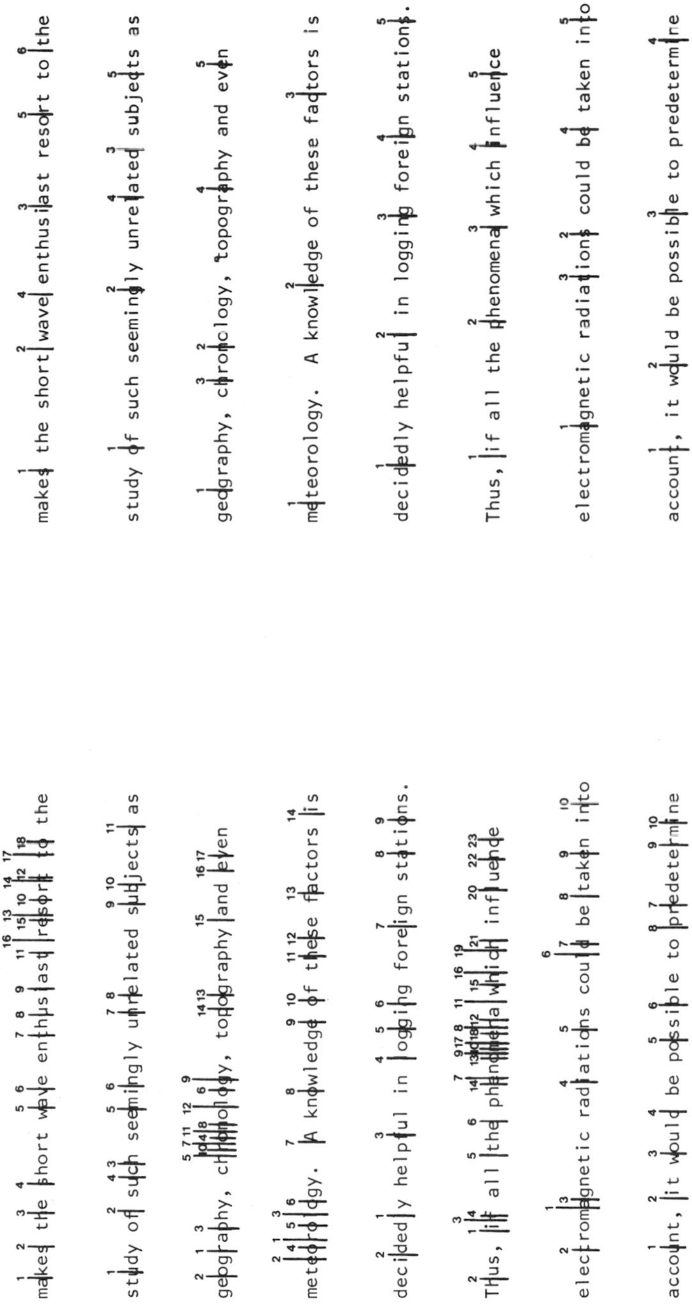
makes the short wave enthusiast resort to the
study of such seemingly unrelated subjects as
geography, chronology, topography and even
meteorology. A knowledge of these factors is
decidedly helpful in logging foreign stations.
Thus, if all the phenomena which influence
electromagnetic radiations could be taken into
account, it would be possible to predetermine
makes the short wave enthusiast resort to the
study of such seemingly unrelated subjects as
geography, chronology, topography and even
meteorology. A knowledge of these factors is
decidedly helpful in logging foreign stations.
Thus, if all the phenomena which influence
electromagnetic radiations could be taken into
account, it would be possible to predetermine

Fig. 4.2. The patterns of fixations made by two readers reading the same passage. The intersection of a vertical line with the line of print indicates a fixation point. The numbers attached to the vertical lines give the order in which the fixations occurred for each line of print. (After [3].)

an essential part of reading and is affected by lighting so it is the aspect that will be considered here.

The first question that arises in considering the influence of lighting on reading is 'which marks need to be visible?' Is it all the letters in a word or is it only a few? This question can be answered by an examination of the eye movements made during reading. Physical methods of eye movement recording have been developed to a high level of sophistication [1, 2], but regardless of the methods used, all those who have studied eye movements during reading have observed the same events. The eye moves in an erratic way along the line of print. Specifically the movement consists of a series of stationary pauses, called fixations, with rapid jumps, called saccades, between them. Normally the movement is in one direction but occasionally saccades indicating a reversal of direction occur [3, 4]. Typical patterns of fixations made whilst reading are shown in Fig. 4.2 [3], the numbers at the top of each indicated fixation giving the order of occurrence. These patterns of eye movements demonstrate that reading does not involve the reader in fixating each letter separately, or even each word. This conclusion is reinforced by the measured speed of reading. When asked to read letters presented individually the best that can be done is about three to four letters per second, which, since English words have an average length of six letters, implies a reading speed of about 35 words per minute. In fact experienced readers can read at a rate of about 300 words per minute [5]. It can be concluded that in reading conventional prose the individual letters are not fixated. In which case how does the reader manage? The answer is first that some off-axis recognition is possible and second that an experienced reader has an expectation about which letters are likely to follow others, and about the structure of sentences. With such capability and knowledge it is possible for the reader to pick up information from only a limited number of points and still construct the meaning of the text. Further, because language is so strongly structured there are no points that are critical for reading. This means the reader does not always have to look at the beginning of the word or the end. Rather the location of fixation points can be anywhere within or between words (see Fig. 4.2). At each fixation point the observer can identify a few letters on either side and this is usually quite sufficient. But this should not be taken to mean that everybody will read in the same way. Figure 4.2 shows the differences in fixations made by two readers on the same passage. One reader had an extensive technical vocabulary and made few fixations. The other had a limited technical vocabulary and words such as 'chronology' and 'phenomena' required numerous fixations. Obviously the latter reader read the passage much more slowly that the former.

To summarise, the conventional reader operates by recognising groups of letters or words from the limited set that the message being read leads him to expect. As a general rule, the more familiar the material being read the fewer the clues he needs and hence the fewer the fixation points; the less familiar the material the more sampling the reader has to do and hence the more frequent the fixations. How the eye movements made during reading are controlled is still a topic of interest [6], but the question that arises here is what implications does this method of operation have for the importance of

'So I'm discovering,' I replied.   I felt full, relaxed,

secure and happier than I had been for a long time.   This

had been an apocalyptic day and the future looked promising.

Faustus crossed one leg over the other, tipped his chair

slightly and clasped his fingers behind his head

Fig. 4.3.   A passage in which one line has the bottom half missing. You should experience little difficulty in reading the passage.

visibility. The answer would seem to be that perfect visibility is not necessary. People can read even when some of the letters are missing or mis-shapen. Figure 4.3 demonstrates this. It shows a sentence where the bottom half of a line is missing. You should experience little difficulty in reading it. Thus reading is possible with poor quality material but there can be little doubt that it is a lot easier with good quality material. How the different characteristics of the material and the lighting influence reading is the topic of interest here.

Studies of the effects of different characteristics of print and of different lighting conditions on various aspects of reading have used a wide range of techniques [7, 8, 9]. To investigate the visibility of individual letters in different typefaces, Tinker used as a measure the extent to which they could be identified when presented for a very short time (usually less than 100 ms) [10]. Other people have measured the distance at which letters in different typefaces, of different sizes and contrasts, can be distinguished [11], and the extent to which they can be put out of focus and still be recognised [12]. These methods have been widely used but as should be apparent from the nature of reading considered above, the link between the visibility of individual letters and reading complete messages is by no means precise. A more relevant approach for examining the latter process is to measure the speed of reading of a set passage. An important feature of this approach is that it needs some form of control of comprehension. Reading at speed is easy if no understanding is being sought, but it is hardly reading at all. Paterson and Tinker [13] introduced comprehension into their experiments by asking the subject to read a short passage in which one word spoiled the meaning. The subject had simply to cross out the offending word. This gives some control of comprehension and enables speed of reading to be used as a simple measure of ease of reading, but it has disadvantages. Zachrisson [14] has pointed out that such a test puts the rapid skimmer of text at a disadvantage even though he may be equally as good as a more careful reader in comprehending the essential meaning. Poulton [15] has criticised the technique because it involves only a minimum level of comprehension. He has suggested that comprehension rate (a comprehension score, derived by asking questions on points of fact contained in the reading passage, divided by the time taken to read it) would be a much more sensitive and meaningful measure. Poulton [15] has also pointed out that where people are familiar with the material to be read or are searching for particular types of information, e.g. a telephone number, rapid skimming across the text is a more usual method of finding the information desired. For studying this sort of activity Poulton uses a technique whereby the reader is given a particular word to find, and to indicate he has found it he has to write down the word following it in the text. In considering the effects of various variables on reading, experiments using all these techniques will be examined. In evaluating the results it is important to remember the above limitations. One other point should also be appreciated. Virtually all the studies of reading performance have involved short reading periods, usually only a few minutes. Whether this short time limit is important is debatable.

4.2.3 Factors Affecting the Ease of Reading of Words
4.2.3.1 The Material
Many different features of printing can be expected to influence the ease of reading. For example, the typeface of the print, its quality, its contrast, its size and its layout are all likely to be important. Further, many of these factors are likely to interact. This means that the number of possible combinations to be studied is enormous. It is not a practical proposition to study all combinations, so the approach most widely used has been to examine each factor individually. Tinker [8] provides a good survey of existing research, and particularly of early work. Poulton [16] gives an interesting criticism of the same work.

The first aspects of printing to be considered are the typeface and its quality of reproduction. Spencer *et al.* [17] measured the number of specified words found within one minute by people reading passages of text. The passages were printed in four different typefaces with the printed image degraded in different ways. Figure 4.4 shows the four typefaces, printed at normal stroke thickness. These typefaces were chosen as being markedly different, but representative of those used in modern printing practice.

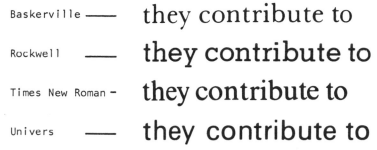

Fig. 4.4. Photographic enlargements of the typefaces used by Spencer *et al.* printed at normal stroke thickness [17].

Figure 4.5 shows the results obtained in terms of the number of words found in one minute of reading against the level of print quality. This latter measure is expressed in terms of the decrease or increase in stroke thickness from the normal stroke thickness for the 9 point† type used. Such changes in stroke thickness commonly occur when printing is copied. It can be seen that there is a marked decrease in reading performance for either a reduction or an increase in stroke thickness. Further, the extent of the decrease is different for different typefaces. At normal stroke widths there is little difference in reading for the different typefaces, but when the stroke width is reduced the Baskerville typeface becomes much more difficult to read than the others, probably because it has the narrowest stroke width under normal printing. When there is an increase in stroke width the Baskerville typeface is the most easily read and the Rockwell typeface is the most difficult, probably because of a tendency for the individual letters to join up and for the spaces in the letters to fill in. It can be concluded that for increasing or decreasing stroke width the different typefaces are likely to have different degrees of readability, but for good quality reproduction, i.e. normal stroke width, there is little to choose between them. This insensitivity to differences in typeface when reading good quality printing is

† A standard typographical measurement used to indicate body size of type, 1 point = 0·35 mm.

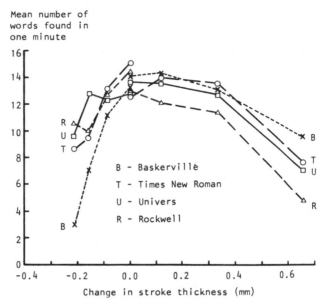

Fig. 4.5.   Mean number of words found in one minute's reading of passages printed in different typefaces, plotted against level of print quality, expressed as the increase or decrease in stroke thickness. (After [17].)

supported by other studies [8, 18, 19]. Therefore it seems likely that, for good quality printing, differences in typeface have to be radical before significant effects on reading occur.

One aspect of the print quality controlled in this investigation was the contrast. The investigation was done with high contrast printing. The question that needs to be considered now is the effect of different contrasts on ease of reading. Spencer et al. [20] measured the number of specified words found in one minute by people reading passages set in 9 point print but with different contrasts. Somewhat surprisingly they did not find any reduction in performance until the contrast of the print dropped below 0·1. Of course this is for good quality printing of reasonable size. The results discussed in Chapter 3 suggest that if smaller sizes had been used contrast would have been more important. Nonetheless these results do imply that slight variations of contrast are unimportant for reading until low contrasts are reached.

Another factor controlled in the above experiments was nominal print size. One feature of size that relates to typeface is the use of capital or lower case letters. Tinker and Paterson [21] measured the speed of reading passages set in the same typeface in 10 point print, but one passage was all in capitals the other was in lower case. Interestingly the passage in lower case print was read more quickly. This result has been confirmed by other experimenters [22, 23]. For example, Poulton found that teleprinter output in capital and lower case was comprehended significantly more rapidly than all upper case output [23]. At first these may seem unexpected results, because in Chapter 3 size of critical detail was shown to be an important factor in determining visual performance. However, letter size is not

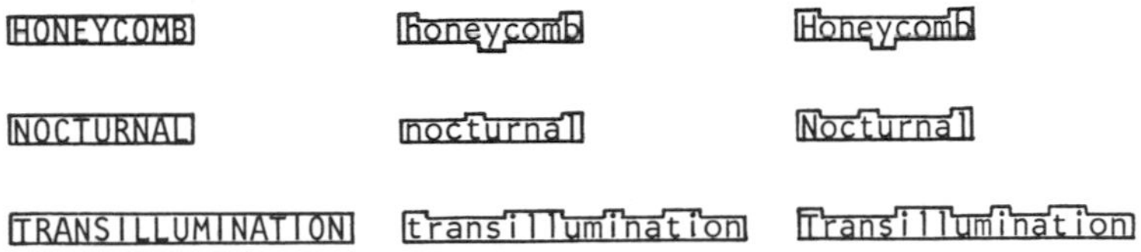

Fig. 4.6. Words in all capitals, all lower case and mixed. The 'signature' of each word is indicated by the line around the word.

everything. Passages printed all in capitals take up more space than when printed in lower case. This fact alone will tend to diminish the speed of reading of all capital print. Familiarity is also important. Readers are more familiar with text in lower case than in all capitals. Further, they may use the capital letters at the start of sentences in a passage set in lower case as clues to the structure of the passage. Finally, it is important to note that words printed in lower case and capitals, and in lower case only, have a characteristic form or 'signature'. Figure 4.6 shows a number of words printed in capital and lower case, combined and separately. The utility of letters coming above and below the 'average' line to form a unique shape to the word is obvious, as is the lack of 'signature' when the word is in capital letters only. It is interesting to note that this 'signature' is the same for various typefaces. This may go some way to explain the apparent insensitivity to different typefaces when reading good quality print.

All this is interesting but what about the effects of size as such? Figure 4.7 shows the same passage printed in different type sizes. Using the same lower case typeface, Paterson and Tinker [24] measured the speed of reading with different sizes of print. They found that 10 point size was the optimum with smaller or larger print being read significantly more slowly. However, in a later experiment [13] they found that 11 point type was read significantly more quickly than 10 point. Poulton found that skimming of newspaper text was carried out more quickly for a 9 point text than an 8 point one [25]. The implication would seem to be that the optimum print size is somewhere

4 pt.

At one time it was something of a status symbol for a city to have more than one evening newspaper. Now London is the only city south of the border to
have more than one.

6 pt.

At one time it was something of a status symbol for a city to have more than one evening newspaper. Now London is the only
city south of the border to have more than one.

8 pt.

At one time it was something of a status symbol for a city to have more than one evening
newspaper. Now London is the only city south of the border to have more than one.

10 pt.

At one time it was something of a status symbol for a city to have more than
one evening newspaper. Now London is the only city south of the border to
have more than one.

12 pt.

At one time it was something of a status symbol for a city to
have more than one evening newspaper. Now London is the
only city south of the border to have more than one.

14 pt.

At one time it was something of a status symbol for a
city to have more than one evening newspaper. Now
London is the only city south of the border to have
more than one.

Fig. 4.7.   The same passage presented in the same typeface in different sizes, ranging from
4 point to 14 point in 2 point steps.

around 10 point, but that it is a rather flat optimum. Nonetheless there is
little doubt that small print (less than 6 point) [24, 26] and large print
(greater than 14 point) [24] is read more slowly than 10 point. Certainly the
pattern of eye movements found with 6 point and 14 point print contains
more fixations and more reversals than are found when reading 10 point
print [27]. A plausible explanation for the existence of this optimum size is
that for the very small print there is difficulty in distinguishing the form of
the words. For the large print it is likely that the individual words cover such

a large area that the complete form of the words cannot be seen at one fixation.

Two other factors that affect speed of reading of print are the length of line of print and the separation between successive lines. Paterson and Tinker examined the effect of line length and spacing between lines for different sizes of type from 6 to 12 point [13]. For all type sizes a 2 point separation between lines was best for reading. With this spacing the optimum line length covered a wide range, the only consistent pattern being that very short lines, less than 14 pica,† were undesirable. From this information Tinker [8] was able to specify a safety zone consisting of limits of spacing of lines and length of lines for different sizes of type. As would be expected these limits were fairly broad. For example, for 10 point type, lines from about 6–13 cm with 2 point spacing would be satisfactory.

The type of results discussed above have been obtained at various times over the last 60 years and can be said to be well established. However, in recent years there has been a shift of interest away from the details of the printing as such, to a more holistic approach. Poulton [28] examined the effect of typeface, size and format of print on the comprehension of scientific journals. Gregory and Poulton [29] studied the advantages of 'justification' of print, i.e. making each line the same length by varying the separation between letters or breaking words. They found that unjustified lines were easier for poor readers. The final stage of this process is illustrated by a paper on the coding of lists and bibliographies [30]. This attempts to identify the optimum way of writing bibliographies from 10 alternative forms, by measuring the speed of finding specific information from the list. The experiment is concerned not with speed of reading in general, but rather with finding information quickly for the specific task of reading through lists. The most important features for rapid search were those that made a clear distinction between successive entries and between the first word of each entry and the rest of the entry.

To summarise, the effects of many different features of print on the speed of reading have been extensively studied and the resulting recommendations can be said to be well established by practice. Presently, interest tends to be concentrated on the layout of information rather than the details of the print in which it is presented. There are still arguments about whether single column or double column presentation is better [31] and how to present tables in text [32], but the general impression given is one of completeness. Spencer [33] provides a useful summary of knowledge in this

---

† Pica is another typographical measure, 1 pica = 4·23 mm.

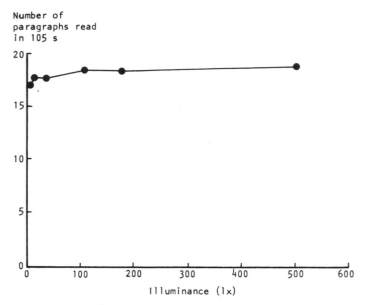

Fig. 4.8.   Mean number of 30-word paragraphs read correctly within 105 s, plotted against illuminance. (After [35].)

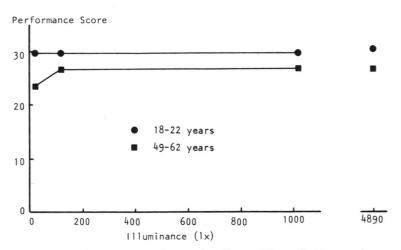

Fig. 4.9.   Mean performance scores for proof reading at different illuminances, by young (18–22 years) and older (49–62 years) subjects. The performance score was derived from the accuracy of proof reading with a bonus for speed. The best possible performance score was 35. (After [36].)

field, covering several points of detail, c.g. margins, page sizes, boldness of print, etc., which for space reasons have not been included here. Another publication of practical value is a case study of the design of the journal *Applied Ergonomics* [34]. The summary table given in that paper could be used as the basis for a checklist for typographic design.

4 2.3.2 The Lighting

It is interesting to contemplate that many of the studies of the way in which various features of print affect the ease of reading contain no mention of the lighting conditions provided. This is in spite of the lip service often paid to the importance of lighting. However, Tinker did examine the effect of illuminance on the speed of reading high contrast, 10 point print [35]. The results are shown in Fig. 4.8. It can be seen that there is little improvement above 30–100 lx. These results were obtained with young subjects reading for short periods. Figure 4.9 shows some results for people of different ages proof reading good quality mimeographs of typed material [36]. Only the older age group shows a marked improvement in performance with increased illuminance, and this ceases at 120 lx. Although this is a difference between age groups, the quality of the material is important for both age groups. Besides the good mimeographed material, similar passages were

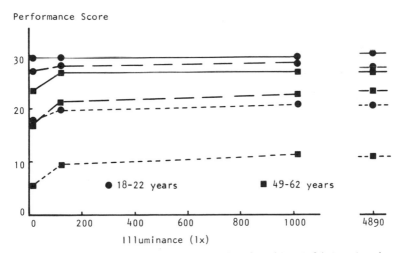

Fig. 4.10. Mean performance scores for proof reading of good (——), fair (— —), and poor (— — — —) quality material at different illuminances by young (18–22 years) and older (49–62 years) subjects. The performance scores were derived from the accuracy of proof reading with a bonus for speed. The best possible performance score was 35. (After [36].)

proof read from moderate and poor quality mimeographs. The results are shown in Fig. 4.10. There is a marked reduction in performance with the decrease in quality for both age groups, but the improvement in performance with increased illuminance is greater for the poor quality materials. It should be noted that this is a particularly strict test of the effect of illuminance on reading performance. It uses a task, proof reading, which

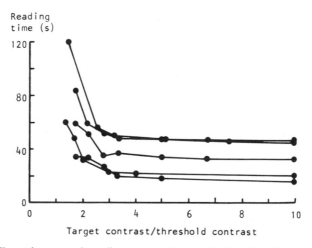

Fig. 4.11.   Time taken to read two-line passages from an Italian scientific journal, plotted against the ratio, target contrast/threshold contrast, for five readers. (After [37].)

calls for much closer inspection of individual letters than usual. Thus the results probably represent a very sensitive measurement of the effect of illuminance. It seems likely, and the evidence in Fig. 4.8 confirms it, that people are less sensitive to illuminance for conventional reading of print of good quality.

Although illuminance has been the most widely examined lighting variable there are other important factors. One of these is the extent of veiling reflections. When veiling reflections occur their effect on most print is to reduce the contrast. Ronchi and Neri [37] measured the speed of reading of a passage of Italian from a scientific journal for different contrasts of the material, at a constant adaptation luminance of 50 cd m^{-2}. The reduction in contrast was achieved by placing a luminous veil across the task. The times taken to read the passages for the five subjects used are shown in Fig. 4.11 plotted against the ratio of target contrast to threshold contrast, a measure similar to Visibility Level. The results show a marked, non-linear increase in the reading time with decreasing contrast ratio, i.e. with reduced task contrast. Similar effects of veiling reflections have been reported by De Boer [38] but this' was for viewing a Landolt ring chart rather than reading script.

All the above studies examined the effect of reduced contrast where it was unavoidable. Unfortunately for simplicity, another study involving extracting information from a telephone directory in the presence of different amounts of veiling reflections [39] failed to show any effect of veiling reflections on task performance. This can be ascribed to the ability of the subject to manipulate the position of the material so as to move the veiling reflections away from where she wished to read.

In a similar study Hopkinson [40] reports some observations of people attempting to read from a book placed on a polished table. On either side of the book were reflected images of the incandescent light sources used in lighting the room. Whilst this did not affect the performance measured over a short period, the subjects considered the reflected images most distracting and irritating, which presumably could affect performance.

Another aspect of lighting that might influence reading is the colour properties of the light source. However, there is little evidence for this, certainly where the material to be read is achromatic, i.e. the usual black print on white paper. Smith and Rea [41] measured the time taken for people to do a numerical verification task, in which the subject had to compare one list of figures with another and detect any differences. The task was done under fluorescent, high pressure sodium and high pressure mercury discharge lamps. The expected increase in performance with increase in illuminance was found but there was no significant difference between the performance for the different light sources.

Finally, it is worth mentioning the possibility that lamp flicker might affect ease of reading. Poulton *et al.* [42] measured the comprehension achieved after reading a passage for a fixed time under flickering illumination. No significant effects of lamp flicker on comprehension rate were found.

To summarise, the lighting variables shown to affect the reading of conventional black-on-white printed material are those expected: illuminance and veiling reflections. The actual values chosen for the illuminance necessary for reading are very much a matter of judgement, involving such factors as the form of material to be read, the level and ease of performance required, the duration of the work involved and the age of the people concerned. There is less equivocation over veiling reflections. The presence of strong veiling reflections over the task will reduce the contrast of the material, which may affect the level of performance. Even if specular reflections do not cover the task but are adjacent to it, they can be distracting and may affect performance indirectly. Veiling reflections should be avoided.

4.2.4 Other Aspects Involved in Reading

The vast majority of reading work is done with black print on white paper in the form of books or papers, with the information conveyed in words. But there are other ways of presenting and obtaining information which could still be called reading. For example, the information can be given on coloured papers and/or in coloured inks. Information can be expressed in letters or numbers. Information can be presented on visual display units. The way in which these departures from conventional reading material affect ease of reading will now be considered briefly.

4.2.4.1 Colour Print and Papers

For black and white printing the contrast of the ink on the paper is simply a matter of the luminances of the ink and the paper. When coloured inks and/or paper are used, colour contrast is also involved. Coloured ink and/or paper is frequently used in advertising material. Tinker and Paterson [43] measured the speed of reading of passages printed in one of 10 different colour combinations in 10 point print. The number of paragraphs read in a set time for each of the different colour combinations was compared with the number read by the same people for the same size black print on white paper. The results are given in Table 4.1. None of the different colour combinations allowed faster reading than black on white.

Table 4.1: Numbers of paragraphs read for different combinations of coloured ink and paper [43]

Colour of ink and paper	Mean number of paragraphs read in 105 s	Mean number of paragraphs read in 105 s for black ink on white paper
Green on white	16·75	17·26
Blue on white	16·06	16·61
Black on yellow	15·77	16·37
Red on yellow	16·31	17·09
Red on white	16·20	17·65
Green on red	16·20	17·91
Orange on black	14·93	16·94
Orange on white	13·75	16·62
Red on green	12·46	17·38
Black on purple	11·00	16·66

The results are explicable in terms of luminance contrast alone. The maximum luminance contrast was achieved by the black on white print and the least was by the black on purple print. To what extent this dominance of luminance contrast would have been reduced if specific colour combinations had been chosen which lay on opposite sides of the CIE 1976 uniform chromaticity scale diagram is unknown, but it does not seem unreasonable that colour would then have had a more marked effect. Nonetheless two conclusions are clear. First that print that has a high luminance contrast is likely to be more easily read than coloured print. Second that simply using colours without considering the luminance contrasts and colour contrasts they might produce can make reading very difficult. As for the effect of light sources with different colour properties, this seems likely to be a minor factor, at least for the range of nominally 'white' light sources used in interiors. There is no doubt that the use of sources with good colour properties can increase the colour contrast presented but this is unlikely to be important unless luminance contrast is low.

4.2.4.2 Black on White or White on Black
Since luminance contrast is a factor in making print easy to read, it is worth while considering if there is any difference between positive and negative contrast as regards reading. Positive contrast is when the print is lighter than the paper, negative contrast is when the print is darker than the paper. From threshold contrast measurements there is evidence to suggest that the negative contrast is more easily detected, at least at low luminances [44]. Of more relevance is the study of the speed of reading of white and black print [45]. White print on black paper (positive contrast) was read about 10% slower than black print on white paper (negative contrast). Against this may be put the failure to find any significant difference between positive and negative contrast materials as they affected the speed of searching for specified words in texts [20]. Such results suggest that any differences between the effect of positive and negative contrast on ease of reading is likely to be small.

4.2.4.3 Words and Numbers
It was emphasised in the discussion on the nature of reading that the information provided by printed words is highly redundant. It is not necessary to discriminate the individual letters to read, indeed printing can be read when half of each letter is missing provided the basic outline of the word is retained. This is because the arrangement of letters in a given

language tends to follow a few patterns; they are certainly not random. This is not true for numbers. There is rarely any way in which the reader can anticipate which number will follow another when reading a bank statement, for example. This implies that for numbers each individual digit will usually have to be identified. Eye movement recordings made while reading numbers and words support this view [46]. Less than half the number of digits were perceived during a fixation pause as were perceived for letters and words, and the average fixation pause for numbers was 40 % longer than for words. This pattern of eye movements implies that speed of reading for numbers will usually be much less than for words. It also suggests that the quality of the material and the lighting will be more important for numbers than for words. In a sense we are back to the old problem of the visual component of the task. For reading words, the visual component is small because the need to discriminate detail is small. For reading numbers, or any other material where the usual linkage between successive letters is broken, lighting will be more important because the need to discriminate detail is greater.

4.2.4.4 Visual Display Units

In recent years information in offices has been increasingly presented on visual display units. In spite of the failure to find any signs of fatigue after six hours of reading in tests with projected microfilm [47], there are many reports of eyestrain, headaches, etc., associated with the use of visual display units [48]. The problems of legibility with visual display units are basically those of conventional reading, but in a more extreme form. The characters shown are frequently more blurred, with the letters poorly defined, and the screen sometimes tends to shimmer if not flicker. Further, the information presented can be moving up the screen as it is being read, is often numerical rather than verbal, and frequently has to be considered as quickly as possible. This last might suggest that most work involving reading from visual display units is stressful, even before considering the details of the way the information is presented. To add to these complications there is the fact that many installations of visual display units are poorly integrated with the rest of the room lighting. Hultgren and Knave [49] report a study of an office where people had complained about the visual display units. They found that much of the trouble was due to an excessive difference between the dark screen of the display unit and the brightness of the rest of the room, since this produced discomfort glare, and reflections of high luminances in the room on the surface of the screen, i.e. a form of veiling reflection. In addition to reducing the contrast of the material to be read, such reflections can mislead the system into focusing at the wrong distance. Guidance on how to avoid such difficulties is available [50].

4.2.5 Conclusions

The study of the relationship between printed material and ease of reading has a long history. Studies of the effects of different typefaces, sizes and layouts have led to general agreement as to the most suitable characteristics for printing when ease of reading is important. Research is no longer concerned with the effects of these components of printing but rather with identifying the best layout of information for specific activities. For example, recent work includes studies of the value of even or uneven right-hand margins for comprehension in reading [29], the effect of positioning tables in text in different ways [32], and the important features of lists and bibliographies that enable the reader to find the desired information quickly [30]. However, there are some components of printing which have not received the necessary attention. Print quality is one of these. Blurring and ghosting of print makes reading difficult, and yet such effects are only just beginning to be studied. A related area which deserves attention is the use of visual display units. These units are becoming increasingly common as an alternative to the printed word, and yet frequently the quality of the writing displayed on the screen is poor. The current problems are those of poor definition, instability of the image, and interaction between the lighting of the room and the display. Studies of the influence of these factors on the ease of reading of visual display units would be valuable.

As for lighting, the position is rather unsatisfactory. Illuminance and veiling reflections have been shown to influence ease of reading but the evidence is largely restricted to good quality, printed material. It would be interesting to examine the effect of lighting on the ease of reading of various forms of poor quality material. For such material lighting would be expected to play a more dominant role than it apparently does for good quality material read by people with good vision.

Finally, it is worth noting one aspect of these studies of reading which causes considerable unease. This is the fact that most of the studies of the effect of different conditions have involved very short work periods. The question is whether the recommendations derived from these studies can be applied to much longer periods of reading. There are two experiments which have failed to show any detrimental effects on speed of reading for long periods [47, 51], but these were using reasonable lighting conditions and materials produced according to good practice. This suggests that if the recommendations derived from short period studies are followed then reading can be done for long periods without trouble. This conclusion leaves open the possibility that small departures from optimum conditions are more important for long durations than for short periods. The importance of deviations from optimum conditions, particularly as regards the possibility of interactions between variables, could usefully be considered for visual display units.

4.3   COLOUR JUDGEMENT

4.3.1 Introduction
The perception of colour is a vital element in many industrial and commercial activities. Apart from the obvious examples of paint and dye production, colour is often important in general manufacture. It is quite common for a product which is designed to appear as a unity to be made from a number of components. The colours of all the components have to match within specified tolerances. Other situations where colour is important occur in shops and hospitals. The freshness of many foodstuffs, such as meat and fish, is at least partly judged by their colour. Similarly, variations in skin colour are often good indicators of a person's state of health.

Before considering the way in which lighting influences the judgement of colours, it is necessary to point out that not all colour judgement work is the same. Two distinct functional aspects of such work can be identified: accuracy and discrimination. Accuracy in colour judgement work involves the appearance of a colour relative to a standard. Discrimination does not demand a standard colour but only the ability to distinguish between colours. Thus discrimination requires only that we are able to separate colours. It is not concerned with the appearance of the colours. Although these two aspects of colour judgement work can be considered separately, it should be realised that most practical activities require both. For example, in manufacture, the lighting should be such that it enables small differences in the colours of components to be discriminated but it should also ensure that the appearance of the whole is what the designer of the product intended.

It is also worth while emphasising the difference between relative and absolute colour judgements. Relative colour judgement occurs where there is available a surface which defines the desired colour, so side-by-side comparisons are possible. This is the situation in much industrial colour judgement work. Absolute colour judgement occurs when the comparison of the object being inspected is with some remembered criterion so a side-by-side comparison is not possible. This is typical of the judgements people make about foodstuffs. The important difference between these two cases lies in the precision of judgement that can be achieved. Side-by-side comparisons allow much finer judgements to be made.

This section will examine the effects of lighting on colour judgement within the framework of the recommendations for lighting such work. It is important to emphasise that only discrimination and accuracy will be considered. The preferred appearance of colours and even preferred colours are not usually involved in colour judgement work and will not be considered here.

4.3.2 Factors Influencing Colour Judgement
The perceived colour of a surface is determined by three factors (a) the light source used to illuminate it, (b) the size and reflectance characteristics of the surface and the nature of its surroundings and (c) the state and organisation of the visual system used to observe it. In simple physical terms the colour of a surface is determined by the spectral composition of the light reflected from the surface. This is known as the spectral emission of the surface, and can be defined as the product of the spectral composition of the light incident on the surface and the spectral reflectance of the surface. This is satisfactory for the physics of it, but for evaluating the effect of lighting on perceived colour it is necessary to consider the visual system as well. This can be done by introducing the CIE Colorimetric System [52], which assumes a standard observer with defined spectral sensitivities (see Chapter 1). This system can be used to locate a colour relative to other colours, either in a colour space [53] or on a chromaticity diagram [52], and by doing this give some indication of the appearance of the colour.

Although the CIE Colorimetric System covers many of the factors that influence perceived colour it does not involve them all. The size of the object is important as is the size, colour and luminance of the surround against which it is seen. Further, the fact that some people have spectral sensitivities different from those of the standard observer is ignored. Thus there can be a marked difference between the measurement of a colour by the CIE Colorimetric System and the perception of that colour by a particular individual. Colour measurement is undertaken to ensure that defined colours can be reproduced and under specified conditions will look the same. In spite of the developments in recent years [54], the information the CIE Colorimetric System gives about the appearance of a colour is still rather imprecise. Although this situation is theoretically less than ideal the CIE Colorimetric System has been found to work reasonably well in practice. Virtually all studies of the effects of lighting conditions on colour judgements have used some form of the CIE Colorimetric System to describe the effect of lighting on the colours. This is apparent in the recommendations that exist for lighting work requiring colour judgement. For example, quantitative recommendations defining light sources suitable for colour judgement work in the graphic arts are given in terms of the CIE Colorimetric System [55]. Some other recommendations, e.g. task illuminance, are also quantified but much advice is given in qualitative form. The rest of this section will be concerned with the factors that affect the performance of colour judgement work and the nature of their effects.

4.3.3 Spectral Emission of Light Sources

Light sources, both natural and artificial, have different spectral emissions (see Chapter 1). In that one of the main factors determining the colour of a surface is the spectral composition of the incident light, different light sources may cause a surface to appear to be different colours. Hence different light sources can render colours accurately or inaccurately, and facilitate or hinder discrimination of colour differences. A useful measure to distinguish between light sources as regards their effects on colour judgement work is the colour gamut (see Chapter 1). The basis of this measure is to take a number of different surfaces with different spectral reflectances, then to calculate the chromaticity co-ordinates of the colours and to plot them on an appropriate uniform chromaticity scale diagram [56]. At present the most appropriate diagram is the CIE 1976 UCS diagram. This is approximately perceptually uniform, so the separation of the points on the diagram should be proportional to the ease of discrimination of differences between the colours, i.e. the further apart they are on the diagram the more easily they can be distinguished. The colour gamut is produced by joining up the points on the CIE 1976 UCS diagram. Obviously the colour gamut produced depends on the surfaces used. Those conventionally used are the eight CIE test colours [57] developed for calculating the CIE Colour Rendering Index (of which more later), but there is no reason why other colours should not be substituted or added provided the colours used are specified and are consistent between comparisons. The important point about the colour gamut of a specific light source is that, for a given set of known surfaces, the size and shape tell us a great deal about the precision of colour discrimination the light source allows, and in which hue regions it is particularly good or bad. Figure 4.12

shows the colour gamut calculated for the CIE test colours for incandescent and fluorescent lamps. An inspection of Fig. 4.12 will reveal that the incandescent and the conventional high efficacy fluorescent lamp (White) both allow a reasonable degree of colour discrimination, but with different sensitivities in different hue regions. However, the biggest colour gamut occurs under the Northlight lamp which is designed for use where precise colour judgements are required. This is what would be expected, but it is comforting to find a measure which reflects the differences beween light sources. Incidentally the same approach shows that the tinted glasses sometimes used in windows transmit light which has a much better colour gamut than most conventional fluorescent lamps [58]. This suggests that colour discrimination which uses daylight is not likely to be too affected by the presence of tinted glass in the windows.

So far the only aspect of the colour gamut that has been considered is how it relates to the discrimination of colours. This ability is simply described by the separation of points on the CIE 1976 UCS diagram. However, by introducing a desired appearance for surface colours by using specified chromaticity co-ordinates on the CIE 1976 UCS diagram to form a standard gamut, the colour gamut can be used to assess the accuracy with which colours are rendered by the light source. The departure of the chromaticity co-ordinates of the surfaces under the lamp being studied from their standard positions gives a measure of this accuracy. Figure 4.13 shows such a plot for the White and Northlight fluorescent lamps with respect to the CIE standard illuminant D65.† Figure 4.13 demonstrates the difference between accuracy and discrimination. The colour gamut for the White fluorescent lamp shows that whilst discrimination is possible the colours are distorted relative to D65, but the Northlight fluorescent lamp provides good discrimination and accurate reproduction relative to D65.

The colour gamut approach is a very descriptive way of assessing the effect of a light source on the colours of a surface. It would be nice if it were used in practice. Unfortunately it is a common belief that practical men always want a single number measure. To meet this requirement the colour gamut has been reduced to the area enclosed by the line joining the points on the CIE 1976 UCS diagram. This area is called the gamut area [56]. Further, when a particular light source is taken as a standard, the relative gamut area can be used as a colour discrimination index [59]. Using a gamut area may have advantages in practice but it inevitably involves

---

† A defined relative spectral power distribution representing daylight with a correlated colour temperature of 6504 K.

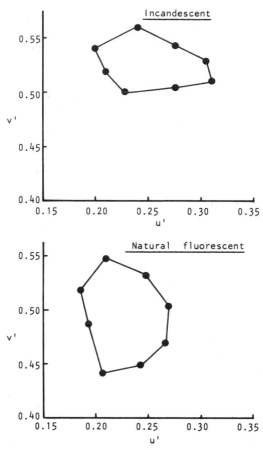

Fig. 4.12.   Colour gamuts on the CIE 1976 UCS diagram for an incandescent lamp (2856 K)
and for White, Natural and Northlight fluorescent lamps.

throwing away the valuable information contained in the shape of the
colour gamut. The loss of information is shown in Fig. 4.12, where the
flattened colour gamut under an incandescent lamp suggests that
discrimination would be poor between yellows and blues but good between
reds and greens, whilst the opposite is indicated by the colour gamut for the
white fluorescent lamp. These very different light sources have similar
gamut areas.

It should be noted that there have been a number of objections to the use
of colour gamut [58]. The two main objections are that sensitivity to colour
differences changes with the state of adaptation of the observer, and that the

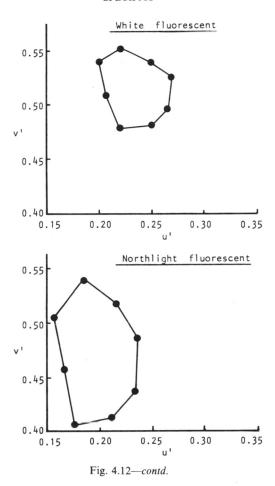

Fig. 4.12—*contd.*

CIE 1976 UCS diagram is not perceptually uniform. However, Pointer [60] has shown that changes in just-noticeable colour differences are small for adaptation to a wide range of colour temperatures. As for the uniformity argument, the CIE 1976 UCS diagram may not be completely uniform but it is the best there is. In principle, any other, more perceptually uniform, colour diagram could be used to define colour gamut, but those which are available are not easy to use. One remaining objection is that the colour gamut only tells us about the differences between colours and will not reveal when colours have been distorted. This is undoubtedly true when the colour gamut is used alone but, as discussed above, when a standard gamut is

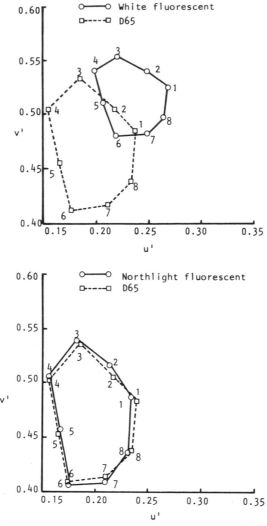

Fig. 4.13.   Positions of the eight CIE test colours on the CIE 1976 UCS diagram when lit by White and Northlight fluorescent lamps and by CIE standard illuminant D65.

introduced accuracy can be described as well. Thus it would appear that colour gamut is a very suitable measure for quantifying the accuracy with which any specific lamp renders colours. However, a different quantity is commonly used: the CIE general Colour Rendering Index [57] (see Chapter 1).

The general Colour Rendering Index is obtained by calculating the co-ordinates in colour space of the eight CIE test colours under the light source of interest, i.e. the test light source, and under a reference light source. The difference between the positions for each test colour for the test and reference light sources is then obtained and scaled, so that perfect agreement between the positions gives a value of 100. This value is called the special Colour Rendering Index for each test colour. For the general Colour Rendering Index the eight special Colour Rendering Indices for the eight test colours are averaged.

Table 4.2: General Colour Rendering Indices for different light sources

Light source	General Colour Rendering Index (Temperature indicates the colour temperature of the reference light source)
High pressure sodium discharge (SON)	29 at 2 000 K
High pressure mercury discharge (MBF)	36 at 4 100 K
White fluorescent (MCF)	56 at 3 450 K
High pressure mercury discharge with metal halide additives (MBIF)	78 at 4 200 K
Tri-band fluorescent (MCF)	83 at 4 100 K
Natural fluorescent (MCF)	84 at 4 000 K
Artificial Daylight fluorescent (MCF)	93 at 6 500 K
Northlight fluorescent (MCF)	93 at 6 750 K
Incandescent (2 700 K)	99 at 2 700 K

Table 4.2 gives the general Colour Rendering Indices (CRI) for some widely used lamps. The general CRI produces a limited amount of information on the accuracy with which colours are reproduced. By averaging the differences across test colours the pattern of discrepancies for different hues is lost. Further, no account is taken of the direction of the differences between the positions for the reference and test light sources. A difference in one direction and a difference in the opposite direction can both give the same value of CRI. Similar objections could be made about the use of the colour discrimination index and gamut area. The basic problem for all these measures is that information is inevitably lost in trying to describe a phenomenon as detailed as relative colour appearance by a single number. The practical problem is to balance the advantages of a single number index, in terms of acceptability, against the gain in information that may be achieved by a more complicated presentation. For the ultimate in precision in assessing the effect of different light sources, the

CIE colour difference equations in colour space [53] can be used. These make it possible to examine the effect of a light source for either discrimination or accuracy but only when the exact colours to be examined can be specified. As an intermediate stage special Colour Rendering Indices for generally important colours such as skin, are sometimes used [57].

However, lighting recommendations for general use have to be applicable to a wide range of conditions, which includes a wide range of unspecified colours. For this reason the suitability of a light source for colour judgement work is almost invariably described in terms of the general CRI. It is therefore of interest to examine how well changes in CRI relate to the performance of colour judgement work. One attempt to do this has been evolved using the Farnsworth–Munsell 100 Hue Test [61]. This test consists of four boxes of coloured discs, successive discs differing from their neighbours in hue but all being of approximately the same perceived chroma and lightness. The discs in each box are presented to the subject in a random order. The subject is then asked to re-order the discs until they show a steady progress around the hue circle. In all, there are 85 discs and departures from the correct order are scored in a way that takes account of the number of errors and the extent of the misplacement. In the experiment to be discussed, the 100 hue test was done under a number of different light sources each producing the same illuminance on the test [62]. The mean errors plotted against CIE general Colour Rendering Index (CRI) and against gamut area calculated for the CIE test colours are shown in Fig. 4.14. The results indicated by a cross were obtained at a different laboratory from those indicated by a square, using different sets of discs and different people. It can be seen that although the absolute level of mean error score is higher for one laboratory than for the other the slope of errors made against CRI or gamut area is similar in both. The important thing to note in Fig. 4.14 is that there is a marked reduction in errors made in ordering the set of discs as the CRI or the gamut area increases. Further, the results for the different lamps do show an expected pattern. The highest mean errors occur for the high pressure sodium discharge lamp and the fewest errors for the Artificial Daylight fluorescent lamp, a lamp specifically designed for colour judgement work [63]. These results confirm that different commercially available light sources do influence the accuracy with which colours can be judged. They also show that there is little difference between CRI and gamut area in predicting the level of errors found. This lack of different occurs because, for the lamps studied, there is strong correlation between CRI and gamut area (0·92). As an indication of the different conditions provided by the lamps, Fig. 4.15 shows the

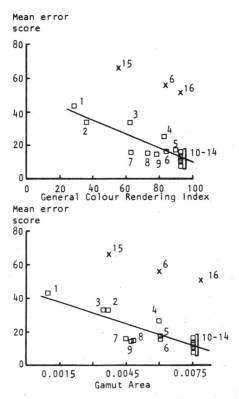

Fig. 4.14.   Mean error scores on the Farnsworth–Munsell 100 hue test for different light sources plotted against the general Colour Rendering Index and the gamut area (after [62]). Light sources: 1, high pressure sodium discharge; 2, high pressure mercury discharge; 3, Homelite fluorescent; 4, Tri-band fluorescent; 5, Kolor-rite fluorescent; 6, Natural fluorescent; 7, Daylight fluorescent; 8, Plus-White fluorescent; 9, high pressure mercury discharge with metal halide; 10–14, Artificial Daylight fluorescent; 15, White fluorescent; 16, Northlight fluorescent.

positions of the 85 disc colours on the CIE 1976 UCS diagram when illuminated by a high pressure mercury discharge lamp and by a Natural fluorescent lamp. The generally closer spacing of the discs under the high pressure mercury discharge lamp, and hence the increased probability of errors in ordering, is obvious.

It can be concluded that, as a general rule, the higher the CRI and the larger the gamut area the better will be the colour judgements made. It must be emphasised that this is a general rule. For specific activities rather different advice may be given. This is neatly exemplified in a study of the

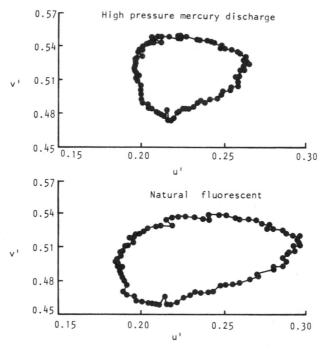

Fig. 4.15. Positions on the CIE 1976 UCS diagram of the 85 colours in the Farnsworth–Munsell 100 hue test when lit by high pressure mercury discharge and Natural fluorescent lamps. (After [62].)

lighting required for hospitals [64]. In this work a variety of hospital staff made a number of colour judgements representative of those made in hospitals. The test material included dermatological lesions, examples of cyanosis, different blood samples, bacteriological cultures and cases of early jaundice. Colour judgements of these samples and cases were made under a number of different fluorescent lamps. The procedure used was a paired comparison in which an observer compared the effect of each light source with the effect of the light source presented immediately before. The decision required was whether the light source presented was better or worse for making the colour judgements than the light source which immediately preceded it. All possible pairs of lamps were examined in this way, the final score for each lamp being based on the number of times each lamp was preferred. The illumination of the tasks under all the light sources was 260 lx. The interesting result for the present discussion was that the rank ordering of the different light sources was the same for all the colour judgements made except for the signs of early jaundice. As might be expected the rank orderings of light sources for all the colour judgements except jaundice follow the change in Colour Rendering Index—the higher the CRI the better the light source for making the colour judgements.†

For the detection of early jaundice a rather different pattern was found. In this case a light source with a low CRI was preferred over one with a high CRI value. The reason for this is that the detection of early jaundice depends on the recognition of a yellowness, which also means an absence of blueness. To provide for this the light source must contain a significant quantity of blue radiation and little red radiation. This requirement is hardly a recipe for good colour judgement for all hues, which suggests that distortion of the skin colours aids the identification of jaundice. All the other medical judgements involved a range of colours, which presumably is why the general rule about light sources with high CRI values being best for colour judgement work applies.

It can be concluded that the spectral emission of a light source does influence the performance of colour judgement work. There exist a number of measures which can be used to quantify the effect of different spectral emissions on the colours of surfaces, the measures varying in the amount and type of information they give. The most general recommendation for lighting for colour judgement work is to use a light source with a high CIE general Colour Rendering Index [55, 67]. This is good general advice. It applies if accurate reproduction of colours and good discrimination between colours is required, provided always that the standard with which the light source of interest is to be compared is one of those recommended for use with the CIE colour rendering system [57]. Since these standards cover a wide range of colour temperatures, and have a continuous spectral distribution, most of the situations of practical concern are dealt with. If it is only discrimination which is important then gamut area can be used as a guide to suitable light sources. These measures are satisfactory as general guides, but if only a few specific colours are of interest then it is worthwhile using the CIE colour difference equations [53] to investigate the effects of different light sources.

4.3.4 Illuminance

Another factor which is included in the recommendations on lighting for

---

† In fact this work uses the NPL–Crawford system of colour rendering specification [65]. This is based on combining the differences in spectral emission in six wavelength regions, between a reference light source and a test light source. However, both the NPL–Crawford system and the CIE Colour Rendering Index rank light sources in the same order [66].

good colour judgement is the illuminance on the task [55, 63, 67]. Obviously there must be a minimum illuminance, because below adaptation luminances of about $3 \, \text{cd} \, \text{m}^{-2}$ colour vision starts to fail as the rod photoreceptors take over from the cone photoreceptors. It is also necessary to consider the Bezold–Brucke effect, as this demonstrates that increasing adaptation luminance changes the perception of hue. The exact nature of the change is that with increasing luminance reds appear yellower and blue greens appear bluer. There is therefore a need for a recommended luminance for colour judgement work so as to avoid the differences in colour judgement caused by large variations in adaptation luminance. The recommendations are given in the form of illuminances, the values recommended being much higher than those necessary to avoid entering the mesopic region of adaptation. The American National Standards Institute [55] recommends 2200 lx. British Standard 950 [63] recommends about 2200 lx. The IES Code [67] suggests 1000 lx as being desirable. Unfortunately, there is little support for these values. Bedford and Wyszecki [68] measured just noticeable differences in wavelength for two 12 min arc discs with centres separated by 24 min arc. Figure 4.16 shows the results obtained for two observers over a range of wavelengths at different retinal illuminations. It is apparent that wavelength discrimination is markedly worse at the extremes of the visible spectrum; but the important point here is that there are no significant differences in wavelength discrimination for the different retinal illuminations. It was only when the stimulus field size was very small (two 1·5 min arc discs, separated by 40 min arc) that any significant reduction in wavelength discrimination with decreased retinal illumination occurred.

Brown [69] measured the effect of adaptation luminance on sensitivity to colour differences by asking two observers to match the two parts of a bipartite circular field subtending 2°. He expressed the accuracy of these matches in terms of ellipses about the chromaticity co-ordinates of the test colour on the CIE 1931 chromaticity diagram. The conclusion reached was that sensitivity to differences in chromaticity is constant until the adaptation luminance drops below about $3 \, \text{cd} \, \text{m}^{-2}$.

These studies suggest that colour discrimination is not strongly dependent on adaptation luminance once photopic vision is operating fully. However, both these studies involved a few observers making careful judgements using a colorimeter, which is rather a long way from real life. One step closer is a Farnsworth–Munsell 100 hue test being done by many observers. Cornu and Harlay [70] used this test to examine the effect of illuminance on the errors made. The mean error scores for the different hue

Just noticeable
difference
in wavelength (nm)

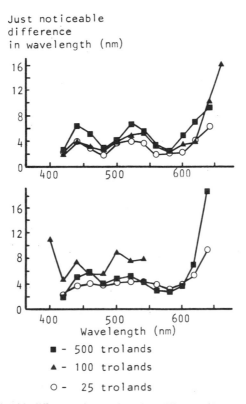

■ - 500 trolands

▲ - 100 trolands

O - 25 trolands

Fig. 4.16. Just noticeable differences in wavelength at different retinal illuminations for two observers comparing monochromatic wavelengths presented in two 12 min arc discs separated by 24 min arc. (After [68].)

regions, obtained from 30 subjects, for a fluorescent and an incandescent lamp are shown in Fig. 4.17. For the fluorescent lamp the illuminances of 300 lx and above did not produce significantly different scores although the scores at 10 lx were significantly worse. For the incandescent lamp the difference in the scores obtained at 300 and 4000 lx was significant. It can be seen from Fig. 4.17 that the difference in mean scores between the two illuminances for the incandescent lamp mainly occur at the extremes of the spectrum. The lack of significant effect of illuminance above 300 lx for the fluorescent lamps was confirmed in the experiment using the Farnsworth–Munsell 100 hue test which has already been discussed [62]. There was no significant effect of illuminance over a range of 300 to 1200 lx except for some older subjects, but Cornu and Harlay used young subjects.

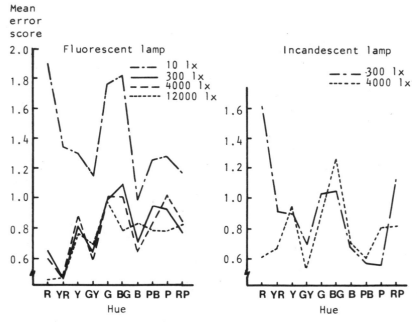

Fig. 4.17.　Mean error scores in 10 different regions of the Farnsworth–Munsell 100 hue test lit to different illuminances by a 'blanc brilliant de lux' fluorescent lamp and an incandescent lamp. The different regions of the Farnsworth–Munsell 100 hue test are indicated on the abscissae by their dominant hue: R = red, YR = yellow/red, Y = yellow, GY = green/yellow, G = green, BG = blue/green, B = blue, PB = purple/blue, P = purple, RP = red/purple [70].

These results suggest that unless colours at the extremes of the visible spectrum are involved, an illuminance of about 300 lx is sufficient for good colour judgement work. Certainly the effects of illuminance are much less marked than the differences between light sources with different spectral emissions. There is one caveat to this conclusion. These results have usually been obtained from young subjects examining reasonably sized coloured fields for short durations. Where prolonged work by older people and/or very small (< 10 min arc) fields are involved then the benefit of a higher illuminance may be more apparent.

4.3.5 Samples and Surrounds
Two problems in applying the results obtained with colorimeters to practical situations are the restricted test field, which is typically about 1–2°, and the dark surround. In practice, colour matching and

discrimination is done with samples larger than 2° and with anything but dark surroundings. The results described above suggest that larger sample sizes would be advantageous for colour judgement work but so far the nature of the surroundings has not been considered.

Brown [71] measured sensitivity to colour differences for different sizes of test colour field and different surround field colours by asking his two observers to make colour matches. Again his results were expressed in terms of the deviation of the match from the chromaticity co-ordinates of the test colour. Brown found that fields of colour subtending 12° produced greater sensitivity to colour differences than 2° fields which covered only the foveal region, which again emphasises the value of having a large size of sample. A further advantage of the large 12° field was that the sensitivity to colour difference was nearly independent of the nature of the surround, but when the 2° field was used the precision of colour matching was reduced if the surround field was a different colour. Brown also notes that the matching was much more difficult to do when the luminance of the surround field was greater than that of the test field. These results indicate the importance of the surroundings if accurate colour judgements are to be made for small objects. The advice given [55, 63, 67] is that the surroundings for colour judgement work should be of medium reflectance and low chroma. Given that this advice has to be applicable to a wide range of colours, it does not seem unreasonable.

4.3.6 Conclusions
This brief review of the way in which lighting affects colour judgements, and the evidence on which some of the recommendations for lighting such work are based, demonstrates the validity of the recommendations, with the possible exception of illuminance. There must be some doubt about the need for illuminances of the order of 1000 lx unless fine discrimination on small samples at the extremes of the visible spectrum is required.

However, it should always be remembered that the recommendations are intended for general use. In other words they are to be used where it is necessary accurately to reproduce, and/or discriminate, a wide range of colours, or where no knowledge of the colours to be examined is available. This is important because there are situations involving particular colours, and/or light sources, where the recommendations on the most important factor, i.e. the spectral emission of the light source, can be misleading. This is because the general CRI is based on the eight CIE test colours. All these test colours have spectral reflectances which only change slowly with wavelength. However, colours can occur which have rapid changes in spectral reflectance. This will make their appearance very dependent on the spectral emission of the light source in the region where the rapid changes occur. Until recently it was generally believed that a uniform spectral emission was necessary for a light source to have a high CRI value, but fluorescent light sources which disprove this have now become available. These sources have a very uneven spectral distribution and yet still give high CRI values. The spectral emission from these light sources is concentrated in three wavelength regions which approximate to the peak sensitivities of the three signal channels believed to be involved in colour vision. The problem they raise is what value to place on the CRI number assigned to these light sources. High CRI values are undoubtedly obtained for these light sources but such lamps markedly change the appearance of some colours from their appearance under daylight [72]. At the moment there is no answer to this question, the best advice is to be aware that some of these tri-band, or prime colour lamps, as they are called can cause distortion of some colours. Really, the only complete method of testing the suitability of a light source for a specific colour judgement task is a field trial. If this is not possible, then the next best approach is to assess its effect on the specified colours of interest by use of the colour difference equations in a CIE 1976 colour space.

The above has also shown up the essentially arbitrary nature of the measures used to describe the effects of light sources on the appearance of colours. The most widely used, the CIE general Colour Rendering Index contains limited information because it is averaged information. The colour gamut approach is better in this respect, in that it potentially gives much more information in a simple form, but is not widely used as yet. Even so the colour gamut only considers the hue and saturation attributes of perceived object colours, and lightness is ignored. The CIE 1976 colour spaces and the associated colour difference equations are superior in that they involve all three attributes: hue, saturation and lightness. However, the calculations involved in colour difference computation are complex. It is to be hoped that with the wider availability of small computers, use of the colour difference equations will become more practicable. Nonetheless it should be noted that even the colour difference equations do not cover all the factors involved in the appearance of surfaces. Surfaces can have a grey content, can be polished, matt or metallic, as well as having hue, saturation and lightness. There are no measures available to account for these perceptions. Even if there were, there would be the practical aspect to consider, namely the value of being able to handle all these perceptions. There is little doubt that it would be useful to extend the use of the colour gamut, and possibly positions in colour space, but the extra computation involved in moving beyond that would have to be justified.

4.4 VISUAL INSPECTION

4.4.1 Introduction
Three distinct types of visual inspection work can be identified: (a) searching an object or an array of objects for defects, (b) measuring an article to see if it is within preset tolerances, and (c) monitoring a known arrangement of information, e.g. a control panel. This section is concerned with the first of these types of inspection, although some of the conclusions may also be applicable to the monitoring activity. The inspection of objects to check for any defects is a part of the production process in many industries but the way in which lighting can aid this inspection has been studied in detail for only a few [73, 74]. This is unfortunate because there is no universally applicable method of lighting for inspection. Each inspection task is different because the nature, pattern and importance of the defects in each are different. The best way to light a particular inspection task will depend on the nature of the material and the defects to be identified.

In view of these comments it may seem perverse to include a section on visual inspection in a chapter which is intended to describe the relationship between light and work for those activities which have been studied in detail. There are two reasons why it is not. The first is simply that there are only a limited number of techniques for using lighting to reveal different types of defect. Therefore inspection lighting for different tasks usually involves different combinations of the same basic techniques. The second is that much work has been done on some of the operational components of visual inspection for other purposes, for example, the factors that influence the efficiency of visual search have been extensively studied for military applications such as aerial reconnaissance [75]. By combining the knowledge that has been gained from such different but related activities and the limited number of suitable lighting techniques available, it is hoped to produce some understanding of the relationship between lighting and visual inspection work.

4.4.2 The Nature of Visual Inspection
The visual inspection process involves two separate but successive components. The first component is the search for and identification of any

defects. In this the observer scans the object looking for defects and concludes the search when the defects have been identified or there is no more time available for searching. The second component is the decision on what to do about the defects. Small defects may be acceptable but if they are not, different types of defect may require different courses of action. For example, in knitwear, poor stitching can be rectified in the machine shop but a flaw in the cloth will require referral to the suppliers. Lighting can only directly affect the first of these components, although, if it enables more defects to be detected, it may also influence the second.

Studies of eye movements whilst searching for defects have revealed the common pattern of fixation and saccade. The observer searches by a series of fixation pauses with rapid saccadic eye movements between them. The first question that arises about this pattern of search is whether the fixations are random or systematic in location. The answer is that, for inspection work, they are systematic to some degree [76]. Even when the inspector is given a completely uniform field to examine, the pattern of eye movements shows a tendency to concentrate on the centre of the field and avoid the edges. However, completely uniform fields are not the only situation occurring in visual inspection work. Often an inspector will be faced with a structured field. For structured fields the pattern of eye movements can be divided into two phases, an initial orientation phase, followed by a search phase based on information obtained in the orientation phase [77, 78]. The balance between these two phases will depend on the inspector's knowledge of the objects being searched. If it is new to him then the orientation phase will be long but if he is inspecting a familiar object the orientation phase will be almost negligible. Thus the pattern of fixations used during a search will be determined largely by his experience of the objects to be searched. In practice, inspectors can be considered as experienced and hence as having expectations about where defects are likely to occur. These locations will be the ones searched. This should not be taken to mean that all inspectors will have the same pattern of fixations. After all, different inspectors may have had different experiences and hence may have different expectations about the locations of defects. Figure 4.18 shows the fixations made by two inspectors examining men's briefs held on a frame. There are some clear differences in the areas examined [76].

Figure 4.18 shows a search pattern involving many fixations; but many fixations are not always necessary. Experienced radiologists can apparently detect abnormalities in chest X-ray plates with a success rate of 70 % when viewing time is restricted to 200 ms, i.e. within one fixation pause. However, such skill is rare and implies a high level of practice (and has a 30 % failure

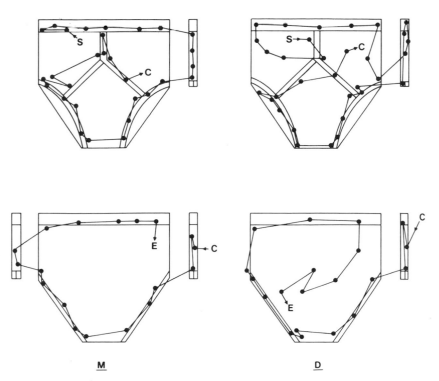

Fig. 4.18. The pattern of fixations made by two inspectors examining men's briefs held on a frame. S = start of scan path, C = end of scan of front and one side, rotation of frame and continuation of scan across back and sides, E = end of search. Inspector M examines only the seams whilst Inspector D examines the fabric as well. (After [76].)

rate) [78]. A more usual search pattern is for the inspection to continue for several fixations. It has generally been found that when this happens there is a tendency for an increase in fixation time to occur as the search time lengthens. Megaw and Richardson [79] found that in searching a matrix of items, if the subjects missed the faulty item on the first scan they increased their fixation times on the second scan.

This mention of time introduces the main quantity used as a measure of the efficiency of visual search: the search time. This is defined as the total time taken from the start of the search to the identification of the target. The question that needs to be considered is the extent to which different factors influence the search time. It is obvious that the area to be searched is relevant but there are a number of other important interacting factors. In order to examine these factors it is convenient to start with a very simple situation, i.e. a very uniform field to be examined in which any blemish is a defect. This situation is representative of some inspection tasks for sheet materials, e.g. glass and some textiles. The characteristic feature of all searching for defects is that detection of the defect takes place when it is away from the visual axis. It is extremely unlikely that the inspector will have the good fortune to fixate directly on the defect. It is much more likely that he will fixate close to it and the defect will be first detected as 'something

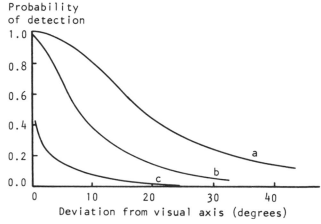

Fig. 4.19.   The probability of detection of targets of (a) contrast = 0·058, size = 19 min arc; (b) contrast = 0·08, size = 10 min arc; (c) contrast = 0·044, size = 10 min arc; within a single fixation pause, plotted against deviation from the visual axis. Each curve can be used to form a visual detection lobe for each target by assuming radial symmetry about the visual axis.

odd' in the periphery and will then be fixated so that the 'something odd' can be recognised. The essential act for rapid inspection is off-axis detection of a departure from the uniform field. In comparison the recognition of a defect once it is directly fixated is usually easy. This situation has been used in one of the more popular models of the search process [80, 81]. The idea behind the model is that each defect has associated with it a visual detection lobe, i.e. a surface centred on the visual axis which defines the probability of detecting the defect at various deviations from the visual axis within a single fixation pause. Figure 4.19 shows some appropriate probability data for detecting targets of different sizes and contrasts. From such results it is possible to calculate a visual detection lobe for each target by assuming radial symmetry about the visual axis. The data in Fig. 4.19 show such visual detection lobes have a maximum at the visual axis but the probability of detecting the defect decreases as the defect is located further off axis. In practice the boundary of the visual detection lobe is usually taken as the deviation from the visual axis for which the defect reaches its threshold condition, i.e. the probability of detection is 50%.

Clearly, different defects will have different visual detection lobes. A large-area, high-contrast hole in some sheet material will have a large visual detection lobe whilst a small-size, low-contrast hole will have a small lobe.

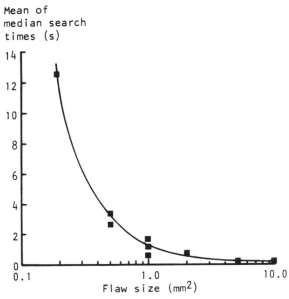

Fig. 4.20. Mean of median search times for detecting a single flaw in a glass sheet, plotted against the flaw size (after [82]).

The size of the visual detection lobe is important because provided the interfixation distance is related to it and the total search area is fixed, the total time taken to cover the search area is inversely proportional to the size of the visual detection lobe. This implies that the search time should be much less for defects with a large visual detection lobe than for defects with a small lobe. Figure 4.20 shows this for the inspection of 610 × 610 mm glass sheets, each containing a single flaw randomly positioned [82]. The larger the flaw size the larger the visual detection lobe and the shorter the search time. Visual detection lobes can be measured by psychophysical procedures or estimated from threshold performance data available for peripheral vision [83]. In either case they appear to be useful for predicting search times for the identification of defects in uniform fields of view.

However, for many inspection tasks the defect appears not in a uniform empty field but in a cluttered field, i.e. one in which many different items are present. This is typical of inspection of small items on a production line. The important distinction between this situation and an empty field is that, whereas in an empty field the detection of anything is the essential act, in a cluttered field the task becomes one of picking out the defect from the many features that can be detected. The additional aspects of this task that affect visual inspection are the extent of the clutter in the field, the ease with which the defect can be discriminated from the clutter and the information the inspector has about the features of the defect.

Drury and Clement [84] have shown that search time is strongly influenced by the amount of clutter in the area to be searched; the greater the amount of clutter the longer the search time. But amount of clutter is not the only important factor affecting search time. For a given amount of clutter the density of clutter is also important; the greater the density, the shorter the search time. Another important factor is the extent to which the items forming the clutter are similar to the defects; the greater the similarity the longer the search time.

Howarth and Bloomfield [85] have suggested a simple equation, based on a random search pattern, which can be used to predict search times. The basic form of the equation is

$$\bar{t} = t_s \times \frac{A}{a}$$

where $\bar{t}$ = mean search time, t_s = mean fixation time, A = total search area, and a = the area around the line of sight within which the target can be detected in a single fixation. For a given article to be inspected A will be constant and t_s reasonably so, but a will depend greatly on the nature of the defect and the background against which it is seen. If t_s and A can be considered as constants then the equation reduces to

$$\bar{t} \propto \frac{1}{a}$$

so the problem resolves itself into measuring a. For an empty field it does not seem unreasonable to identify the visual detection lobe with a. For a cluttered field Howarth and Bloomfield [85] have shown that a can be related to what they called the 'discriminability' of the defect. To do this they used an inspection task in which the inspector had to examine a set of similar items and to pick out those that were different. Figure 4.21 shows the results obtained for an experiment with discs, the mean search time being

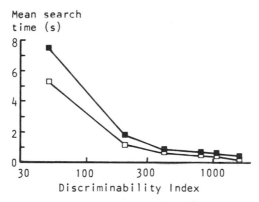

Fig. 4.21. Mean search times for two observers locating a disc target amongst an array of larger discs, plotted against the discriminability of the target disc. The discriminability index is the square of the difference in diameter of the target and non-target discs. (After [86].)

plotted against the square of the discriminability of the defects. In this case discriminability is defined as the difference in diameters of the target and non-target discs. Figure 4.22 shows a similar result [86, 87] but for distinguishing rectangles and squares from other squares. This time the discriminability is defined as the difference between the square roots of the areas of the targets and the non-targets. Bloomfield [86, 88] has also shown that search time is related to discriminability for a low contrast disc in an empty field, where discriminability is the difference beween the diameter of the disc and the smallest disc just visible foveally. This suggests there might be a link between the visual detection lobe concept and the discriminability of the defect in a cluttered field. In fact a link has been shown to exist [89]. The area around the visual axis within which a difference between one disc located amongst others of a different size can be detected on 50 % of occasions in a single fixation, i.e. the visual detection lobe for the difference, has been found to be proportional to the square of the discriminability as measured by the expression $(d_b - d_t - d_0)$; where d_b = the diameter of a non-target disc, d_t = the diameter of the target disc, and d_0 = the smallest difference in diameter that can just be detected foveally.

This result suggests that the various measures of discriminability considered can be regarded as measures of effective visual detection lobes, but now the lobe is determined not only by the target but also by the other items amongst which it is seen. Engel has shown how such effective visual detection lobes can be measured and has demonstrated that they are related to the probability of finding a target in a cluttered field within a fixed time;

the larger the effective visual detection lobe, the greater the probability of finding the target in the time [90, 91].

All the above results indicate that the discriminability of the defect is important, which implies that search time will be strongly influenced by the specification of what constitutes a defect. Williams [92] studied search times for finding a specified item from a display of 100 items, which could vary in

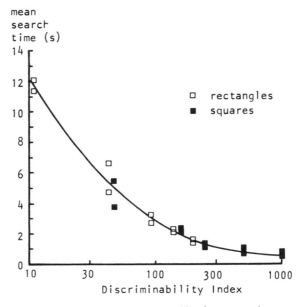

Fig. 4.22.   Mean search times for two observers searching for rectangular or square targets amongst an array of square non-targets, plotted against an index of discriminability. The discriminability index is given by $(\sqrt{A_1} - \sqrt{A_2})^2$ where A_1 and A_2 are the areas of the targets and non-targets respectively. (After [86].)

size, shape, colour and the information they contained. In fact the information within each item was a two-digit number. The inspector was asked to locate a particular number, the number being specified either alone or together with various combinations of the size, colour and shape of the item it was in. For example, the inspector could be asked to 'find the number 45 which is in a large, blue square', or to 'find the number 45 which is in a square', or simply to 'find the number 45'. The mean search times for the different target specifications are given in Table 4.3. It can be seen that the longest search times occurred when the number alone was specified, the shortest occurring when the colour and size of the item in which it lay were also specified. This is not simply a case of the more factors that are specified the fewer the items that have to be inspected. The point is that some parts of the specification are much more important than others. Table 4.3 shows that whenever the colour of the item in which the desired number lay was specified, short search times were achieved. Specifying the colour reduces the items where the number might be to 20% of the total but so does

Table 4.3: Mean time to find target for different target specifications [92]

Target specification	Mean time (s)
Number only	22·8
Number and shape	20·7
Number and size	16·4
Number, size and shape	15·8
Number and colour	7·6
Number, colour and shape	7·1
Number, colour, size and shape	6·4
Number, colour and size	6·1

specifying the shape and yet the latter has much less effect on search time. A similar observation applies to the specification of size. It is clear from Table 4.3 that specifying different aspects of the defect to be sought can have a marked influence on search time, a result confirmed by Bartz [93] in his study of the effect of different typefaces on the time taken to find a location on a map. The interesting question now is how this comes about. Fortunately, Williams [92] also measured the eye movements made during the search. As might be expected he found that when the item containing the required number was specified by number and colour, fixations were made predominantly on items of that colour. When the number and shape were specified there was very little change in the pattern of fixations from that which occurred when the number alone was given. This difference in pattern of fixations can be explained if the items of the specified colour have a large enough visual detection lobe so that fixation on one item with specified colour allows off-axis detection of an adjacent item of the same colour. In these conditions the pattern of eye movements during the search can be efficient and the resulting search time short.

The search situations considered, a uniform field and a cluttered field, have demonstrated the framework of an approach to understanding visual search. However, in practice, the visual search component in visual inspection is likely to be much more complicated than the situations considered so far. For a start, experienced inspectors are likely to have stereotyped search patterns which may or may not be suitable for the distribution of defects but which are unlikely to be random [76]. Next there is the point that inspectors are rarely searching for a single type of defect. Usually there are many types of defect. The effect of this is not as clear as might be expected. In some circumstances increasing the range of defects to be found does not affect search time [79], in other conditions it produces an increase [94, 95]. A further complicating factor is that for some types of inspection work the inspector will have to keep a count of the numbers of specific types of defects occurring. This too will tend to increase search time [95]. The main effect which compensates for these potential causes of an increase in search time is experience. In a search task which involved looking for specified alphanumeric characters from amongst others [95] the effect of 30 days' practice was to halve the search times for the more difficult conditions. From the subjects' comments, this seems to have occurred because they established a mental set of the targets they were looking for. This may have influenced the range over which the targets could be detected, i.e. increased their visual detection lobes. Certainly the specified letters appeared to 'pop out' of the display.

Mention of the effect of experience introduces the other more cognitive component of an inspection task. Having located the defect the inspector then has to decide what to do about it. In the real world this is very much influenced by social, organisational and psychological factors [96, 97]. For example, the levels at which a defect is ignored or sent for rectification, or the article is scrapped, are sometimes determined by the state of the market for the product or by relationships between the production workers and the inspectors. Further, the importance of a particular defect can vary with the role of the inspector within the organisation. Pope *et al.* [98] have shown that the dimensions on which inspectors judge the quality of knitwear depends on their role in the organisation. Each role will give greater importance to the defects that affect its area of responsibility. It should be apparent that the factors influencing the decision on what to do about an identified defect are complex and diffuse and have little to do with lighting.

From this brief discussion of the nature of visual inspection it can be concluded that successful inspection calls for much more than just the ability to see a defect. In other words, visual inspection is more than visual search. Inspection is often done at a set speed, which limits the time available for searching, and ultimately involves the judgement of a number of different and sometimes opposing factors. Lighting has a part to play in visual inspection but it is a limited part. Other factors such as the time

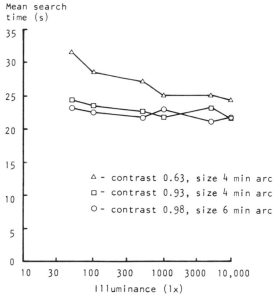

Fig. 4.23. Mean search times for locating a specified integer number from a random array of
100 such numbers, plotted against illuminance. (After [100].)

allowed to inspect an item and the manner of presentation are also
important. Megaw [99] gives an interesting survey of these factors.

4.4.3 The Effect of Lighting on Visual Inspection

The function of lighting for inspection is to take the defects that are present
and reveal them. From the previous discussion it is apparent that a
reasonable basis for doing this is to try to increase the visual detection lobe
of any defect. Since the visual detection lobe is determined by the off-axis
discriminability of the defect relative to the background, be it plain or filled
with similar items, any aspect of lighting which aids this discrimination will
be useful. The most widely applicable aspect is the illuminance on the
object. Increases in illuminance produce increases in luminance and hence
increases in contrast sensitivity and visual acuity, both of which aid
discriminability. Further, increasing illuminance should be the most
general approach, because it works by changing the operating state of the
visual system and hence is independent of the particular defect. An example
of this approach is the experiment by Muck and Bodmann [100] who had
people searching a display of 100 integer numbers scattered across a table
top. Figure 4.23 shows the mean search times for numbers of different size
and contrast. It can be seen that as the illuminance increases search time

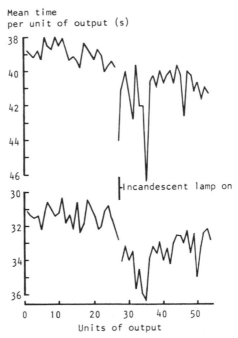

Fig. 4.24.   Mean time taken by two workers to inspect and pack cartons of 25 shotgun
cartridges before and after an incandescent lamp was switched on. (After [101].)

decreases. It should also be noted that there are differences between the
search times for the high contrast and the medium contrast numbers even
for the same size and illuminance conditions. This can be explained by the
observation that the medium contrast numbers would be expected to have a
smaller visual detection lobe, which means that many more fixations would
be required to cover the same search area. Hence the mean search times
should always be much longer for the low contrast than for the high
contrast targets; which indeed they are.

Although illuminance is effective in this case it should be noted that it is
not always so in practice. In an early study of inspecting and packing
shotgun cartridges [101], the result shown in Fig. 4.24 was found. This
displays the times taken for the inspection and packing of cartons of 25
cartridge cases during the period immediately before and after lighting was
switched on in the afternoon of a winter working period. The sudden
reduction in speed of inspection with the onset of the lighting from a single
incandescent lamp overhead is obvious. The important point to note is that
the onset of this lighting almost certainly increased the illuminance on the
task but this caused a worsening of performance. The inspectors stated that

the electric lighting produced an element of reflected glare from the brass caps of the cartridge cases and the cases were less uniformly lit. Thus in this case the increased illuminance was provided in such a way that the defect became more difficult to see.

Another area where illuminance is likely to be ineffective is for revealing form in three-dimensional objects. For such objects the distribution of light is more important. The effect of distribution is well demonstrated in a paper by Faulkner and Murphy [102]. Figure 4.25 shows two views of the same material, the upper being lit by general diffuse illumination, the lower by spot-lighting aimed at grazing incidence across the surface. The increase in the visibility of the loose thread caused by the highlights and the shadows produced by the grazing incidence beam is apparent. Effectively the highlights and shadows mean the defect is larger and has a higher contrast, which in turn means its visual detection lobe will be greater. Faulkner and Murphy list 17 different methods of lighting for inspection. Their methods can be classified into three types: (a) those that rely on the distribution of light relative to the material, as described above; (b) those that rely on some special physical property of the light emitted which interacts with the material being inspected, e.g. ultraviolet radiation for detecting the presence of some types of impurities in a product; and (c) those that call for the projection of a regular image onto or through the material being studied. This last method is shown in Fig. 4.26. Any distortion of the grid when it is viewed through the beaker indicates a defect in the glass. It should be appreciated that all these techniques operate by increasing the size and/or contrast of the defect, either directly or by using it to affect some other object. Table 4.4 gives a comprehensive list of methods for inspection lighting for materials of various types [103].

All this is for inspection of defects which are of sufficient size that they can be seen with the naked eye. However, there are an increasing number of activities in which the objects and defects to be inspected are so small that they have to be magnified for them to be seen at all. This introduces a special problem, namely the magnification to be used. The greater the magnification the bigger is the size of the defect and hence the easier it should be to detect. However, a limit to this process occurs because the magnification also increases the search area. This suggests that there might be an optimum magnification to use for each size of defect. Smith [104] has demonstrated that such an optimum does occur. Figure 4.27 shows the time taken per correct inspection for examining (7×7) or (8×8) arrays of circles, each array containing from 0 to 3 imperfect circles as defects. The imperfection consisted of a $1 \cdot 2$ min arc gap in the circle. These arrays of

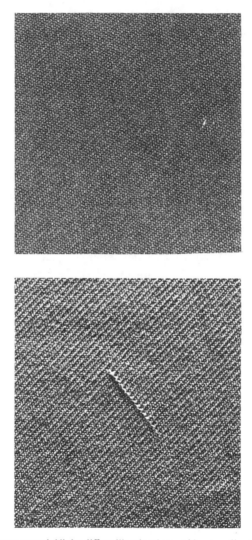

Fig. 4.25.   The same material lit by diffuse illumination and by very directional lighting. The increase in visibility of the loose thread under the directional lighting is dramatic. There is no difference in magnification for the two photographs. (After [102].)

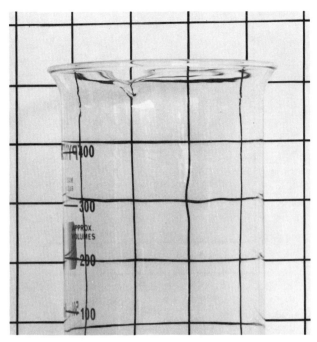

Fig. 4.26. A defect in the transparent glass beaker is revealed by the distortion in the grid seen through the beaker.

circles were inspected at various levels of magnification and illuminance but even at the maximum magnification an array did not exceed the field of view of the microscope. Figure 4.27 shows there is a clear optimum for this task when the magnified gap size subtended at the eye is between 10 and 15 min arc. This optimum is independent of the size of the array although the difference between arrays was small. It is also interesting to note that here the effect of increased illuminance is to decrease the time time taken for a correct inspection. The high and low labels on Fig. 4.27 refer to the illuminances that were considered suitable and insufficient respectively.

By now it should be apparent that there are a number of techniques available for lighting inspection tasks, each technique being suitable for a different type of defect. Unfortunately the typical inspector's work involves looking at many different types of defect simultaneously. This implies that different lighting may be required for different defects. Gillies [74] provides a good example of a lighting system designed for a particular inspection job. The problem was to inspect a heavy, shaped, glass sheet product for faults such as cracks and bubbles in the glass. Figure 4.28 shows the inspection

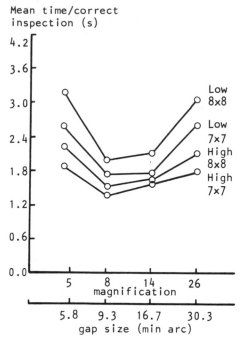

Fig. 4.27. Mean time/correct inspection for examining arrays of 7 × 7 or 8 × 8 circles for circles with gaps, at different levels of magnification and illuminance [104].

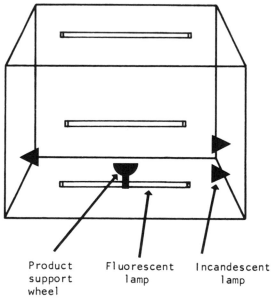

Product support wheel   Fluorescent lamp   Incandescent lamp

Fig. 4.28. A schematic diagram of the inspection lighting for a shaped glass product. (After [74].)

Table 4.4:   Classification of visual tasks and lighting techniques (after [103])

Part I—Flat surfaces

| Classification of visual task | | Example | | Lighting technique | |
General characteristics	Description	Lighting requirements	Luminaire type†	Locate luminaire‡
A. Opaque Materials 1. Diffuse detail and background				
(a) Unbroken surface	Newspaper proof reading	High visibility with comfort	2 or 3	To prevent direct glare and shadows (A)
(b) Broken surface	Scratch on unglazed tile	To emphasise surface break	1	To direct light obliquely to surface (C)
2. Specular detail and background				
(a) Unbroken surface	Dent, warps, uneven surface	Emphasise unevenness	5	So that image of source and pattern is reflected to eye (D)
(b) Broken surface	Scratch, scribe, engraving, punch marks	Create contrast of cut against specular surface	3 or 4 or 5 when not practical to orient task	So detail appears bright against a dark background So that image of source is reflected to eye and break appears dark (D)
(c) Specular coating over specular background	Inspection of finish plating over underplating	To show up uncovered spots	4 with colour of source selected to create maximum colour contrast between two coatings	So reflection of source image is towards the eye (D)

Table 4.4—contd.

Classification of visual task	Example	Lighting requirements	Lighting technique	
General characteristics	Description		Luminaire type†	Locate luminaire†
3. Combined specular and diffuse surfaces				
(a) Specular detail on diffuse light background	Shiny ink or pencil marks on dull paper	To produce maximum contrast without veiling reflections	3 or 4	So direction of reflected light does not coincide with angle of view (A)
(b) Specular detail on diffuse dark background	Punch or scribe marks on dull metal	To create bright reflection from detail	2 or 3	So direction of reflected light from detail coincides with angle of view (B)
(c) Diffuse detail on specular light background	Graduations on a steel scale	To create a uniform, low-brightness reflection from specular background	3 or 4	So reflected image of source coincides with angle of view (B or D)
(d) Diffuse detail on specular dark background	Wax marks on car body	To produce high brightness of detail against dark background	2 or 3	So direction of reflected light does not coincide with angle of view (A)
B. Translucent Materials				
1. With diffuse surface	Frosted or etched glass or plastic, lightweight fabrics, hosiery	Maximum visibility of surface detail	Treat as opaque, diffuse surface—See A.1	
		Maximum visibility of detail within material	Transilluminate behind material with 2, 3, 4 (E)	
2. With specular surface	Scratch on opal glass or plastic	Maximum visibility of surface detail	Treat as opaque, specular surface—See A.2	
		Maximum visibility of detail within material	Transilluminate behind material with 2, 3 or 4 (E)	

C. Transparent Materials				
Clear material with specular surface	Plate glass	To produce visibility of details within material such as bubbles and details on surface such as scratches	5 and 1	Transparent material should move in front of 5 then in front of black background with 1 directed obliquely. 1 should be directed to prevent reflected glare
D. Transparent over Opaque Materials				
1. Transparent material over diffuse background	Instrument panel	Maximum visibility of scale and pointer without veiling reflections	1	So reflection of source does not coincide with angle of view (A)
	Varnished desk top	Maximum visibility of detail on or in transparent coating or on diffuse background		
		Emphasis of uneven surface	5	So that image of source and pattern is reflected to the eye (D)
2. Transparent material over a specular background	Glass mirror	Maximum visibility of detail on or in transparent material	1	So reflection of source does not coincide with angle of view. Mirror should reflect a black background (A)
		Maximum visibility of detail on specular background	5	So that image of source and pattern is reflected to the eye (D)

Table 4.4—contd.
Part II—Three-dimensional objects

| Classification of visual task | | Example | | Lighting technique | |
General characteristics	Description	Lighting requirements	Luminaire type†	Locate luminaire†
A. Opaque Materials 1. Diffuse detail and background	Dirt on a casting or blow holes in a casting	To emphasise detail	2 or 3 or	To prevent direct glare and shadows (A)
			1	In relation to task to emphasise detail by means of highlight and shadow (B or C)
			2 or 3 as an ultraviolet source when object has a fluorescent coating	To direct ultraviolet radiation to all points to be checked
2. Specular detail and background (a) Detail on the surface	Dent on silverware	To emphasise surface unevenness	5	To reflect image of source to eye (D)
	Inspection of finish plating over underplating	To show up areas not properly plated	4 plus proper colour	To reflect image of source to eye (D)
(b) Detail in the surface	Scratch on a watch case	To emphasise surface break	4	To reflect image of source to eye (D)
3. Combination specular and diffuse (a) Specular detail on diffuse background	Scribe mark on casting	To make line glitter against dull background	2 or 3	In relation to task for best visibility. Adjustable equipment often helpful. Overhead to reflect image of source to eye (B or D)

(b) Diffuse detail on specular background	Micrometer scale	To create luminous background against which scale markings can be seen in high contrast	3 or 4	With axis normal to axis of micrometer
B. Translucent Materials				
1. Diffuse surface	Lamp shade	To show imperfections in material	2	Behind or within for transillumination (E)
2. Specular surface	Glass enclosing globe	To emphasise surface irregularities	5	Overhead to reflect image of source to eye (D)
		To check homogeneity	2	Behind or within for transillumination
C. Transparent Materials				
Clear material with specular surface	Bottles, glassware—empty or filled with clear liquid	To emphasise surface irregularities	1	To be directed obliquely to objects
		To emphasise cracks, chips and foreign particles	4 or 5	Behind for transillumination. Motion of objects is helpful (E)

† *Key to luminaire type*
1 = narrow, concentrated beam, e.g. spot lights, fluorescent luminaires with narrow light distribution.
2 = wide spread beam from high-luminance small-area luminaires, e.g. incandescent or high pressure discharge lamps in diffusing luminaires.
3 = wide spread beam from moderate-luminance large-area luminaires, e.g. fluorescent luminaires with wide light distribution.
4 = uniform moderate luminance luminaire (variation in luminance < 2:1), e.g. fluorescent lamps behind a diffusing panel.
5 = uniform moderate luminance luminaire (variation in luminance < 2:1) with pattern superimposed.

Table 4.4—*contd.*

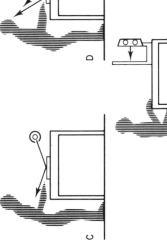

‡ *Key to location of luminaires*

installation. It involves three fluorescent lamps and three incandescent lamps in a black box. The product is held on the support wheel and the various lamps are used in the following ways. The overhead fluorescent light is used to reveal any surface defects on the upper face. By tilting the product the reflection of the overhead light can be made to traverse the whole surface. Any fault will appear as a dark point on the image. The underneath fluorescent lamp is used for the examination of the face and radius of the product by transmitted light. The product is rotated on the support wheel so that each part passes over the lamp. This process facilitates the detection of bubbles in the glass. The incandescent lamps are used in the examination for cracks in the side and radius of the product. As the edge passes the incandescent lamp any cracks sparkle. Finally, the back wall fluorescent lamp is used to detect surface defects with the product held so that the light is transmitted through the product face.

This is rather a complex example but it does illustrate the way in which lighting can be used to aid visual inspection. Just providing more light may make some contribution, but a specific installation designed to reveal the defects is often a lot better. Dekoker and Frier [105] reached a similar conclusion for the inspection of metal sheets.

4.4.4 Conclusion

The essence of good inspection lighting is to make conspicuous those features which make the defects different from the rest of the product. Unfortunately, it is apparent from the above discussion that there is no one method of lighting which will do this for all types of defect. Lighting can both help and hinder visual inspection. The best procedure when considering lighting for visual inspection is, to first establish the nature of the defects that need to be identified, and then to decide on the type and arrangement of lighting that is most likely to increase their visual detection lobes without discomfort to the inspector. Then equipment that provides the type of lighting required can be constructed. Much advice is available on the best inspection lighting for different materials (see Table 4.4 and refs [102, 106, 107]). This advice is correct but it should be remembered that a defect is not a single entity. Defects in a product can have a number of components. For example, a scratch on the surface will have a size, a contrast and a texture, and possibly will reveal a different colour. The important aspects of the defect are those that make the defect different, e.g. a scratch different from other marks on a surface. Given that appropriate lighting conditions can be identified, then lighting can make a contribution to easing the difficulties of visual inspection; but it should not be supposed that it will solve all the problems of inspection work [108, 109].

4.5 DRIVING

4.5.1 Introduction

About one quarter of all road miles are travelled during the hours of darkness. Roadway lighting and vehicle lighting are designed to make such journeys safer, quicker and more comfortable. In practice little is known about the effects of such lighting on journey times but the effects on road safety and, in a restricted sense, on comfort, have been extensively studied. That safety whilst driving at night deserves special consideration is borne out by the fact that in developed countries the number of road fatalities at night is likely to equal or exceed the number occurring by day, even though the traffic density is typically only a third of the daytime value [110]. It is the effect of roadway and vehicle lighting on safety whilst driving at night that will be considered here.

4.5.2 The Visual Component of Driving

Before examining the effects of roadway and vehicle lighting in detail, it is necessary to consider the nature of the driver's task. In essence this consists of assessing visual information and manoeuvring the vehicle appropriately. When the driver fails in either of these aspects of his work an accident is likely to happen. An indication of the importance of visual information to driving performance can be gained from the data collected in 'on the spot' investigations of 2036 accidents in a rural area [111, 112]. Figure 4.29 is derived from this work. It shows the percentage of accidents in which, in the opinion of the investigating team, each listed factor was involved. The percentages do not sum to 100% because road accidents rarely have a single cause, usually a combination of causes is involved. Even so, Fig. 4.29 declares that in 95% of the accidents studied the road user contributed to the accident. Further, 44% of the accidents involved what is termed perceptual error. Included under perceptual error are distractions and lack of attention, incorrect interpretation, misjudgement of speed and distance, and a category intriguingly called 'looked but failed to see'. Not included under perceptual error are a lack of skill or care on the part of the driver, and any reduction in the driver's ability such as can be produced by drugs or alcohol.

A more detailed examination of the class of perceptual error can be used to show the visual requirements of driving. The first two contributory causes of accidents, distraction and lack of attention, both stem from the need to make decisions rapidly. Frequently drivers have only one or two seconds in which to decide whether to make a particular manoeuvre. This time limit would matter less if all parts of the retina had the same capability

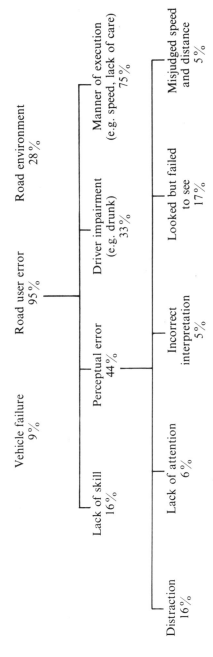

Fig. 4.29. Apparent contributory factors in road accidents [111, 112].

for discrimination but, as was discussed in Chapter 2, they do not. Discrimination is best, and hence the identification of objects is easiest, in the central area of the retina, the fovea, but declines rapidly as the object moves away from the visual axis. In the few seconds available the driver will only be able to fixate and accommodate on a few places. Eye movement studies [113] have shown that drivers make about three fixations per second. The relevance of this is that any distraction or lack of attention can result in some of these fixations not being used to examine critical locations, and hence important information being missed. The question now is 'how does a driver know where the critical locations are before he looks at them?' Obviously experience of both general driving and knowledge of the particular piece of road is very important. There is certainly evidence that eye movement patterns are different for experienced and inexperienced drivers [114]. But experience is not the only factor. Another important feature is the conspicuity of any particular object. The more conspicuous a feature is, the more likely it is to be detected by peripheral vision and then to be examined foveally. Conspicuous objects are those that stand out from the background. High conspicuity can be produced by relative motion of an object against a background and can be facilitated by having a high luminance contrast between the object and the background (e.g. motor cycles ridden with headlights on during the day), or by flashing lights (as on ambulances). However, it should be noted that although conspicuous objects are, by definition, highly visible, visible objects are not necessarily conspicuous. Conspicuity is also affected by the complexity of the background against which the object is seen. If all motorists drove with headlights on during daytime they would all be highly visible but all would not necessarily be conspicuous.

When attention is directed to a location by the conspicuity of the object there, it is unlikely that any difficulty will be found in identifying the object. However, when experience suggests the driver examine a particular location, then the visibility of any particular object there needs to be considered. The visibility of the object will depend on such attributes as its luminance, size, shape, highlights, movement and colour, the background against which it is seen and the state of adaptation of the driver. Much effort has been put into measuring the minimum values of some of these factors that are visible under different conditions. Figure 4.30 shows a relationship between luminance and size found when tail-lights, disc obstacles and pedestrian dummies are 'just visible' under no road lighting and 'no glare' conditions. It can be seen that the log luminance of the obstacle plotted against the log visual area of the obstacle gives a nearly

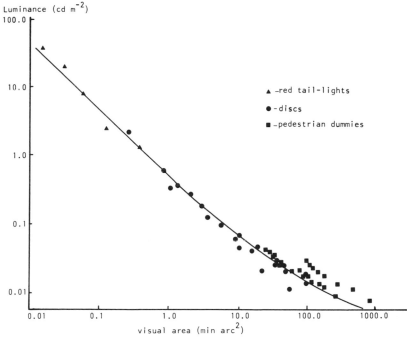

Fig. 4.30. The relationship between luminance and visual area for an object to be 'just visible' under no road lighting/no glare conditions. (After [115].)

straight line; the smaller the obstacle the greater the obstacle luminance has to be before it is just visible [115].

The results shown in Fig. 4.30 are very much what would be expected from the performance characteristics of the visual system discussed in Chapter 2. If such visual characteristics are an important aspect of driving, then people with poor vision would be expected to have more accidents. Burg [116] measured the static and dynamic visual acuity, the field of vision, the extent of misalignment in the two eyes and the rate of recovery from glare for 17 500 drivers, who between them had had over 5200 accidents in the previous three years. He was able to find only very weak correlations between the measured visual abilities and the drivers' accident records. This unexpected result can be explained in two ways. First it is possible that drivers with worse visual abilities are aware of their capabilities and drive within them. Certainly younger drivers tend to have more accidents even though they also tend to have better eyesight. Second there is the point that the visual abilities measured were very simple. Davidson [117], in an extensive review, has confirmed that simple visual abilities are at best weakly correlated with accident occurrence. However, he does suggest that some more complex measures of visual ability show promise of being more closely related to driver performance. This is to be expected because driving involves high level perception. The driver often has to detect the presence and/or movements of a number of different obstacles, recognise them and their characteristic patterns of movement, and then relate them to each

Fig. 4.31.   A road layout which was frequently misinterpreted by drivers. The major road actually bears right but the minor road continues straight on with the same width and tree line as the major road. Accidents were common at this site until the road layout was changed.

other, all in a short time before making a manoeuvre. It is the failure of this co-ordinating aspect of the driver's task that appears, in Fig. 4.29, as a cause of accidents under the heading 'incorrect interpretation'. Incidentally it is important to realise that road directions can be misinterpreted as well as dynamic traffic situations. Figure 4.31 shows a major road going right with a minor road continuing the line of the major road. An assumption that the major road went straight ahead, and a consequent decision to overtake, could lead to disaster.

Whichever explanation of the poor relationship between the simple visual abilities of drivers and accidents is correct, and probably both are involved, one thing is clear; that for people with sufficient visual capacity to drive at all, their driving is strongly influenced by factors other than their basic visual abilities. This conclusion is supported by the existence of a 'looked but failed to see' category of accident cause (Fig. 4.29). A typical example of this is a motorist who pulls out from a side road straight into the path of a motorcyclist on the main road. The driver looked along the main road but failed to see the motorcyclist. It is not simply that the motorcyclist has a low visibility, if the driver had been told the motorcyclist was there he could have detected him easily. The problem is also one of expectancy. The driver is searching for a car or lorry, he is not expecting a motorcycle and therefore does not see one. To overcome this phenomenon of expectancy, strong signals are necessary, i.e. conspicuity has to be increased. It can be seen from Fig. 4.29 that 'looked but failed to see', makes a by no means negligible contribution to accidents.

The final category of causes of accidents is misjudgement of speed and distance. These judgements are both involved in two manoeuvres that produce many serious accidents, turning across traffic and overtaking. Both these manoeuvres require the driver to move from his correct side of the road and go across or into the opposing traffic. To do this safely it is often necessary to make accurate judgements of how far off an approaching vehicle is and at what speed it is coming. Experimental evidence [118] suggests that people are not very good at making either of these judgements, particularly when the judgement has to be made at a long distance and the approaching vehicle is moving directly towards the driver so the change in size and the perceived relative motion are small.

By now it will be appreciated that the driver's task is a complex one. Within a very limited time he has to interpret what is likely to happen on the road ahead. To do this he has developed a series of expectations of other drivers' behaviour and of what are appropriate locations to examine. He will be faced with objects of different degrees of visibility and conspicuity, and will have to make judgements for which his visual system is not very well suited. It is in this context that roadway and vehicle lighting has to operate.

4.5.3 Roadway Lighting

Roadway lighting is provided in urban areas and on main traffic routes mainly to enable a driver to see further and better than he could using vehicle lighting alone. A number of different approaches have been used to examine the effectiveness of roadway lighting. The most direct examines the relationship between roadway lighting and accident occurrence. However, this has not been widely used because of the practical difficulties it presents. The accident researcher has no control over the events he seeks to study. Accidents happen infrequently and to different drivers in different circumstances. Each accident has a number of different factors influencing it, e.g. driver, traffic flow, construction of road, type of lighting if any, surroundings, weather, etc. Any attempt to identify the overall effect of one aspect of one factor (e.g. the luminance distribution of roadway lighting) on accidents therefore calls for the collection of considerable amounts of data either for a long time or from a large number of sites or both. An alternative approach, namely to do tests of driver performance under different lighting conditions, has been used more frequently. These tests take the form of observing driver behaviour or of measuring the visibility of obstacles. These latter tests are useful but it must be remembered that they relate to only a small part of the driver's task. Another approach used has been to make subjective appraisals of roadway lighting. This can be done for individual factors or for complete installations. In either case it should be remembered that the information gathered concerns the observers' opinions of how well they think they can see or drive, rather than any objective measurement of how well they can see or drive. The difference between a subjective opinion and an objective performance can be large. Nonetheless, subjective appraisals can be useful in identifying important areas of the field of view and the effect of those variables that influence discomfort rather than performance.

Different aspects of roadway lighting have been examined by different methods, either separately or in combination. In considering the effectiveness of roadway lighting, the first question that comes to mind is 'does roadway lighting have any effect on accident occurrence at all?' The answer is positive, Tanner [119] showed that improving roadway lighting produced a marked reduction in the number of accidents occurring at night, typically about 30%. Similar results have been found by other researchers in a number of countries, so there can be little doubt about the general effect [110, 120, 121]. However, it should be noted that even this reduction will not make accident rates during night and day simply proportional to the traffic flows during these times. This is not only because roadway lighting is usually a poor substitute for daylight but also because road user populations are different during night and day.

Although there is little doubt that some form of roadway lighting will reduce the number of accidents at night, the question now concerns what form the lighting should take. Recently the CIE has published a set of recommendations for roadway lighting [122]. Five criteria are given for good roadway lighting. These relate to the average road surface luminance, the uniformity of luminance, the extent of discomfort glare, the extent of disability glare and the directional guidance provided to the driver. Each of these recommendations will now be examined in turn.

The first criterion, of average road surface luminance, has some experimental evidence to support the view that it is an important factor in accident occurrence. Figure 4.32 shows a plot of the ratio of night casualty accidents to day casualty accidents against what is called light level for 34 sections, totalling 81 miles, of urban traffic routes in Sydney, Australia [123]. Light level is simply the luminous flux emitted in the lower hemisphere of the roadway lighting lanterns, divided by the carriageway

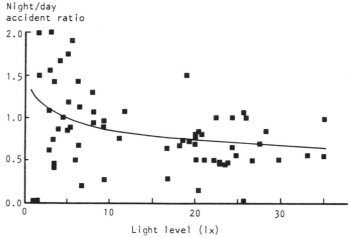

Fig. 4.32. Night/day casualty accident ratios plotted against the light level. The light level is the luminous flux emitted in the lower hemisphere of the roadway lighting lanterns divided by the carriageway area. The curve is the best fit through the points. (After [123].)

area. It should be related to the illuminance on the carriageway unless the luminous intensity distribution of the lanterns is very unusual. Therefore it should also be related to the average road surface luminance, although the link may be approximate since different road surfaces have different luminance factors. Although there is a wide scatter of results in Fig. 4.32 there is a statistically significant link between light level and the night/day casualty accident ratio. Specifically, the link looks like a law of diminishing returns, so that installing even a low light level will give a marked reduction in night accidents, thereby reducing the night/day accident ratio; but further increases in light level produce much smaller, if any, changes in the number of night accidents. Light levels of the order of 10–15 lower hemisphere lumens per square metre of carriageway, which is the level at which the curve starts to flatten, imply an average carriageway luminance of about 1 cd m^{-2}.

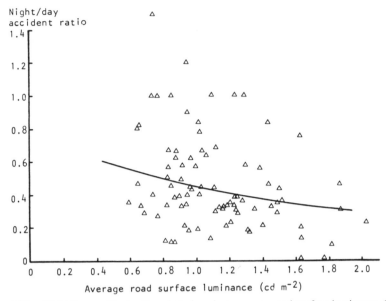

Fig. 4.33.   Night/day accident ratios plotted against average road surface luminance. The curve is the best fitting exponential through the data, after weighting each ratio for the number of accidents to which it relates. (After [124].)

Further evidence of the importance of average road surface luminance comes from the work of Hargroves and Scott [124]. They collected statistics of road accidents known to the police and measured the average road surface luminance for 89, 1 km lengths of single carriageway road, each with a $50 \, km \, h^{-1}$ speed limit. Figure 4.33 shows the night/day accident ratios for accidents occurring over a three-year period plotted against average road surface luminances in dry conditions. The curve shown is the best fitting exponential curve through the data, with the night/day accident ratios weighted to give greater importance to those sites where accidents occurred most frequently. Again a law of diminishing returns is apparent such that increasing the average road surface luminance tends to reduce the night/day accident ratio. The average road surface luminance recommended by the CIE for this class of road is either 1 or $2 \, cd \, m^{-2}$, depending on the surroundings [122]. Thus both these studies of road surface luminance and accident ratios offer some support to the CIE recommendations. However, the wide scatter in the data shown in Figs 4.32 and 4.33 also demonstrates that factors other than average road surface luminance are important in road safety.

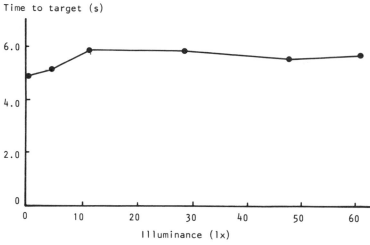

Fig. 4.34. Mean time to target for unwarned drivers detecting an obstacle on the carriageway plotted against the horizontal illuminance on the carriageway [125].

Gallagher *et al.* [125] used a driver performance test to measure the effect of road surface luminance. They placed a human size obstacle in one lane of a three lane urban road which could be lit to different illuminances and different uniformities. Unsuspecting drivers were observed and the distances at which they first took avoiding action was recorded, as was the speed at which they were travelling. The speeds and distances were combined to form a measure of the time to target when avoiding action was taken. The mean times to target are plotted against horizontal illuminance on the carriageway in Fig. 4.34. It can be seen that there is no improvement in performance above a horizontal illuminance of 10 lx, which is similar to the results for overall effect of light level on night/day accident ratio shown in Fig. 4.32. Further support comes from a paper by Fischer [126]. He reported that observers rated average road surface luminances of $1.4 \, \text{cd m}^{-2}$ as good for a number of different roadway lighting installations in the Netherlands. Finally, indirect support for average road surface luminances of about $1 \, \text{cd m}^{-2}$ comes from some observations by De Boer [127] on the conditions when drivers turned on vehicle lights on unlit roads. The results of his observations are shown in Fig. 4.35. It can be seen that as the average road surface luminance falls below $1 \, \text{cd m}^{-2}$ the percentage of drivers turning on their headlights increases rapidly. Clearly road surface luminances below about $1 \, \text{cd m}^{-2}$ are considered insufficient for safe driving.

Gallagher *et al.* [125] also examined the effect of the uniformity of illuminances across the carriageway but no significant differences in time to target were found. This is probably because the importance of uniformity is that non-uniform lighting provides places where a potential hazard may be hidden. If the obstacle is not in such a place then the non-uniformity is of little concern. Narisada [128] also examined the effect of non-uniformity.

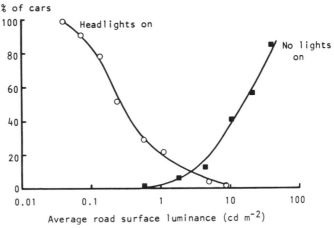

Fig. 4.35.   The percentage of cars without lights or with headlights, on unlit roads at various average road surface luminances. (After [127].)

His criterion was not one of task performance but of simply detecting the location of a square shape with a minimum contrast of 0·25 and a size of 0·2 m × 0·2 m. This size of obstacle has been shown to be considered a moderate hazard by drivers [121]. Using a simulated roadway lighting scene Narisada found that for a given value of road surface luminance, the greater the uniformity, the higher the probability of detecting the obstacle. He also reported that in order to maintain a given probability of detection the average road surface luminance has to be increased if the uniformity is decreased, the relationship being of the form

$$\bar{L} = \frac{\bar{L}_u}{U_{o1}^2}$$

where $\bar{L}$ = average road surface luminance $(cd\,m^{-2})$, U_o = overall uniformity = minimum road surface luminance/average road surface luminance, and $\bar{L}_u$ = average road surface luminance for the set probability of detection when $U_o = 1\cdot0$ $(cd\,m^{-2})$.

There is little doubt that overall uniformity of luminance is important, a value of 0·4 being recommended by the CIE [122]. Fischer [126] also reports observers' ratings of the non-uniformity in a number of roadway lighting installations in the Netherlands. It was found that a luminance uniformity (L_{min}/L_{max}) of at least 0·7, when measured along the axis of the carriageway, was necessary for the degree of uniformity to be rated as good. This is the basis of the lengthwise luminance uniformity recommendation made by the CIE [122].

A further feature of roadway lighting that is thought to affect the visibility of objects is the extent of disability glare from the lanterns. There are several different experimentally derived formulas that express the extent of disability glare in different situations [129, 130, 131]. Probably the most widely used formula is

$$L_v = \sum_n 10 E_n \theta_n^{-2}$$

where L_v = the equivalent veiling luminance, i.e. the luminance of a uniform light veil, which, when superimposed over a target, increases its luminance difference threshold by the same amount as the installation of interest does (cd m^{-2}); E_n = the illuminance at the eye from the nth glare source (lux); and θ_n = the angle between the line of sight and the nth glare source (degrees).

The CIE roadway lighting recommendation [122] limits the effect of the disability glare by restricting the allowed percentage increase in luminance difference threshold occurring under the roadway lighting when the equivalent veiling luminance is allowed for. This criterion is known as the threshold increment, limiting values varying from 10–20%, depending on the class of road. The percentage threshold increment (TI) can be obtained from the formula

$$TI = 65 \frac{L_v}{(\bar{L})^{0·8}}$$

where L_v = the equivalent veiling luminance (cd m^{-2}), and $\bar{L}$ = the average road surface luminance (cd m^{-2}).

The importance of disability glare diminishes rapidly as the angle between the line of sight and the lantern increases. Modern lantern design to national standards has ensured that disability glare from roadway lighting is a relatively minor matter compared with the disability glare produced by vehicle headlights, of which more later.

Disability glare directly affects the visibility of an object. Discomfort glare does not—or rather it has no direct measureable effect. It may well be that discomfort glare affects driver performance by distraction or fatigue but this is a matter of belief rather than proof at present. Nonetheless, criteria for controlling discomfort glare exist, the one used by the CIE [122] being the Glare Mark (see Chapter 7). This measure takes into account the luminous intensity of the lanterns at 80° and 88° from the downward vertical in the plane of the road axis, the average road surface luminance, the number and height of lanterns, and the flashed area of the lantern at 76° from the downward vertical in the plane of the road axis. Cornwell [132] had observers make a subjective appraisal of different sorts of roadway lighting from a car driven through them. It showed that some of these factors are more important than others, specifically the cut-off angle (i.e. the luminous intensity distribution for the lantern) and the flashed area. He suggests that the discomfort glare experienced from roadway lighting is better predicted from these two factors alone. Fischer [126] also obtained ratings of the extent of glare experienced in different roadway lighting installations. He showed a Glare Mark of 7 was necessary before the glare from the installation was considered satisfactory. In spite of this, the CIE recommended Glare Marks in the range of 4–6 depending on the type of road, all of which indicate greater discomfort.

All the above criteria: average roadway luminance, luminance uniformities, threshold increment and Glare Mark are quantitative. However, it should not be assumed that they cover all the important aspects of roadway lighting. The CIE recommendations also mention the role of roadway lighting in providing visual and optical guidance. Visual guidance concerns the effect lighting can have on the visibility of the road itself and on the road markings and other directional aids. Optical guidance refers to the extent to which the layout of the roadway lighting installation itself can indicate to the driver the direction the road is taking.

Another important factor which is mentioned but not quantified in the CIE recommendations is the lighting of the carriageway surround. Such lighting has two purposes. First, it increases the visibility of pedestrians at the edge of the carriageway [133], who incidentally often over-estimate their own visibility to the driver [134]. Second, it gives some extra guidance to the driver. Eye movement recordings [135] taken during driving show that, when manoeuvring round a bend, drivers tend to fixate on lateral placement cues, such as curbs, rather than relying on peripheral vision, which is relied on on straight roads.

Finally, another feature of roadway lighting which has been investigated is the colour of the light source used. For economic reasons most roadway lighting uses either high pressure mercury, or low or high pressure sodium discharge lamps. These lamps differ markedly in their spectral emissions. Several different studies have been made on the effectiveness of these different light sources [136, 137, 138] without any clear conclusions. This would suggest that any effects are small. Part of the problem is that the observer of roadway lighting is usually using mesopic vision. This means that his perception of the scene may be different from its photometric measurement, particularly if, as is usual, the measuring instrument is calibrated to the CIE relative photopic visual efficiency function (V_λ). This problem is magnified by the fact that the light sources used emit a number of strong spectral lines which would tend to emphasise any differences between the CIE V_λ function, the actual spectral response of the measuring instrument and the spectral response of the observer.

All the criteria discussed above are involved in the design of roadway lighting, but one very important factor is not usually considered. The criteria assume a dry road. If the road is wet, marked changes in its reflection properties occur which obviously affect its luminance pattern. Gordon [133] had observers rating the visibility of 20 cm cubes and of people wearing clothes of known reflectance. He found that when the road was wet the visibility of both cubes and people was reduced. The area of road where there is a high probability of seeing the object is smaller for wet roads than dry roads and the reduction of visibility with distance is greater. This suggests that wet road surfaces are more critical surfaces for roadway lighting than dry road surfaces. Considerable efforts are now being made to extend the design principles for dry roads to wet roads [139] although Gordon doubts if this is possible. Incidentally, it is interesting to note that Gordon's observers did not produce consistent differences in ratings of satisfaction for wet and dry roads, although the visibility of obstacles was much changed. A similar lack of difference in ratings of various features of roadway lighting for wet and dry conditions is reported by Mainwaring and Stainsby [140].

This demonstrates a common problem of simple appraisals. Presumably the observers were rating whether the roadway lighting was reasonable for the conditions experienced rather than on some absolute basis.

The above discussion of roadway lighting has dealt with the quantitative and qualitative criteria. If they are all met then it seems likely that the roadway lighting will be satisfactory. However, to meet them all is likely to be an expensive business, so it is useful to know their relative importance. Cornwell [132] asked his observers to rate the quality of the visibility provided by the installations as well as the quality of the road luminance, the luminance uniformity, the glare limitation and the visual guidance. He was then able to examine the relationship between these ratings. For dry roads he found the rated quality of visibility was most influenced by the road luminance followed by the visual guidance, luminance uniformity and glare, in that order. For wet roads he found that the rated quality of visibility was most affected by the uniformity. The next most important factor was the road luminance followed by the extent of visual guidance offered and the glare conditions. This increased importance given to non-uniformity in wet conditions does not seem unreasonable given that one of the effects of a wet road surface is to increase the variation in luminance across the surface. The CIE recommendations [122] are consistent with these findings. They suggest that, where circumstances make a reduction in roadway lighting standards necessary, average road surface luminance and overall uniformity should be maintained in preference to Glare Mark and lengthwise uniformity.

By now it should be apparent that the specific criteria adopted for roadway lighting are a matter of judgement as regards both level and importance. Only at the coarsest level is there evidence that roadway lighting affects accident occurrence, most of the detailed criteria being based on their effect on the visibility of objects.

The distribution of visibilities occurring under any specific roadway lighting installation has been used in attempts to devise a single figure index of merit for an installation. The first such measure was suggested by Waldram [141]. His measure, called revealing power, is the percentage of small objects with pedestrian-like reflection characteristics which will be seen when scattered uniformly over the carriageway. Lambert et al. [142] have extended this idea by combining the frequency distribution of all probable object luminances with the frequency distribution of all the luminances of the carriageway. These luminances are used to calculate luminance differences which are compared with the minimum luminance difference necessary for detection. The results of these comparisons are used to produce a figure of merit which is the probability of detection of objects occurring on the carriageway.

A slightly different approach has been adopted by Blackwell and Blackwell [143] and Gallagher [144]. Both attempt to construct a measure of visibility, which is really a measure of how far the target is above threshold. Blackwell and Blackwell use a visibility meter to measure the equivalent contrast of any particular object, and then express this as a ratio of the threshold contrast; i.e. they determine the Visibility Level of the object (see Chapter 3). For the figure of merit, the distribution of Visibility Levels that occur across the carriageway for the object is calculated.

Gallagher uses a similar measure of the extent to which a target is above the threshold, called the Visibility Index, but he uses the actual physical contrast of the object rather than its equivalent contrast as the numerator. This is simpler, but because of the variation in luminance of three-dimensional objects, it seems less likely to be consistently related to the subjective visibility of the object. At present there is considerable controversy in the world of roadway lighting about whether it is useful to express the performance of a lighting installation in terms of the visibility it achieves, which is one of the reasons for its existence, or in photometric terms, which are the measures used in its design. There is something to be said for both sides. Although providing visibility is only one aspect of roadway lighting, it is an important one. It therefore seems reasonable that roadway lighting designs should be evaluated by the distribution of visibilities they produce even though they are designed in photometric terms. Certainly Gallagher [144], using a similar method to that described earlier of observing the behaviour of unwarned drivers when an object is placed on the carriageway, has shown that the time at which drivers take avoiding action is closely related to Visibility Index (and Blackwell avers that Visibility Index is simply a constant multiple of Visibility Level). It may well be that these visibility measures can be related more closely to driver performance than simple photometric measures because visibility is one of the factors the driver bases his judgement on. To balance this, Economopoulos [145] has shown that the distance at which a 0.2×0.2 m object of reflectance 0.11 can be detected on the road by a driver whilst driving relates directly to the average road surface luminance, and that, for any specific object, increasing road surface luminance gives increased visibility, as measured by a quantity similar to the Visibility Index. Basically this is an argument about means rather than ends. Ultimately the aim is to relate roadway lighting conditions to road safety and/or traffic flow rate. It seems unlikely that either simple photometric measurements or the visibility measurements can reveal any precise relationship. Other factors are of equal, if not greater, importance. Rather than arguing about one aspect of driving in detail, it would seem more fruitful to concentrate research activities on identifying which features of the roadway scene are used by drivers in making particular manoeuvres, e.g. by eye movement recordings, and then attempting to identify how different forms of roadway lighting affect the necessary judgements.

4.5.4 Vehicle Lighting
The examination of vehicle lighting can conveniently be divided into two

parts: (a) lighting that is designed primarily to indicate the presence or give information about the movement of a vehicle, (b) lighting that is there principally to enable the driver to see. The former category, e.g. tail lights, side lights and indicator lights, are usually small and of low luminous intensity. The latter, e.g. headlights, are somewhat bigger and of such high luminous intensity that disability glare occurs when two vehicles meet. The former lights are items to be detected, the latter are the means of making obstacles on the carriageway visible.

The conditions necessary for small objects to be visible in roadways have been extensively studied. Hills [115] has produced a predictive model of the factors influencing the visibility of small, round or square areas of light. Figure 4.30 shows a plot of the luminance required for different objects of different visual areas to be just visible when no roadway lighting and no disability glare from oncoming vehicles is present. The results shown were all obtained in field conditions. The scatter in the data is remarkably small considering the different objects used. Based on this curve Hills has developed a series of similar curves for a wide range of different background luminances (Fig. 4.36). In this case the ordinate is the increment of the object luminance necessary for it to be seen against the background luminance. Different values of background luminance enable the effects of different levels of roadway lighting to be examined. Hills has shown that by using these curves he can plausibly predict the experimental results of other workers [146, 147]. This gives the use of Fig. 4.36 some justification but closer examination of the predictions that arise would be worthwhile. Nonetheless the usefulness of such information should be apparent. For a known luminance of background and target, e.g. a rear light, the visual area of the target necessary for it to be just visible can be derived. Then for any specific physical size the distance at which the light will be just visible is available. By juggling the visibility distance thought necessary for safety and the background luminance against which the target will be seen, it is possible to specify the necessary size and luminance for the target to be just visible. Further, from a number of other studies, Hills suggests a series of factors which, when applied to the luminance increment, change the criteria of assessment from 'just visible' to other subjective criteria such as 'just obvious'. However, it has to be remembered that the observers in these experiments were expecting to see an obstacle. Roper and Howard [148] have shown that an unexpected obstacle is only seen at half the distance at which an expected obstacle is detected. Such information forms the background upon which national and international standards of vehicle lighting are based [149], but it is only a background. The people who

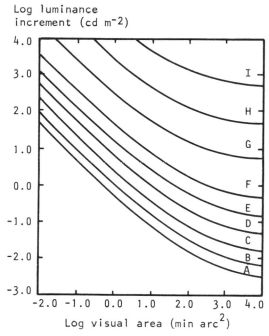

Fig. 4.36. Relationships between luminance increment and visual area at different background luminances, for small targets to be just visible under night driving conditions [115]. Each curve is for one background luminance as follows: A, $0.01\,cd\,m^{-2}$; B, $0.1\,cd\,m^{-2}$; C, $0.32\,cd\,m^{-2}$; D, $1.0\,cd\,m^{-2}$; E, $3.2\,cd\,m^{-2}$; F, $10\,cd\,m^{-2}$; G, $100\,cd\,m^{-2}$; H, $1000\,cd\,m^{-2}$; I, $10\,000\,cd\,m^{-2}$.

determine the standard have to make judgements of the right criteria of visibility to adopt and the distance at which the target should be seen. They are also constrained by what is technically possible, hence the difference in standards for the rear lights of cars [149] and bicycles [150].

Having examined indicator lights, the other part of vehicle lighting needs to be considered. Headlights are important for objects where luminance depends on the reflection of light rather than for such self-luminous objects as rear lights. However, as the data for pedestrian dummies shown in Fig. 4.30 fit the same line, and these data were obtained when driving with dipped headlights and no road lighting, it seems reasonable to suggest that the data shown in Fig. 4.36 can also be used to predict the luminance increment, size or distance for such an object to be visible in the stated conditions, and hence the necessary luminous intensity of headlights. There is one condition attached to these data which is important—no disability glare was present from the headlights of approaching vehicles. In modern day traffic conditions this is somewhat unrealistic.

The extent of disability is quantified by the equivalent veiling luminance, which is the luminance of a uniform veil which changes the luminance difference threshold as much as the actual lighting of interest does. It does not seem unreasonable to suggest that the effect of disability glare on the characteristics necessary for an obstacle to be visible can be obtained from Fig. 4.36 by adding the equivalent veiling luminance to the background luminance, and adjusting the luminance increment and size appropriately. But, before this is done, it should be noted that the formula for equivalent veiling luminance given earlier does not hold for small deviations from the line of sight. These are quite usual when facing headlights. Hartmann and Moser [151] have shown that for angles less than $1.5°$ from the line of sight, the loss of visibility associated with disability glare is much greater than would be predicted, probably due to neural interaction occurring in the retina in addition to scatter in the optic media. For such small angles the suggested formula [115] for equivalent veiling luminance is

$$L_v = \sum_n 9.2 E_n \theta_n^{-3.44}$$

where L_v, E_n and θ_n are as defined earlier.

The effectiveness of different headlight intensities for detecting obstacles in the presence of an approaching vehicle has been studied several times. One of the earlier studies was that of Jehu [152]. He measured the distance at which an observer could recognise which of two standard obstacles, a 45 cm diameter circle or a 30 × 50 cm rectangle, both of reflectance 0·07, were present in the carriageway. The results, which are for one subject only, are shown in Fig. 4.37. Seeing distance is plotted against disability glare intensity, i.e. the luminous intensity of the opposing headlights, towards the observer, for a beam luminous intensity towards the object. The dotted line shows the conventional stopping distance for the 50 km h^{-1} speed used. From such a diagram it is theoretically possible to calculate the likelihood of identifying a standard obstacle throughout the full passage of one car approaching another. The visibility distance changes because the glare intensity changes as the cars approach each other, the extent of the change depending on the luminous intensity distributions of the beams and the relative positions of the cars. Mortimer and Becker [153], using computer simulation and field measurements, have shown that visibility distance for two opposing cars diminishes as the cars close and then starts to increase

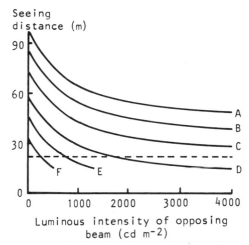

Fig. 4.37. Seeing distance for distinguishing between circular and rectangular obstacles on the carriageway from a car travelling at 50 km h⁻¹ for different headlight and opposing beam luminous intensities. The headlight luminous intensities towards the obstacles are: A, 20 000 cd; B, 10 000 cd; C, 5000 cd; D, 2000 cd; E, 1000 cd; F, 500 cd. The obstacles were placed 3 m to the side of the opposing beam and 3 m behind it. The dotted line indicates the conventional stopping distance for the speed used. (After [152].)

rapidly. The separation at which the visibility distance is minimum depends on the relative luminous intensity distribution of the headlights, the relative positions of the two cars and the obstacles to be seen, and the physical characteristics of the obstacle itself. This study assumes the same luminous intensity distribution for both cars. Helmers and Rumar [154] used cars with headlights adjustable to different high beam luminous intensities. Subjects were driven towards a parked car with its headlights on and asked to indicate when they saw a range of obstacles in the road ahead. The obstacles were flat grey 1·0 × 0·4 m rectangles with a reflectance of 0·045. The results were presented in the form of visibility distance, i.e. the distance at which the obstacles were seen. It was found that when both the observer's vehicle and the opposing vehicle had about the same peak luminous intensity, there was little increase in visibility distance above a luminous intensity of 50 000 cd, the visibility distance being about 60–80 m. However, if the approaching car had a beam of luminous intensity three or more times greater than the observer's car, then the visibility distance was reduced by about one-third compared with that when both vehicles had equal luminous intensities. Figure 4.38 shows these results and describes the pattern of visibility distance change as the cars approach. The actual luminous intensity values to be provided in headlights depend on the distances at

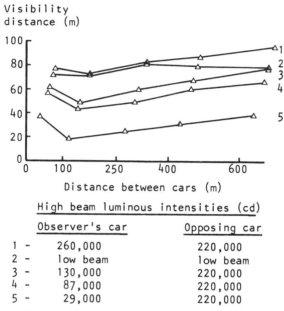

Fig. 4.38. The visibility distance for detecting the presence of a 1·0 m × 0·4 m rectangle of reflectance 0·045 in the presence of opposing headlights, plotted against the distance between the observer's car and the opposing car, for different combinations of headlight luminous intensity. (After [154].)

which it is necessary to detect any obstacles, and on the nature of the obstacles to be detected. For a speed of the order of 120 km h⁻¹ a normal stopping distance in good conditions is about 100 m.

Helmers and Rumar [154] show that for their small, dark grey obstacle, a conventional headlight installation with a high beam luminous intensity of about 80 000 cd gives a visibility distance of about 220 m when no opposing beam is present. However, for two opposing cars with equal luminous intensity headlights, the visibility distance is reduced to about 60–80 m, which is less than the stopping distance. As is evident from Fig. 4.38, increasing the beam intensities equally makes little difference, but when the opposing car has a higher luminous intensity, then the visibility distance reduces even more. It is obvious that driving at high speeds against opposing traffic approaches an act of faith.

There exist national and international standards that relate to vehicle headlights and which are based on this type of information. The standards specify luminous intensity distributions in terms of minimum and maximum luminous intensities in specified directions. For dipped

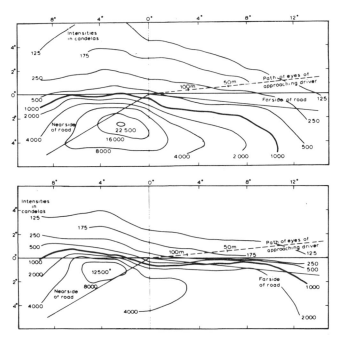

Fig. 4.39. Typical iso-candela diagrams of (upper) Anglo-American dipped headlight beam
and (lower) European dipped headlight beam. (After [160].)

headlights there are two broad classes of distributions, the Anglo-American
beam [155] and the European beam [156]. Differences between them are
slight, the principal feature being that the Anglo-American beam provides a
somewhat higher luminous intensity along the lane directly ahead of the car
(Fig. 4.39). Extensive field studies by Watson [157, 158] have shown that
there is little to choose between the two beams, the Anglo-American beam
giving greater recognition distances on left-hand curves and tops of hills for
nearside obstacles, and the European beam giving greater recognition
distances for offside objects on straight roads and on right-hand curves.

    However, one aspect in which the two beam distributions do differ is
discomfort glare. Schmidt-Clausen and Bindels [159] developed an
expression relating to the discomfort felt with headlights of different
luminous intensity distribution. Figure 4.40 shows the isoglare assessment
surfaces for different illuminances at the eye (glare illuminance) and
different adaptation luminances, with the glare source at different angles
from the line of sight. As might be expected, as the adaptation luminance
increases, the extent of discomfort diminishes; as the angle from the line of

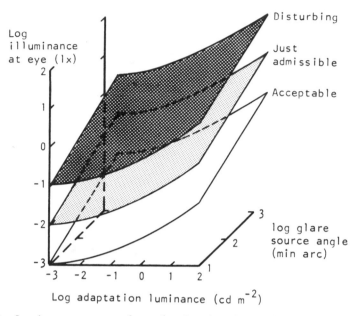

Fig. 4.40.   Iso-glare assessment surfaces plotted against the illuminance at the eye, the adaptation luminance and the angle between the glare source and the line of sight. (After [159].)

sight increases, so the sensation of discomfort reduces; and as the glare illuminance increases, so the discomfort increases. These results are interesting but as far as is known discomfort glare is not considered in any standards of vehicle lighting. Disability glare which affects performance and can also be uncomfortable is considered much more important. However, discomfort glare ought to be involved in preference for beam distribution and in dipped beam behaviour. From studies involving these factors it can be predicted that a pair of European beam headlights on an unlit road will give slightly less glare than a pair of Anglo-American beams [160].

Another aspect of headlights is colour. It is mandatory in France for vehicles to be equipped with yellow headlights, whilst the rest of the world seems to manage on white. An exhaustive review of the relative merits or otherwise of yellow or white headlights for driving [161] concludes that there is no significant difference in their effects.

It can be concluded that the requirements for vehicle lighting are fairly well understood. The problems that arise are basically those of application.

The most pressing problem is how to control the disability glare experienced when meeting another car, since this seriously reduces visibility distances. Headlights producing polarised light have been suggested many times but little progress has been made in their application [160].

4.5.5 Interaction Between Roadway Lighting and Vehicle Lighting

Roadway lighting and vehicle lighting have been considered separately in this section because that is the way the research has usually been conducted. Even so, it is sometimes possible to extend the effects of vehicle lighting into lit roads. Figure 4.36 can be applied to the visibility of rear lights, direction indicators, etc., under both vehicle and roadway lighting conditions. Thus it is possible to indicate at what distance such objects become visible in these conditions. As for disability glare, this occurs on both lit and unlit roads but the general rule is that the roadway lighting will diminish the effect of disability glare from headlights because the adaptation luminance is increased. The same applies to discomfort glare. In the presence of roadway lighting the sense of discomfort created by headlights will be diminished because the background luminance is greater. The real problem with the interaction of roadway lighting and vehicle lighting comes in their combined effect on the visibility of obstacles which are not self-luminous. The problem arises because the design approach to roadway lighting is basically to light the road surface uniformly as seen by the driver. This provides a bright backcloth against which obstacles in the road can then be seen in silhouette. The assumption is that the obstacles will be in silhouette because it is their vertical surfaces which are seen by the driver whereas the roadway is a horizontal surface. This principle works to some degree, although the inevitable inter-reflections, the differences in reflectances and the three-dimensional nature of many obstacles on roads diminish its purity. However, once headlights are introduced into the scene a dramatic change occurs. Headlights are good at lighting vertical surfaces, not so good at lighting horizontal surfaces. Thus, between them, roadway lighting and headlights are capable of lighting all surfaces of obstacles and hence may sometimes reduce their visibility rather than increase it. Ketvirtis [121] measured the visibility distance of three different objects on lit, dual carriageway roads with the test vehicle using dipped headlights and experiencing opposing traffic. The results obtained are shown in Fig. 4.41. As the background luminance from the roadway lighting increases above about $0.2\,\mathrm{cd\,m^{-2}}$ there is a decrease in visibility distance followed by an increase. This discontinuity in visibility distance deserves closer study, particularly with more specular objects.

Rather less dramatic, but still a vexed question, is the attitude to driving with dipped headlights under good roadway lighting, the idea being that the headlights are then primarily to indicate presence rather than to allow the driver to see better. At present dipped headlights tend to produce too much disability glare under conventional roadway lighting conditions. This means they cause problems in seeing for other drivers and pedestrians as

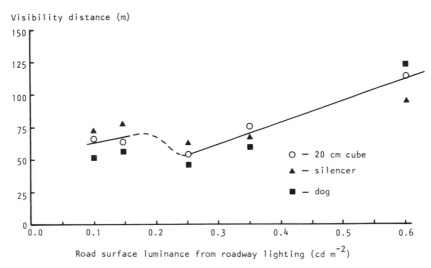

Fig. 4.41. Visibility distances for a 20 cm cube (reflectance = 0·05), a car silencer (reflectance = 0·12) and a poodle dog (reflectance = 0·04) plotted against the road surface luminance produced by the roadway lighting. The visibility distances were obtained from a vehicle using dipped headlights and facing opposing traffic. (After [121].)

well as indicating the vehicle's presence. The alternative of driving on sidelights alone is not thought to be satisfactory because the luminous intensity is insufficient to attract attention under roadway lighting, particularly in a crowded street. The best solution would appear to be the proposed town beam [162] which is produced by dimming the present dipped headlight to give a straight ahead luminous intensity of 80 cd. Although these three alternatives exist, the choice and enforcement of almost any one of them would be better than the present situation. It is a common experience to meet a stream of oncoming traffic with headlights being used by the majority but with a few sidelights-only vehicles amongst them. It is extremely easy for other drivers and pedestrians to look but not to see these vehicles.

4.5.6 Conclusion

Many people have spent considerable amounts of time and money examining the effect of vehicle lighting and roadway lighting on the performance of drivers. The current position is that much is known about the effects of such lighting on obstacle visibility and rather less about the effects on driver behaviour. Vehicle lighting can be divided into that which indicates the presence of a vehicle, e.g. rear lights, and that which enables the driver to see the road ahead, e.g. headlights. Data are available for predicting the conditions for which indicating lights become just visible. The visibility of objects for different headlight luminous intensities, with and without opposing traffic, has been extensively studied, the results indicating that disability glare from oncoming traffic is the major problem. As for roadway lighting, the effects of road surface luminance, uniformity, disability and discomfort glare, and visual guidance are all considered. There is evidence to show that night/day accident ratios decrease with increasing road surface luminance but the other factors have only been examined for their effect on obstacle visibility and driver comfort. The information obtained has been used in drawing up recommendations on conditions necessary for good roadway lighting [122]. The major area waiting to be explored is the influence of wet roads on the effectiveness of roadway and vehicle lighting. Almost invariably wet roads make seeing conditions more difficult. Important though this problem is, the future of roadway and vehicle lighting needs to be considered on a wider basis. Measurements of the visibility of obstacles have been useful but it is about time that other aspects of drivers' tasks were considered. Some studies of what drivers look at when making different manoeuvres would be a start. It might then be possible to identify what sort of information they were seeking and to develop ways of lighting those areas so that the information was easier to see. Studies using realistic three-dimensional objects with their different patterns of luminance under different conditions of roadway lighting and headlights would also be worthwhile. They might lead to a new approach to roadway lighting rather than the silhouette philosophy.

Finally, a realisation that understanding and standards are one thing but practice is another, can be very conducive to getting problems into perspective. Yerrell [163] reports a set of roadside measurements of headlight luminous intensity in Europe. There was a very large spread of intensities. Such a spread indicates that fine changes in patterns of beam distribution are likely to be lost in practice. Using the data on speed collected at the same time, Yerrell calculated the probability of drivers actually being able to distinguish Jehu's targets [152] at 60 m, and being

Table 4.5:   Calculated percentage of situations where the
target is seen at 60 m and the vehicle is capable of stopping
before reaching it [160]

Site	With glare from opposing traffic at 60 m	In the absence of opposing traffic
United Kingdom	35 (trunk)	56 (trunk)
	22 (M-way)	35 (M-way)
Belgium	17	18
Netherlands	33	42
Germany	17	18
France	17 (M-way)	22 (M-way)

able to stop within that distance. His results are shown in Table 4.5. They indicate a great deal of faith on the part of drivers moving at night on dipped beams. Perhaps the cause of accident reduction might be better served by examining why such behaviour occurs rather than arguing further over the precise design characteristics of roadway lighting.

4.6   SUMMARY

This chapter examines the effect of lighting conditions on four specific activities, namely reading, making colour judgements, inspecting for defects and driving. The form of the examination for each of these topics is first to consider the nature of the activity and then to discuss the aspects of lighting that are important to its performance.

For reading, the aspects considered are the illuminance on the print, veiling reflections and light source colour properties. The importance of these variables is interwoven with the typeface used, the print size, contrast and quality as well as layout variables, so these factors are also examined.

The discussion of making colour judgements is more general. A distinction is drawn between discrimination of colours and accurate reproduction of colours. The measures used to characterise these two aspects and the effects of lighting on them are discussed at length. The variables examined are the spectral distribution of the light source, the illuminance on the sample, the size of the sample and the nature of the surroundings.

Visual inspection work has two components, identifying the defect and deciding what to do about it. Lighting principally affects the former. The factors influencing detection are considered in terms of visual detection lobes and discriminability for defects on plain and cluttered backgrounds. From this the principles to be followed to improve visual inspection by lighting are derived.

The visual element of driving is essentially an information collection and interpretation task. The effects of roadway lighting and vehicle lighting on accident rates and the visibility of obstacles are discussed.

Conclusions on the effects of lighting for each of these activities are given at the end of the appropriate sections.

REFERENCES

1. Young, L. R. and Sheena, D. Survey of eye movement recording methods, *Behaviour Research Methods and Instrumentation*, **7**, 397, 1975.
2. Laycock, J. The measurement and analysis of eye movements, in *Search and the Human Observer*, eds J. N. Clare and M. A. Sinclair, Taylor and Francis, London, 1979.
3. Buswell, G. T. *How adults read*, Supplementary Educational Monograph 45, University of Chicago, 1937.
4. Tinker, M. A. Recent studies of eye movements in reading, *Psychological Bulletin*, **65**, 215, 1958.
5. Kolers, P. A. Experiments in reading, *Scientific American*, **227**, 814, 1972.
6. Monty, R. A. and Senders, J. W. *Eye Movements and Psychological Processes*, Lawrence Erlbaryn Associates, Hillsdale, New Jersey, 1970.
7. Huey, E. B. On the psychology and physiology of reading, *Amer. J. Psychol.*, **11**, 283, 1900.
8. Tinker, M. A. *Legibility of Print*, Iowa State University Press, 1963.
9. Burnhill, P., Hartley, J. and Davies, L. Typographic decision making: layout of indexes, *Applied Ergonomics*, **5**, 35, 1977.
10. Tinker, M. A. The relative legibility of the letters, the digits and of certain mathematical signs, *J. Gen. Psychol.*, **1**, 472, 1928.
11. Roethlein, B. A. The relative legibility of different faces of printing types, *Amer. J. Psychol.*, **23**, 1, 1912.
12. Burtt, H. E. and Basch, I. Legibility of Bodoni, Baskerville Roman and Cheltenham type faces, *J. Appl. Psychol.*, **7**, 237, 1923.
13. Paterson, D. G. and Tinker, M. A. *How to Make Type Readable*, Harper, New York, 1940.
14. Zachrisson, B. *Studies in the Legibility of Printed Text*, Alonquist and Wiksell, Stockholm, 1965.
15. Poulton, E. C. The measurement of legibility, *Printing Technology*, **12**, 72, 1968.
16. Poulton, E. C. A note on printing to make comprehension easier, *Ergonomics*, **3**, 245, 1960.
17. Spencer, H., Reynolds, L. and Coe, B. *The effect of image degradation and background noise on the legibility of text and numerals in four different type-faces*, Readability of Print Research Unit, Royal College of Art, London, 1977.
18. Pyke, R. K. *Report on the legibility of print*, HMSO, London, 1926.
19. Poulton, E. C. Letter differentiation and rate of comprehension in reading, *J. Appl. Psychol.*, **49**, 358, 1965.
20. Spencer, H., Reynolds, L. and Coe, B. *The effect of image/background contrast and polarity on the legibility of printed materials*, Readability of Print Research Unit, Royal College of Art, London, 1977.

21. Tinker, M. A. and Paterson, D. G. Influence of type form on speed of reading, *J. Appl. Psychol.*, **12**, 359, 1928.
22. Breland, R. and Breland, M. K. Legibility of message headlines printed in capitals and lower case, *J. Appl. Psychol.*, **28**, 117, 1944.
23. Poulton, E. C. Rate of comprehension of an existing teleprinter output and of possible alternatives, *J. Appl. Psychol.*, **52**, 16, 1968.
24. Paterson, D. G. and Tinker, M. A. Studies of typographical factors influencing speed of reading. II. Size of type, *J. Appl. Psychol.*, **13**, 120, 1929.
25. Poulton, E. C. Skimming (scanning) news items printed in 8 point and 9 point letters, *Ergonomics*, **10**, 713, 1967.
26. Poulton, E. C. Skimming lists of food ingredients printed in different sizes, *J. Appl. Psychol.*, **53**, 55, 1969.
27. Paterson, D. C. and Tinker, M. A. Influence of size of type on eye movements, *J. Appl. Psychol.*, **26**, 227, 1942.
28. Poulton, E. C. *Effects of printing types and formats on the comprehension of scientific journals*, MRC Applied Psychology Research Unit, Cambridge, Report 346, 1959. (A condensed version appeared in *Nature*, **184**, 824, 1959.)
29. Gregory, M. and Poulton, E. C. Even versus uneven right hand margins and the rate of comprehension in reading, *Ergonomics*, **13**, 427, 1970.
30. Spencer, H., Reynolds, L. and Coe, B. Typographical coding in lists and bibliographies, *Applied Ergonomics*, **5**, 136, 1974.
31. Hartley, J., Burnhill, P. and Fraser, S. Typographical problems of journal design, *Applied Ergonomics*, **5**, 15, 1974.
32. Burnhill, P., Hartley, J. and Young, M. Tables in text, *Applied Ergonomics*, **7**, 13, 1976.
33. Spencer, H. *The Visible Word*, Lund Humphries, London, 1968.
34. Poulton, E. C., Warren, T. R. and Bond, J. Ergonomics in journal design, *Applied Ergonomics*, **1**, 207, 1970.
35. Tinker, M. A. The effect of illumination intensities upon speed of perception and upon fatigue in reading, *J. Educ. Psychol.*, **30**, 561, 1939.
36. Smith, S. W. and Rea, M. S. Proof-reading under different levels of illumination, *J.I.E.S.*, **8**, 47, 1978.
37. Ronchi, L. and Neri, M. Speed of reading versus target contrast, *Atti della Fondazione Giorgio Ronchi*, **29**, 957, 1974.
38. De Boer, J. B. Performance and comfort in the presence of veiling reflections, *Ltg Res. and Technol.*, **9**, 169, 1977.
39. Boyce, P. R. *Veiling reflections: an experimental study on their effects on office work*, Electricity Council Research Centre Memorandum 1230, Capenhurst, UK, 1979.
40. Hopkinson, R. G. A preliminary study of reflected glare, in *Architectural Physics: Lighting*, ed. R. G. Hopkinson, HMSO, London, 1963.
41. Smith, S. W. and Rea, M. S. Relationships between office task performance and ratings of feelings and task evaluation under different light sources and levels, *Proc. CIE 19th session, Kyoto*, 1979.
42. Poulton, E. C., Kendall, P. G. and Thomas, R. J. Reading efficiency in flickering light, *Nature*, **209**, 1267, 1966.
43. Tinker, M. A. and Paterson, D. G. Studies of typographical factors influencing speed of reading. VII. Variations in colour of print and background, *J. Appl. Psychol.*, **15**, 471, 1931.
44. Patel, A. S. and Jones, R. W. Increment and decrement visual thresholds, *J. Opt. Soc. Amer.*, **58**, 696, 1968.
45. Paterson, D. G. and Tinker, M. A. Studies of typographical factors influencing speed of reading. VI. Black type versus white type, *J. Appl. Psychol.*, **15**, 241, 1931.
46. Terry, P. W. The reading problem in arithmetic, *J. Educ. Psychol.*, **12**, 365, 1921.
47. Carmichael, L. and Dearborn, W. F. *Reading and Visual Fatigue*, Greenwood Press, Westport, Connecticut, 1947.

48. Cakir, A. The incidence and importance of eyestrain amongst VDU operators, *Eyestrain and VDU's symposium*, *The Ergonomics Society, Loughborough, UK*, 1979.
49. Hultgren, G. V. and Knave, B. Discomfort glare and disturbances from light reflection in an office environment with CRT display terminals, *Applied Ergonomics*, **5**, 2, 1974.
50. Cakir, A., Hart, D. J. and Stewart, T. F. M. *The VDT Manual*, Inca-Fiej Research Association, Darmstadt, West Germany, 1979.
51. Anderson, I. H. and Meredith, C. W. The reading of projected books with special reference to rate and visual fatigue, *J. Educational Res.*, **41**, 453, 1948.
52. Commission Internationale de l'Eclairage. *Colorimetry*, CIE Publication 15, 1971.
53. Commission Internationale de l'Eclairage. *Official recommendations on uniform colour spaces, colour difference equations and psycho-metric colour terms*, CIE Publication 15, Supplement 2, 1978.
54. Hunt, R. W. G. The specification of colour appearance. I. Concepts and terms. II. Effects of changes in viewing conditions, *Colour Research and Application*, **2**, 55 and 109, 1977.
55. American National Standards Institute. *Viewing conditions for the appraisal of colour quality and colour uniformity in the graphic arts*, ANSI PH2.32, New York, 1972.
56. Pracejus, W. G. Preliminary report on a new approach to colour acceptance studies, *Illum. Engng*, **62**, 633, 1967.
57. Commission Internationale de l'Eclairage. *Methods of measuring and specifying colour rendering properties of light sources*, CIE Publication 13.2, 1974.
58. Lynes, J. A. *Colour discrimination and heat rejecting window glasses*, Plymouth Polytechnic, School of Architecture, Report 10/73, Plymouth, UK, 1973.
59. Thornton, W. A. Colour discrimination index, *J. Opt. Soc. Amer.*, **62**, 191, 1972.
60. Pointer, M. R. Colour discrimination as a function of observer adaptation, *J. Opt. Soc. Amer.*, **64**, 750, 1974.
61. Farnsworth, D. The Farnsworth–Munsell 100 hue and dichotomous tests for colour vision, *J. Opt. Soc. Amer.*, **33**, 568, 1943.
62. Boyce, P. R. and Simons, R. H. Hue discrimination and light sources, *Ltg Res. and Technol.*, **9**, 125, 1977.
63. British Standards Institution. BS 950, *Artificial daylight for the assessment of colour*, 1967; and Amendment 1, 1968.
64. Joint Committee on Lighting and Vision. *Spectral requirements of light sources for clinical purposes*, MRC Memorandum 43, HMSO, London, 1965.
65. Crawford, B. H. Colour rendering, tolerances and colour rendering properties of light sources, *Trans. Illum. Engng Soc. (London)*, **28**, 60, 1963.
66. Crawford, B. H. The colour rendering properties of illuminants: the application of psycho-physical measures to their evaluation, *Brit. J. Appl. Phys.*, **14**, 319, 1963.
67. Illuminating Engineering Society. *IES Code for Interior Lighting*, IES, London, 1977.
68. Bedford, R. E. and Wyszecki, G. W. Wavelength discrimination for point sources, *J. Opt. Soc. Amer.*, **48**, 129, 1958.
69. Brown, W. R. J. The influence of luminance level on visual sensitivity to colour differences, *J. Opt. Soc. Amer.*, **41**, 684, 1951.
70. Cornu, L. and Harlay, F. Modifications de la discrimination chromatique en fonction de l'éclairement, *Vision Research*, **9**, 1273, 1969.
71. Brown, W. R. J. The effect of field size and chromatic surrounds on colour discrimination, *J. Opt. Soc. Amer.*, **42**, 837, 1952.
72. Colour Measurement Committee of the Society of Dyers and Colourists. The impact of modern lighting on the dyer, *Journal of the Society of Dyers and Colourists*, **92**, 407, 1976.
73. Fox, J. G. Quality control of coins, in *Human Factors in Work Design and Production*, eds H. G. Maule and J. S. Weiner, Taylor and Francis, London, 1977.
74. Gillies, G. J. Glass inspection, in *Human Reliability in Quality Control*, eds C. G. Drury and J. G. Fox, Taylor and Francis, London, 1975.

75. Morris, A. and Horne, E. P. (eds). *Visual Search Techniques*, US National Academy of Sciences, Washington, DC, 1960.
76. Megaw, E. D. and Richardson, J. Eye movements and visual inspection, *Applied Ergonomics*, **10**, 145, 1979.
77. Enoch, J. M. Natural tendencies in visual search of a complex display, in *Visual Search Techniques*, eds A. Morris and E. P. Horne, US National Academy of Sciences, Washington, DC, 1960.
78. Kundel, H. L. and Nodine, C. F. Interpreting chest radiographs without visual search, *Radiology*, **116**, 527, 1975.
79. Megaw, E. D. and Richardson, J. Target uncertainty and visual scanning strategies, *Human Factors*, **21**, 303, 1979.
80. Lamar, E. S. Operational background and physical conditions relative to visual search problems, in *Visual Search Techniques*, eds A. Morris and E. P. Horne, US National Academy of Sciences, Washington, DC, 1960.
81. Bloomfield, J. R. Theoretical approaches to visual search, in *Human Reliability in Quality Control*, eds C. G. Drury and J. G. Fox, Taylor and Francis, London, 1975.
82. Drury, C. G. Inspection of sheet material—model and data, *Human Factors*, **17**, 257, 1975.
83. Barbur, J. L. Visual periphery, in *Search and the Human Observer*, eds J. N. Clare and M. A. Sinclair, Taylor and Francis, London, 1979.
84. Drury, C. G. and Clement, M. R. The effect of area, density and number of background characters on visual search, *Human Factors*, **20**, 597, 1978.
85. Howarth, C. I. and Bloomfield, J. R. A rational equation for predicting search times in simple inspection tasks, *Psychonomic Science*, **17**, 225, 1969.
86. Bloomfield, J. R. Studies in visual search, in *Human Reliability in Quality Control*, eds C. G. Drury and J. G. Fox, Taylor and Francis, London, 1975.
87. Bloomfield, J. R., Howarth, C. I. and Bone, K. E. *Visual search and pattern recognition*, University of Nottingham research report, Nottingham, UK, 1972.
88. Bloomfield, J. R. *Visual search*, Ph.D. Thesis, University of Nottingham, UK, 1970.
89. Bloomfield, J. R. and Howarth, C. I. Testing visual search theory, in *Image Evaluation*, *Proc. NATO Advisory Group on Human Factors*, Munich, ed. H. W. Leibowitz, 1969.
90. Engel, F. L. Visual conspicuity, directed attention and retinal locus, *Vision Research*, **11**, 563, 1971.
91. Engel, F. L. Visual conspicuity, visual search and fixation tendencies of the eye, *Vision Research*, **17**, 95, 1977.
92. Williams, L. C. The effect of target specification on objects fixated during visual search, *Perception and Psychophysics*, **1**, 315, 1966.
93. Bartz, B. S. Experimental use of the search task in an analysis of type legibility in cartography, *J. Typographical Research*, **4**, 147, 1970.
94. Kristofferson, M. W. Types and frequency of errors in visual search, *Perception and Psychophysics*, **11**, 325, 1972.
95. Gould, J. D. and Carn, R. Visual search, complex backgrounds, mental counters and eye movements, *Perception and Psychophysics*, **14**, 125, 1973.
96. Mills, R. and Sinclair, M. A. Aspects of inspection in a knitwear company, *Applied Ergonomics*, **7**, 97, 1976.
97. Thomas, L. F. and Seaborne, A. F. M. The socio-technical context of industrial inspection, *Occupational Psychology*, **35**, 36, 1961.
98. Pope, M. C., Shaw, M. L. and Thomas, L. F. *A report on the use of repertory grid techniques in final inspection*, Centre for the Study of Human Learning, Brunel University, Uxbridge, UK, 1977.
99. Megaw, E. D. Factors affecting visual inspection accuracy, *Applied Ergonomics*, **10**, 27, 1979.

100. Muck, E. and Bodmann, H. W. Die bedeutung des beleuchtungsniveaus bei praktischer sehtatigkeit, *Lichttechnik*, **13**, 502, 1961.
101. Wyatt, S. and Langdon, J. N. *Inspection processes in industry*, MRC Industrial Health Research Board, Report 63, HMSO, London, 1932.
102. Faulkner, T. W. and Murphy, T. J. Lighting for difficult visual tasks, *Human Factors*, **15**, 149, 1973.
103. Illuminating Engineering Society of North America. *IES Lighting Handbook*, 5th edition, IES, New York, 1972.
104. Smith, G. L. Inspector performance on microminiature tasks, in *Human Reliability in Quality Control*, eds C. G. Drury and J. G. Fox, Taylor and Francis, London, 1975.
105. Dekoker, N. and Frier, J. P. Visibility of specular and semi-specular tasks in sheet metal surfaces, *Illum. Engng.*, **64**, 167, 1969.
106. Bellchambers, H. E. and Phillipson, J. Lighting for inspection, *Trans. Illum. Engng Soc. (London)*, **27**, 71, 1962.
107. Frier, J. P. Difficult industrial tasks and how to light them, *Illum. Engng*, **62**, 283, 1967.
108. Schoonard, J. W. and Gould, J. D. Field of view and target uncertainty in visual search and inspection, *Human Factors*, **15**, 33, 1973.
109. Buck, J. R. Dynamic visual inspection: task factors, theory and economics, in *Human Reliability in Quality Control*, eds C. G. Drury and J. G. Fox, Taylor and Francis, London, 1975.
110. Fisher, A. J. Road lighting as an accident countermeasure, *Australian Road Research*, **7**, 3, 1977.
111. Sabey, B. E. and Staughton, E. C. Interacting roles of road environment, vehicle and road user in accidents, *5th International Conference of the International Association for Accident and Traffic Medicine, London*, 1975.
112. Hills, B. L. Vision, visibility and perception in driving, *Perception*. **9**, 183, 1980.
113. Mourant, R. R. and Rockwell, T. H. Mapping eye-movement patterns to the visual scene in driving: an exploratory study, *Human Factors*, **12**, 81, 1970.
114. Mourant R. R. and Rockwell, T. H. Strategies of visual search by novice and experienced drivers, *Human Factors*, **14**, 325, 1972.
115. Hills, B. L. Visibility under night driving conditions: derivation of $(\Delta L, A)$ characteristics and factors in their application, *Ltg Res. and Technol.*, **8**, 11, 1976.
116. Burg, A. *The relationship between vision test scores and driving records: general findings*, Report 67-24, Department of Engineering, University of California, Los Angeles, 1967.
117. Davidson, P. A. The role of drivers vision in road safety, *Ltg Res. and Technol.*, **10**, 125, 1978.
118. Jones, H. V. and Heimstra, N. W. Ability of drivers to make critical passing judgements, *J. of Engineering Psychol.*, **3**, 117, 1964.
119. Tanner, J. Reduction of accidents by improved street lighting, *Light and Lighting*, **51**, 11, 1958.
120. Commission Internationale de l'Eclairage. *Street lighting and accidents*, CIE Publication 8, 1960.
121. Ketvirtis, A. *Road illumination and traffic safety*, Transport Canada, Ottawa, 1977.
122. Commission Internationale de l'Eclairage. *International recommendations for the lighting of roads for motorized traffic*, CIE Publication 12/2, 1977.
123. Skene, P. *The cost effectiveness of upgrading urban street lighting*, M. Eng. Sc. project, School of Transportation and Traffic, University of New South Wales, Australia, 1976.
124. Hargroves, R. A. and Scott, P. P. Measurements of road lighting and accidents—the results, *Public Lighting*, **44**, 213, 1979.
125. Gallagher, V. P., Janoff, M. S. and Farber, E. Interaction between fixed and vehicular illumination systems on city streets, *J.I.E.S.*, **4**, 3, 1974.
126. Fischer, D. European approach to the luminance aspect of roadway lighting, *J.I.E.S.*, **4**, 111, 1975.

127. De Boer, J. A. Strassenleuchtdichte und blendungsfreiheit als praktische massstare fur die gute offentlurcher beleuchtung, *Lichttechnik*, **10**, 359, 1958.
128. Narisada, K. Influence of non-uniformity in road surface luminance of public lighting installations upon perception of objects on the road surface by car drivers, *Proc. CIE, 17th session, Barcelona*, 1971.
129. Stiles, W. S. The effect of glare on the brightness difference threshold, *Proc. Roy. Soc.*, **B104**, 322, 1929.
130. Adrian, W. and Eberbach, K. Uber den einfluss der leuchtenausdehnung bei der physhologischen blendung in der strassenbeleuchtung, *Lichttechnik*, **19**, 67A, 1967.
131. Fry, G. D. A re-evaluation of the scattering theory of glare, *Illum. Engng*, **49**, 98, 1954.
132. Cornwell, P. R. Appraisals of traffic route lighting installations, *Ltg Res. and Technol.*, **5**, 10, 1973.
133. Gordon, P. Appraisals of visibility on lighted dry and wet roads, *Ltg Res. and Technol.*, **9**, 177, 1977.
134. Allen, M. J., Hazlett, R. D., Tacker, H. L. and Graham, B. V. Actual pedestrian visibility and the pedestrian's estimate of his own visibility, *Am. J. of Optometry and Arch. of Am. Ac. of Optometry*, **47**, 44, 1970.
135. Shinar, D., McDowell, E. D. and Rockwell, T. H. Eye movements in curve negotiation, *Human Factors*, **19**, 63, 1977.
136. Eastman, A. A. and McNelis, J. F. An evaluation of sodium, mercury and filament lighting for roadways, *Illum. Engng*, **58**, 28, 1963.
137. De Boer, J. B. Modern light sources for highways, *J.I.E.S.*, **3**, 142, 1974.
138. Buck, J. A., McCowan, T. K. and McNelis, J. F. Roadway visibility as a function of light source colour, *J.I.E.S.*, **5**, 20, 1975.
139. Commission Internationale de l'Eclairage. *Road lighting for wet conditions*, CIE Publication 47, 1979.
140. Mainwaring, G. and Stainsby, R. G. The effect of lantern light distribution on subjective assessments in a street lighting installation, *Trans. Illum. Engng Soc. (London)*, **33**, 98, 1968.
141. Waldram, J. M. The revealing power of street lighting installations, *Trans. Illum. Engng Soc. (London)*, **3**, 173, 1938.
142. Lambert, G. K., Marsden, A. M. and Simons, R. H. Lantern intensity distribution and installation performance, *Public Lighting*, **160**, 27, 1973.
143. Blackwell, D. M. and Blackwell, H. R. A proposed procedure for predicting performance aspects of roadway lighting in terms of visibility, *J.I.E.S.*, **6**, 148, 1977.
144. Gallagher, V. P. A visibility metric for safe lighting of city streets, *J.I.E.S.*, **5**, 90, 1976.
145. Economopoulos, I. A. Relationship between lighting parameters and visual performance in road lighting, *Lichttechnische Gesellschaft, 3rd International Symposium, 'Measures of road lighting effectiveness'*, *Karlsruhe*, July 1977.
146. Dunbar, C. Necessary values of brightness contrast in artificially lighted streets, *Trans. Illum. Engng. Soc. (London)*, **3**, 187, 1938.
147. Moore, R. L. *Rear lights of motor vehicles and pedal cycles*, Road Research Technical Paper 25, HMSO, London, 1952.
148. Roper, V. J. and Howard, E. A. Seeing with motor car headlamps, *Trans. Illum. Engng Soc.* **33**, 417, 1938.
149. British Standards Institution. BS AU 40, *Motor vehicle lighting and signalling equipment. Part 1, Side, rear, parking, marker and stop lamps, and direction indicators*, 1963, Amendment 1, 1967.
150. British Standards Institution. BS 3648, *Cycle rear lamps*, 1963.
151. Hartmann, E. and Moser, E. A. The law of physiological glare at very small glare angles, *Lichttechnik*, **20**, 67A, 1968.
152. Jehu, V. J. A method of evaluating seeing distances on a straight road for vehicle meeting beams, *Trans. Illum. Engng Soc. (London)*, **20**, 57, 1955.

153. Mortimer, R. G. and Becker, J. M. *Development of a computer simulation to predict the visibility distances provided by headlamp beams*, Report UM-HSRI-1AF-73-15, University of Michigan, 1973.
154. Helmers, G. and Rumar, K. High beam intensity and obstacle visibility, *Ltg Res. and Technol.*, **7**, 35, 1975.
155. British Standards Institution. BS AU 40, *Motor vehicle lighting and signalling equipment. Part 3, Headlamps with pre-focus filament lamps. Part 4, Sealed beam headlights*, 1966, Amendment 1, 1976.
156. European Economic Community Regulation E/ECE 324, 1965.
157. Watson, R. L. *A comparison of three different dipped headlight beams in meeting situations*, Road Research Laboratory, Technical Note TN486, Crowthorne, UK, 1970.
158. Watson, R. L. *Further comparison of dipped headlight beams*, Road Research Laboratory, Technical Note TN492, Crowthorne, UK, 1970.
159. Schmidt-Clausen, H. J. and Bindels, J. Th. H. Assessment of discomfort glare in motor vehicle lighting, *Ltg Res. and Technol.*, **6**, 79, 1974.
160. Yerrell, J. S. Vehicle headlights, *Ltg Res. and Technol.*, **8**, 69, 1976.
161. Schroeder, D. A. *White or yellow light for vehicle headlights*, Institute for Road Safety Research, SWOV, Voorburg, Netherlands, 1976.
162. Fisher, A. J. The luminous intensity requirement of vehicle front lights for use in towns, *Ergonomics*, **17**, 87, 1974.
163. Yerrell, J. S. *Headlight intensities in Europe and Britain*, Road Research Laboratory, LR383, Crowthorne, UK, 1971.

5

Uncertainties

5.1 INTRODUCTION

Although a consistent overall pattern of the effect of light on work has been demonstrated and some specific activities have been examined in detail, our understanding of the relationship between light and work is not complete. There remain a number of areas of uncertainty and controversy. These areas arise either because of a misunderstanding of the available evidence, or because the available evidence conflicts, or because there is no firm evidence about effects which are believed rather than known to occur. The aim of this chapter is to examine the more important of these areas, to demonstrate the balance of the available evidence and, where necessary, to suggest means whereby the truth might be revealed.

5.2 THRESHOLD/SUPRATHRESHOLD

One area of almost perpetual disagreement concerning the relationship between light and work is the relevance of threshold performance to everyday work. For many people the exact link between the effects of lighting conditions on threshold performance and on real life tasks is by no means obvious. This is because of the numerous differences between threshold experiments and everyday work. Among these differences are the following.

(a)   The level of performance usually taken as the threshold is 50% correct, which is hardly realistic for everyday work performance.

(b)   The conditions for threshold measurements are very restrictive and
       not usually representative of real work conditions. For example, for
       most threshold measurements the target occurs in a known place at
       a known time for a fixed duration.
(c)   The balance between speed and accuracy is different in real life and
       in threshold experiments. In experiments, speed is usually fixed and
       the threshold is determined by a measure of accuracy. In real life the
       balance between speed and accuracy is fixed by the worker, usually
       at a level of high accuracy so that changes in lighting mainly
       produce changes in speed.
(d)   The activities requested of the observer in threshold experiments,
       usually simple detection, are a long way from the more integrated
       aspects of real work.

These differences are sufficient to explain the tendency to talk about two
discrete areas of work which may be related to lighting conditions in
different ways, namely threshold and suprathreshold. The implication of
this separation is that although threshold studies are interesting to some, it
is only suprathreshold conditions that are relevant for everyday work.
Therefore the first point that needs to be considered is the extent to which
threshold and suprathreshold performance are different, in kind as well as
degree. As soon as an attempt is made to do this another problem appears.
This is the lack of a definition of suprathreshold performance. Obviously
suprathreshold performance is above the 50% level which defines
threshold, but how far above? Is it 100% detection, or is it anything above
50%, or is it some entirely different aspect of performance? In the absence
of a quantitative definition it is necessary to define by common usage. On
examination it is apparent that suprathreshold performance is used
synonymously with everyday performance of everyday tasks. This means
that the question being considered reverts to whether threshold
performance measures are relevant to the types of task performed in
everyday life. To answer this it is necessary to examine the differences
between threshold measures and everyday tasks in detail.
    First, the level of performance will be considered. Although threshold
levels are conventionally defined at 50% performance, higher levels of
performance do occur in threshold experiments. In fact in some conditions
near to 100% performance is obtained. The important point is whether
these different levels of performance are related in some consistent way.
Experience shows that the results obtained in threshold experiments can
usually be fitted by a normal cumulative frequency distribution curve

(Fig. 5.1) [1]. Only two quantities are necessary to define such a curve, the mean value, which is the threshold value, and the standard deviation of the results, which defines the slope of the curve. Different stimulus conditions will have different means and standard deviations but all the curves go in the same direction, higher levels of stimulus being required for higher levels of performance. Thus there is some sort of relationship between more realistic

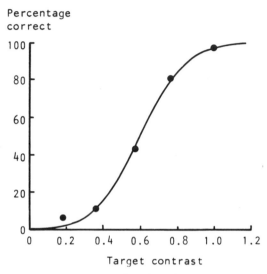

Fig. 5.1. Percentage correct responses, plotted against target contrast. The data was obtained for an observer detecting the presence of a disc appearing for a fixed time in a known place. The disc has a higher luminance than the background so luminance contrast, $C = |(L_t - L_b)/L_b|$ can be greater than unity [1].

levels of performance and threshold levels, which is consistent in direction at least. However, the exact value of the relationship is dependent on the particular conditions.

The second difference is the unrealistic nature of the conditions for many threshold measurements. But there is no necessity for the experimental conditions to be unrealistic. Threshold measurements could be taken for targets appearing at random, in unexpected positions and for different times. Nor should the likely importance of such factors be underestimated. Reduced presentation time is known to increase threshold contrast [2]. It seems only too probable that increasing the uncertainty about when and where a target will appear would do the same. Unfortunately most of the threshold data available are for targets occurring in a known position at a known time. This is because most of the threshold performance results have been obtained by people who were primarily interested in the operation of the visual system, rather than any effects of lighting on task performance. The third difference concerns the balance between speed and accuracy. Shifting the level of performance to be achieved to higher values, e.g. 95 % detection can overcome the problem of the increased accuracy called for in real life tasks. It also seems reasonable to suppose that limiting the time of exposure of the target in a threshold experiment can effectively vary the speed required [3]. This suggestion is based on the assumption that an exposure of 200 ms is equivalent to handling five targets per second. Whilst this assumption is unlikely to be valid at very short times, it does suggest that it is possible to put an element of speed into threshold measurements.

The fourth difference, the nature of the performance requested, is the most difficult to deal with. In many threshold experiments a simple measure of performance is required, so a simple type of activity is usually necessary. In real life many tasks are not simple and do not produce an easily quantifiable output. This difference in the nature and complexity of the work demanded for threshold measures and for the performance of everyday tasks is serious. It cannot be bridged and hence should always be remembered when considering the relevance of threshold measures for everyday tasks.

The overall conclusion from the above discussion is that there is no theoretical reason why the effect of lighting conditions on threshold performance should not be a useful guide to the effect of the same conditions on the performance of some everyday tasks. A hard distinction between threshold and suprathreshold is false. Threshold performance is simply a low level on the performance continuum which stretches from no performance to perfect performance. The vital question when considering the relevance of threshold performance data to everyday work is the conditions under which they were obtained.

If this conclusion is accepted the next point to consider is the value of the vast array of threshold performance data that already exist (see Chapter 2). Unfortunately most of the data available have been obtained under very simple conditions so their value as a predictor for most real tasks is limited. Nonetheless they are often the only data available. In this situation they can be used to suggest the relative importance of different lighting conditions and the direction of the effect they have, but that is about all. If accurate information is needed on the performance of a specific task under different lighting conditions, there is no substitute for a proper experimental investigation of the task itself.

5.3   EASE OF SEEING AND VISUAL PERFORMANCE

A burgeoning controversy surrounds the implications of two measures which attempt to quantify the ease with which targets can be seen under different lighting conditions. Visibility Level (see Chapter 3) is one, the other is the apparent contrast of the target, i.e. the contrast it appears to have. Kulikowski [4] investigated the variation of apparent contrast with luminance by presenting two sinusoidal grating targets to an observer, both with a spatial frequency of $5\,\mathrm{deg}^{-1}$, but with different average luminances. One grating was taken as the standard and the observer was asked to adjust the contrast of the other until both gratings appeared to have the same contrast, i.e. until the apparent contrasts of the two gratings were equal. Kulikowski found that the apparent contrasts of the gratings were equal when the difference (luminance contrast—threshold contrast) of each grating was the same. Further, this relationship was found to hold for a wide range of luminances and for different spatial frequencies. This result implies that the apparent contrast of a target is directly related to the difference between its luminance contrast and its threshold contrast. This in turn suggests that as luminance increases the apparent contrast of a target will increase, since the luminance contrast will stay the same but the threshold contrast will decrease (Chapter 2).

Yonemura et al. [5] performed a similar matching experiment using printed words. Figure 5.2 shows their results in terms of equal apparent contrast contours, i.e. all the connected points are perceived as having the same apparent contrast. The interesting feature of these data is the difference between high and low contrast conditions. For low contrast conditions, as luminance increases a smaller luminance contrast is required to maintain equal apparent contrast. This is what would be expected from Kulikowski's results. However, at high contrasts, as luminance increases the luminance contrast necessary to maintain equal apparent contrast first decreases and then increases. This increase would not be predicted from Kulikowski's results. To confuse matters further a recent study which attempted to replicate the results of Yonemura et al. did not find any upward trend in the equal apparent contrast contour for high contrast, high luminance conditions [6].

It is important that this confusion of data should be resolved. Apparent contrast can be considered as an aspect of how easy a target is to see. In considering only contrast it is more limited than Visibility Level, but it should still indicate the same effect of lighting conditions on ease of seeing. If Kulikowski's conclusion is generally true then both apparent contrast

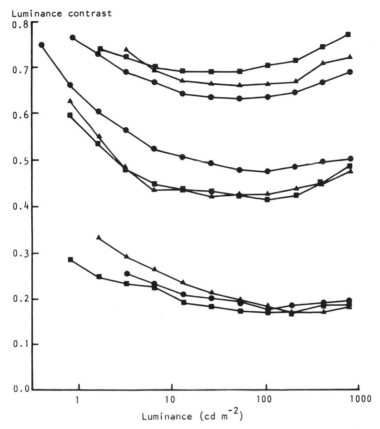

Fig. 5.2.   Equal apparent contrast contours for words in three letter sizes: ●, 6·5 min arc; ■, 4·0 min arc; ▲, 3·2 min arc. (After [5].)

and Visibility Level predict the same effect for increased luminance, i.e. greater ease of seeing. However, if Yonemura *et al.* are correct about what happens to apparent contrast for high contrasts at high luminances then some conflict between apparent contrast and Visibility Level is evident. Specifically, the results of Yonemura *et al.* imply that while increasing luminance makes low contrast targets easier to see, for high contrast targets increases in luminance eventually cause a reduction in the ease of seeing. In comparison, Visibility Level implies that increases in luminance increase ease of seeing for both low and high contrast targets. The question now is 'how might this conflict be resolved?' A critical experiment could be based on the assumption that ease of seeing can be related to task performance. If this assumption seems reasonable then the critical question is 'can a task be found in which a high contrast target produces a decline in performance at luminances where a low contrast target does not?' Figure 5.2 suggests that the luminances in question are several hundred candelas per square metre, which is higher than is normally produced by interior lighting even on white paper. This means the effect of any decline found is unlikely to be of great practical significance. Nonetheless the controversy does raise an important question of principle, namely, whether steadily increasing luminance always produce an increase in ease of seeing or not, and if not under what condition this assumption fails. Until more experimental data which examine the critical region of high contrast tasks at high luminances are available, the question must remain unresolved.

5.4 EFFORT AND FATIGUE

5.4.1 Effort

The measurements used to evaluate the influence on lighting on the performance of tasks generally only involve the output produced. Speed of working, or number of errors made, or some combination of them, are typical measures. But to produce an output some form of input is required. In general this input can be considered as having two components, the information collected from the visual world and the effort made to perform the task. In all the experiments discussed earlier, the variation in input information produced by lighting conditions has been used as the independent variable, and the measured output as the dependent variable. Implicit in this procedure is the assumption that the effort made by the subject is constant over the different experimental conditions. This assumption is not completely unreasonable because in an obviously experimental situation most people are well motivated to succeed, so a consistently high level of effort is likely. Nonetheless it is an assumption, and when considering the effects of lighting in detail, small changes in effort may well become significant. Of particular interest is the possibility that the effect of poor lighting conditions on work output can be masked by an increase in effort made. Such compensation is believed to happen when the effects of temperature on work are examined in the laboratory, because it is very difficult to show any effects on performance even under conditions which are considered to be uncomfortable [7]. It would be interesting to know if compensatory increases of effort occurred with poor lighting conditions as well. This alone is reason enough to examine the variation in effort, as well as output, with lighting conditions; but there is another reason. For some tasks it is difficult to define a suitable output measure. In this situation a simple measure of effort would be very valuable.

The problem is to find a quantity that is a valid measure of effort. Luckiesh and Moss [8] looked long and hard for a suitable measure to use with visual tasks. They examined a range of simple physiological variables, chosen on the basis that such measures could not be voluntarily controlled and would be influenced by any stress caused by the particular lighting conditions. Finger pressure, heart rate and blink rate were all studied, the measurements usually being taken before and after a specified time spent reading. Although all these measures were studied, it was blink rate that was eventually chosen as the most sensitive and consistent. Luckiesh and Moss [8] showed that different blink rates were obtained for reading over a one-hour period under different illuminances, using type faces of different sizes, in the presence or absence of glare, and whilst wearing incorrect or correct spectacles. Further, the worse conditions, i.e. low illuminance, smaller print size, the presence of glare and incorrect spectacles, all produced higher blink rates. Luckiesh and Moss considered blink rate to be a sensitive and reliable indicator of the effort involved in reading [9]. Unfortunately, McPherson [10] failed to replicate the above results and experiments by Tinker [11] showed that there were considerable inter- and intra-individual variations in blink rate during reading. As an example, Tinker observed that under the same lighting conditions some subjects showed considerable decreases in blink rate, others showed large increases, while the blink rate of yet others oscillated. Bitterman [12] found the same variable effects for blink rate. Thus, in spite of its early promise, blink rate cannot be considered a very consistent index of the effort involved in visual work.

The failure of blink rate can be explained by two observations. First, blinking, like headaches, is probably a generalised response to a large number of different circumstances; in which case variations due to a specific condition will be difficult to identify. Second, blinking is not concerned with the major activity involved in much visual work, namely the central processing of information obtained from the visual world. Rather it is connected with the peripheral but necessary activities of protecting, moistening and cleaning the cornea. In order to get closer to the effort involved in a specific visual task, an approach that involves information processing activity would seem more promising.

One approach to measuring the extent of central information processing is the dual task method. This rests on the assumption that man can only process information at a finite rate, i.e. he has a limited information handling capacity. In most situations he uses less than the full capacity, so spare capacity exists. When he makes an effort to increase his level of performance or maintain his performance in worsening conditions, he uses more of the spare capacity. The essence of the dual task approach is to measure the residual spare capacity under different conditions. The method involves giving the subject two tasks to do simultaneously, one being the visual task of interest, the other task being something simple but which causes a continuous load. The two tasks are assumed to take up all or nearly all the information handling capacity. Then, when extra effort is called for to maintain performance on the visual task, the extra processing capacity required will be taken from the second task and will be indicated by a reduction in its performance. This approach has quite a long history (see Welford [13] for a review), and has been used successfully in realistic situations. Brown and Poulton [14] had people driving around the city of Cambridge and at the same time listening to sets of three digit numbers presented at a fixed rate. As well as driving, the driver had to sum the three digits and speak the total directly after each set of digits was given. It was found that in areas requiring complex decisions for driving, e.g. shopping areas, there was a significant increase in the number of errors made in the additions. This is what would be expected from the limited-capacity model of human information processing. Similar results were found for pilots with different levels of skill making a simulated landing [15]. There is thus evidence that the difficulty of the task does affect the proportion of the total capacity which is used and that the variation in the capacity used can be measured by a secondary loading task. This observation led to an experiment, aimed at comparing the sensitivity of performance measurements for a single task with dual task measurements, for evaluating the effects of illuminance on task performance [16]. The visual task used in both single and dual task modes was the detection of the gap orientation in complex Landolt rings with gaps of size 4 min arc and contrast 0·6. The subject had to indicate the orientation of the gap on a push button keyboard. Each Landolt ring was presented for 1 s. The output measure was simply the percentage of correct responses out of the total of 150 Landolt rings presented. In the dual task mode the secondary task was an auditory tone discrimination task, which required the subject to indicate which of two tones was being presented at any one time. Each second, one of the two tones would be presented, the tone being cut off after the subject had made a response until the next presentation was due. Again the measure of performance used was the percentage of correct responses made. The responses for the tone discrimination task were made using foot pedals. It

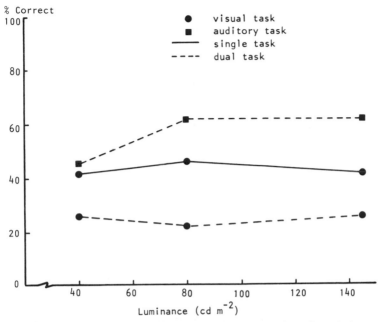

Fig. 5.3.   Percentage of correct responses made for a visual task performed alone, and together with a secondary auditory task, plotted against luminance. There is no significant variation in the percentage correct scores for the visual task in either the single or dual task modes. However, the auditory task in the dual task mode shows a significant ($p < 0.05$) reduction at the lowest luminance [16].

should be noted that there was no conflict in input and output modes for the two tasks, the primary task was eye input and hand output; the secondary was ear input and foot output. Therefore any overlap occurs in the central processing of information. Both single tasks and dual tasks were done under three different task luminances, 40, 80 and 145 cd m^{-2}. The results obtained are shown in Fig. 5.3. It can be seen that there is no significant variation in performance of the visual task alone with luminance; but when the secondary (or auditory) task is added, the performance of the secondary task shows a significant reduction in percentage correct score at the lowest luminance whilst the visual task performance is invariant with luminance. This is an indication that the dual task approach can be more sensitive to changes in task difficulty caused by changes in task luminance than the simple output measures alone. But this is not always so. The dual task method was also used in an experiment examining the effects of age on the performance of a range of tasks under different illuminances [17]. In this case the dual task was not more sensitive than the single task. The value of the dual task approach therefore remains an unanswered question. However, its success in complicated but realistic situations, such as driving, suggests it could be useful as an index of different lighting conditions. For example, it would be interesting to use the dual task approach to evaluate the effectiveness of street lighting. But its use does pose some problems for the experimenter. Rolfe [18] discusses some of these. Amongst them are such matters as the maintenance of performance on the primary task, the priority which the subject gives to one task over the other, and the possibility of time sharing on the two tasks and hence the extent to which the information processing can be considered to compete for space in a single channel. All these are valid questions and require careful consideration before the dual task approach is used. Certainly they suggest that careful experimental design and interpretation of results are necessary when it is used.

Both the simple physiological measures and the dual task approach attempt to provide objective measures of the effort involved in performing the task. There is much to be said for objective measures [19] but subjective measures are also possible; and there seems no reason why the effort made by the subject on a visual task should not be assessed in this way. Figure 5.4 shows some results from an experiment in which this was done [20]. The task was to mark off the gaps of a specific orientation for simple and complex Landolt rings seen under different illuminances. The size of gap (4 min arc) and the contrast (0·6) were the same for both simple and complex rings, so that the visual difficulty of the two tasks was the same, but the area to be searched was greater for the complex ring than for the simple ring. From Fig. 5.4 it can be seen that there are differences in time taken for the two ring types but increasing the illuminance improves the performance for both ring types. Both these effects are statistically significant. The subjects were also asked to rate the effort they made to perform the task on a seven-point rating scale with the ends labelled 'very little' and 'very great'. The pattern of ratings follows the pattern of the time taken in the performance results. The two ring types require different degrees of effort and the effect of increasing illuminance is to reduce the effort required. These differences in effort rating are also statistically significant. It is apparent that consistent results can be obtained by simple subjective ratings of effort.†

It is interesting to note that the above results conflict with the usual

---

† It is worth noting that a simple physiological measure, heart rate variability, was also measured in this experiment but failed to distinguish between the performance on the simple and complex rings or the different illuminances.

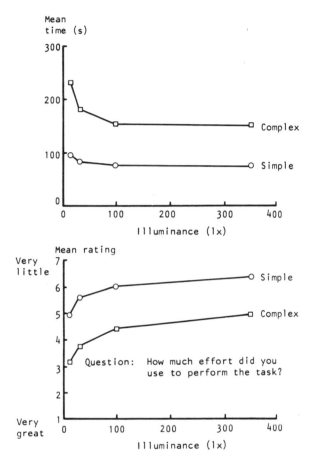

Fig. 5.4.   Mean time taken and mean ratings of effort for doing Landolt ring chart tests with simple and complex rings at different illuminances. (After [20].)

assumption stated at the beginning of this section, namely that effort is constant under different experimental conditions. Rather the results in Fig. 5.4 support Helson's 'par' hypothesis [21]. This states that in most tasks the subject forms his own criterion of a satisfactory level of performance and adjusts his efforts to maintain his performance at that level. The fact that he does not always succeed in this aim is shown by the different levels of performance obtained; but the change in effort ratings with decreasing illuminance suggests that some compensatory changes in effort are being made. The overall conclusion from this brief consideration of the available data is that the effort made by subjects to perform visual tasks under different lighting conditions is unlikely to be constant. There is some evidence to suggest that, in a laboratory context at least, a compensatory increase in effort will be made when lighting conditions on a task worsen, which will tend to diminish the effect of poor lighting on performance. This possibility, and it is little more than that because the above conclusion is based on the very limited amount of data directly relevant to lighting, indicates the value of assessing the effort when studying visual performance. This subject deserves more extensive and careful study than it has been given in the past.

5.4.2 Fatigue

When work is sustained for a long period, effort becomes particularly important. This is because most sustained activities eventually lead to a degree of fatigue, which will normally lead to a decrease in performance unless effort is increased, which in turn will lead to more fatigue and so on. The question of interest here is how lighting conditions affect the occurrence of fatigue during prolonged work. Before attempting to answer this question it is necessary to consider the nature of fatigue. The phenomenon of fatigue can conveniently be divided into two parts, physical fatigue and mental fatigue. Physical fatigue is well understood. Physical work involves muscular contractions. Prolonged use of the same sets of muscles produces physiological changes which either make the muscles incapable of further contractions or stop the central nervous system sending efferent impulses to the muscles. In either case the muscles stop operating. These effects can and have been studied by straightforward physiological methods. There is no doubt they occur, the duration of the work necessary before breakdown depending on the intensity of the work.

By comparison mental fatigue is not well understood. Basically this is because there are no simple physiological measures analogous to those used for physical fatigue available. Indeed, such measures are unlikely ever to be found, given that mental fatigue involves the whole being, not just limited parts of the musculature. A useful analogy to aid understanding of this difference is a computer system. Physical fatigue can be considered as a failure of the hardware. Mental fatigue can be considered as a failure of the software. Computer hardware can only be used in a limited number of ways. Computer software can be used in an almost infinite number of ways. The failure of the hardware in a computer system can easily be identified. Failure of software is much more difficult to establish. Nonetheless software failure does occur as does mental fatigue. It is a matter of common experience that prolonged and difficult mental work leads to feelings of tiredness.

Whilst this computer analogy helps understanding, it does not assist in answering the basic question, 'how do you identify the occurrence of mental fatigue and how do you quantify it?' At the moment there is no simple answer to these questions but the most promising approach appears to be to consider the patterns of task performance that occur when mental fatigue is reported. Three changes of pattern have been identified. The first is a change in the capabilities of the sensory systems. For example, continual measurement of visual acuity over 55 minutes showed a 23 % reduction in visual acuity, which was restored after five minutes rest [22]. This rapid restoration after rest is a characteristic feature of fatigue. The second change is the slowing of sensory-motor performance. This is the effect which is most widely associated with fatigue, as it implies a decline in performance. Typical of work which demonstrates this type of response is a study by Singleton [23], using a serial choice reaction time task. The subject had to move a joystick in one of four directions according to which of four lights was lit, the pace of the work being controlled by the subject. It was found that the time per response gradually lengthened over each test, the lengthening occurring in the time taken to make the decision as to which way to move. Further, the lengthening of the time per response increased with the difficulty of the decision. For tasks in which the subject has to work at a pace set by the task, slowing of response leads to breakdowns of performance in the sense that some events calling for action are ignored or attention is concentrated on a diminished aspect of the task. As an example of this, Davis reports a study of two-hour flights in a simulator under blind flying conditions [24]. It was found that as time progressed, more and more pilots forgot to reset the fuel indicator at 10-minute intervals as required. The third type of change that occurs is an increase in the irregularity of timing. Bertelson and Joffe [25] have shown this effect, again for a serial choice reaction task. The subject was asked to press one of four keys, each one corresponding to one of the numbers, 1–4, displayed. As soon as the subject answered, a new number was presented. The distribution of reaction times changed over the 30-minutes work, with the number of long reaction times markedly increasing. Close examination of the data showed that throughout the working period the reaction times oscillated, gradually increasing until a very long reaction time occurred which was then followed by a rapid recovery till the reaction time approached that achieved at the start of the experiment. These very long reaction times are known as 'Bills blocks', after Bills [26] who first identified them. Anyone who has tried to do a prolonged period of mental arithmetic will know the phenomenon.

The problem now is to examine how lighting affects fatigue. Studies on this subject can be divided into two types, those which have sought a particular form of physical fatigue and those which have tried to identify mental fatigue. The particular form of physical fatigue can reasonably be called ocular fatigue since what has been sought is some form of decrease in efficiency in the ocular motor system, usually in the ciliary muscle which provides accommodation changes, or in the extra-ocular muscles controlling eye movements. The idea behind these measurements is that poor lighting conditions will require additional use of the ocular motor system and this will be revealed by changes in the responses of the system. Collins and Pruen [27] measured the time taken to change accommodation from a far to a near distance, i.e. the accommodation time, before and after working for two hours on a vernier setting task lit to an illuminance of either 10 lx or 300 lx. Only if the results for the two illuminance conditions were taken together was there a significant increase in accommodation time after the two-hours work. There was no significant difference between the changes in the accommodation time after two-hours work for the two illuminances used. Sakaguchi and Nagai [28] also used accommodation time in a study of the effect of reading Landolt ring charts of different sizes and contrasts for up to 80 minutes. Their results show an increase in accommodation time as the duration of the task increased. Further, they showed that the increase was greater for an illuminance of 50 lx than for one of 600 lx and for sodium discharge lamps rather than White tubular fluorescent lamps. Unfortunately these results were obtained from a very few subjects so they need confirmation. Simonson and Brozek [29] measured the speed of eye movements between two fixation points before and after two hours of work involving reading small letters presented in an aperture at half-second intervals. They found that the number of movements made in a set time decreased after two hours of work but the difference between the reductions in movements that occurred for the task illuminances used was not significant. These results are typical of studies of ocular fatigue produced by different lighting conditions. It is relatively easy to show ocular fatigue occurring after prolonged work at a difficult visual task but it is much more difficult to demonstrate any significant differences between lighting conditions. At best the literature on ocular fatigue can be said to show that some lighting conditions can produce ocular fatigue for some tasks sometimes. Certainly there is little indication of the lighting conditions necessary for minimising ocular fatigue.

This unsatisfactory picture is repeated when lighting conditions and mental fatigue are considered. Most of the early work used simple measures of central functioning, particularly the change in critical fusion frequency (CFF) measured before and after work. Grandjean and Battig [30] showed that counting dust particles under a microscope, operating telephone exchanges or finishing needles, all of which are difficult visual tasks, reduced the CFF significantly over the working day. They also showed that the reduction in CFF could be limited by resting during a midday break or by reducing the contrast at the edge of the field of view for the microscope work. Baschera and Grandjean [31] measured CFF for people doing prolonged repetitive work at low, moderate and high levels of difficulty. They found a reduction in CFF over the working period which varied with the degree of task difficulty, the least reduction occurring for the task of moderate difficulty and the greatest with the task of low difficulty. Thus CFF would appear to be related to both fatigue and boredom which makes interpretation of any results difficult. There are also problems with its measurement. Rey [32] discusses the factors that can cause a reduction in CFF when fatigue is not likely to be present, and concludes that the method of determining the CFF, the exposure to flicker during the task, visual defects and drugs can all affect CFF. In these circumstances great care is needed in establishing the conditions of work and the method of CFF measurement before ascribing any change in CFF to fatigue.

Another feature of performance that has been used to examine the effect of lighting conditions on fatigue is the decrement in performance that occurs throughout the test because of slowing of sensory motor response. Simonson and Brozek [29], in the experiment already mentioned, found a significant improvement in average performance as the task illuminance was increased to 1000 lx, but this was followed by a reduction as illuminance was further increased to 3000 lx. At all illuminances there was a worsening of performance over the two-hour period but the reduction was a minimum at 1000 lx. Thus both simple performance and the measure of fatigue, the decline in performance throughout the test, showed the same illuminance to be the optimum.

Another measure of fatigue is the variability of performance arising from the occurrence of blocking. Khek and Krivohlavy [33] developed a task requiring the subject to indicate the direction of the gap in a Landolt ring of size 2 min arc and contrast 0·96 presented for a limited time. The rate of presentation was determined by a preliminary measurement of the maximum rate the subject could handle. When this had been determined the subject had to work for 30 minutes on the task with the Landolt rings

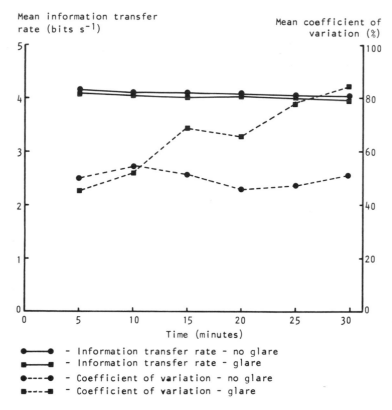

Fig. 5.5.   Mean information transfer rates and mean coefficients of variation for a 30-minute period spent working at a paced visual task with and without a glare source. (After [33].)

presented at his maximum rate. From these results the rate of information transfer and its variability was calculated. The results are shown in Fig. 5.5. Two conditions are presented, with and without a high intensity glare source at 45° to the line of sight. It can be seen that there is little change in the information transfer rate over the 30 minutes but there is a marked increase in variability when glare is present. This is to be expected if, as seems likely, strong glare produces a degree of fatigue.

From this brief review it should be apparent that although fatigue caused by visual conditions has been studied over many years, little progress has been made. The understanding at present is that what is commonly called fatigue undoubtedly occurs but should be considered in two separate forms, physical and mental fatigue.

Physical fatigue, as affected by the lighting conditions, will be made manifest by impaired functioning of the muscles in the ocular motor system. Lighting in which there are poor cues for such ocular motor functions as accommodation and/or fixation, and tasks which require rapid and frequent changes in these functions may lead to ocular fatigue.

Mental fatigue involves the central information processing capability. This makes it much more difficult to comprehend and much less easy to measure. In fact mental fatigue is usually inferred from the occurrence of one or more of a number of symptoms after prolonged work, namely a reduction in the performance capabilities of the visual system, or a decrease in the rate of working, or an increase in the variability of working, or some more detailed breakdown of performance. Most of these symptoms of mental fatigue have been shown to occur after intense work under difficult lighting conditions but no common pattern has been revealed.

In spite of this rather dismal picture it should always be remembered that fatigue does occur and lighting conditions may influence its occurrence. Fatigue in relation to lighting deserves further study but there is little point in such study until suitable measures of fatigue have been identified.

5.4.3 Similarities

Effort and fatigue are similar and inter-related. The inter-relation comes about because greater effort will lead to fatigue occurring earlier. Making the correct level of effort for the expected duration of the task is what is involved in the subject pacing his performance. The similarity is that both are essentially 'internal' to the subject. Only he can tell you the effort he is making and the degree of fatigue he is experiencing in producing a given performance. In spite of the 'internal' nature of the effort and fatigue, attempts to measure them as they are affected by lighting conditions have concentrated on external objective measures. This had led to a very unsatisfactory situation. Much work has gone into seeking simple physiological indices of fatigue and effort, yet the specificity of physiological response to stress [34] indicates that considerable differences between people are to be expected. Further, many of the responses that have been used are general responses to many different stimuli. To hope to obtain a clear and sensitive single measure in these circumstances is to have faith indeed. Even if such a measure could be found there would still remain the problem of interpretation. What does a 10 % reduction in CFF mean? Is it important and how much fatigue does it indicate? This last criticism also applies to the performance measurements that have been used.

Variability of performance and performance decrement can be used to identify the onset of fatigue, but what degree of fatigue does a given decrement indicate? Even the dual task approach, which seems the most satisfactory system for measuring effort, can present difficulties in interpretation. The one aspect all these measures can be used for is to identify the optimum conditions for a given task. They will successfully identify the condition for which minimum fatigue occurs or minimum effort is required, but they will not tell you how much fatigue or effort is involved in a departure from that condition. Thus the information that even a careful investigation of the pattern of performance can give about fatigue and effort is very limited. Yet there is another approach which has only rarely been used for studying lighting conditions. Why not accept that effort and fatigue are essentially 'internal' matters and simply ask the subject about his feelings? There are a number of ways in which the asking can be done, e.g. discussion and ratings, some of which can be easily quantified. In many ways the study of effort and fatigue is similar to the study of thermal comfort. You can try to measure thermal comfort by objective measures of skin temperature or observation of posture but ultimately such objective measures have to be 'calibrated' against a subjective assessment of the comfort of the conditions. It is only then that the objective measures have much meaning. Future studies of the effort made or the fatigue experienced under different lighting conditions would be more useful if they adopted the same approach. Subjective assessments of effort and fatigue should be taken alongside objective measures [19]. It is only by adopting this approach that we can expect to make much progress in understanding the complete relationship between lighting conditions and performance, effort and fatigue.

5.5 INDIRECT EFFECTS OF LIGHTING ON WORK

5.5.1 Introduction
Lighting can affect the performance of work in two ways. It can do so directly by changing the physical characteristics of the task and/or by varying the operating state of the visual system; the combined effect being to alter the visibility of the task. This is what has been considered so far. However, lighting can also influence the performance of work indirectly, i.e. without altering the visibility of the task. This indirect influence of lighting can occur either because of distraction, or because of a change in the general level of activation of the individual, or because lighting affects mood and hence motivation. Unfortunately, whilst this is all plausible and

widely believed there is little reliable experimental evidence that such indirect effects of lighting occur. In order to examine the probability of an indirect effect of lighting on performance occurring, it is necessary to argue from general psychological knowledge, and by analogy from features of the physical environment other than lighting.

5.5.2 Distraction

In considering the possible distracting role of lighting the first point to consider is why some events are distracting and others are not. The clue lies in the word event. An event is a discrete happening, a change in the environment. It is changes in the environment that are distracting: the greater the change the more distracting it is likely to be. On this basis any event which is distracting is likely to be novel, transient, unpredictable and of a different intensity to the rest of the environmental stimuli. These views are supported by the finding that noise in the form of sudden rocket firings influences the efficiency of decision making [35]. Novelty, transience and unpredictability are readily understandable but intensity requires some explanation. Distraction is related to change but the change can be to either higher or lower intensities. Sudden periods of silence in a noisy environment are just as distracting as loud noises in a quiet environment.

There is one other important aspect that influences the strength of the distraction produced by a stimulus—meaning. When the sudden event has meaning any distraction will be enhanced. Illustrative of this are some results from a survey of user attitudes to a landscaped office [36]. Noise was described as distracting, but what was important was not the overall sound pressure level but the overheard conversations. Conversations have meaning, noise usually does not.

The next question to consider is how distraction affects performance. Over the long term the answer appears to be very little, except when the distraction occurs frequently. The reduction in efficiency associated with rocket firings [35] could only be found in the first few seconds after the onset of the noise. This limited effect occurs because distraction only produces a brief deviation of attention, so the only loss is a time loss and in many tasks this can be easily made up. However, if the distraction occurs frequently, the total time loss may become significant; but even then there is some defence. The effectiveness of a distraction decreases with repeated exposure. In a word, a process of habituation occurs [37]. This suggests that simple repetitive mechanical distractions are unlikely to cause serious problems because habituation is easy, but very variable distractions such as overhead conversations may well produce an overall fall in efficiency. In addition to this variation of distraction with different stimuli, it is important to realise that the effectiveness of any given stimulus will depend on the nature of the task being done. The more complex the task, i.e. the greater the extent of continuous thought required for each part of the task, the more susceptible the person carrying it out will be to distraction.

It is now necessary to consider the features of conventional lighting installations which are likely to cause distraction. From the above discussion flickering and/or high luminance light sources seen directly or by reflection are the obvious candidates. Because the rest of the visual world is stable a flickering light source will be a source of distraction, especially as detection of flicker is much better in the far periphery than most other visual functions. As for high luminance light sources, this is simply a matter of differences in intensities of stimulation in the visual field. It can be concluded that some features of conventional lighting are likely to be distracting and hence may affect the performance of some tasks. The extent to which performance is affected will depend on the nature of the task.

5.5.3 Arousal

The conventional model of the way in which stimulation affects people involves the concept of arousal. The idea behind arousal is that at any given time each individual has an identifiable level of wakefulness. When he is asleep he is in a state of low arousal. When he is greatly excited or anxious he is in a state of high arousal. The level of arousal can be identified by various physiological and behavioural measures [38]. The important points for this discussion are that arousal level is determined partly by the stimulation from the environment and that it can be related to human performance. The relationship is expressed by the Yerkes–Dodson Law. Figure 5.6 shows a schematic illustration of this law. As arousal level increases from the level of sleeping, performance on a given task increases; but as arousal continues to increase, performance starts to fall. In other words, performance shows an optimum. Unfortunately different tasks have their optima at different arousal levels, the more complex the task the lower the arousal level for optimum performance. This is an important point. It explains why the same stimulation can enhance performance on one task and cause a deterioration on another. The use of music as an aid to productivity is an example of the practical application of arousal. Music causes an increase in arousal, which for dull repetitive tasks can produce an increase in performance [39]. For more complex tasks it may increase arousal too much, and hence decrease performance. This pattern of the effect of the stimuli being dependent on the nature of the task is similar to that discussed under the heading of distraction. This suggests that distraction can be considered as a special aspect of the more general arousal effect.

All the above discussion has been about increasing arousal but the possibility of decreasing arousal should not be forgotten. Decreased arousal will also cause a reduction in performance for some tasks and an increase for others, depending on the complexity of the task.

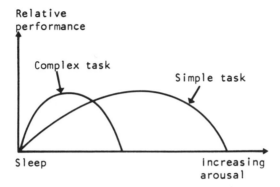

Fig. 5.6.   A schematic illustration of the Yerkes–Dodson law. Increasing arousal from sleep first increases the performance of a task and then decreases it. The optimum performance is reached at different levels of arousal depending on the complexity of the task; the more complex the task the lower the level of arousal for optimum performance.

It is now necessary to consider the aspects of lighting that are likely to change arousal. Since distraction may be considered as an aspect of arousal it seems probable that the features of lighting that cause distraction increase arousal; but, in general, the novelty, transience, unpredictability, intensity and meaning of the lighting changes are all likely to be important. Certainly the effect of intermittent aperiodic noise on the performance of a complex visual psychomotor task is to reduce the level of performance more than does constant or periodic noise. [40].

As for lighting, an experiment which related systematic variation in peripheral lighting in a room to task performance [41] failed to show any significant effects on the simple, constantly lit, clerical checking task used. This may have been because the variable lighting used did not change the level of arousal much, which does not seem unlikely if the subjects were already aroused by taking part in the experiment. Another possibility is that the varying lighting did increase arousal but part of the extra activity generated was directed towards something else and not towards the task. This illustrates one severe difficulty for the practical use of arousal by lighting. You may be able to increase arousal but there is no guarantee as to where the extra activity will be directed. It may be put into complaining about the lighting.

As for decreased arousal, the features of lighting likely to be effective are the opposite of those causing an increase in arousal. Thus uniform, dull, invariant and predictable lighting is likely to decrease arousal and hence to be unsuitable for all areas other than those where very complex tasks are being done.

Although non-uniformity, either spatial or temporal, seems the obvious way of producing arousal through lighting, there is also the possibility that the overall amount of light alone may influence arousal. Certainly light acts as a timer of glandular and metabolic functions [42] and can produce various biochemical changes normally associated with arousal [43]. Further, there is evidence that a high level of continuous noise [44] and a high temperature [45] do produce increases in arousal, although the effect on performance is inconsistent. The problem for lighting is whether the range of luminances likely to be found in conventional lighting is sufficient to produce such changes in arousal. Given the enormous range of luminances over which the visual system can operate, and the efficiency of adaptation, the effect of a uniform illuminance over a range of say 100 to 1000 lx on arousal seems unlikely to be large. In support of this, Sommer and Herbst [46] failed to find any effect of illuminance over the range 200 to 2000 lx on the performance of an audio-typing task.

5.5.4 Motivation

In the same way that distraction can be considered as one aspect of arousal, so arousal can be considered as part of a more general experience—the mood state. One of the most widely held beliefs about lighting is that good lighting puts people in a good mood and will encourage them to work harder. There is no doubt that lighting can give an interior a particular appearance (Chapter 6) and no doubt that mood state can influence performance [47]. The weak link is the consistency and duration of the relationship between appearance and mood. This is exemplified by the work of Kreiger [48]. In a careful study he failed to find any difference in mood produced in two rooms which the observers rated as having significantly different visual appearances. A possible explanation for this failure is that many other factors besides the visual appearance affect mood. This fact, together with the phenomenon of habituation to environmental stimuli, make any consistent effect of visual appearance on performance improbable.

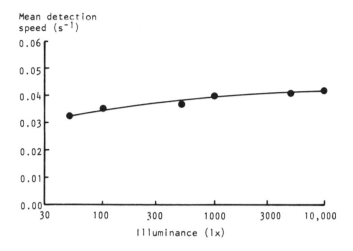

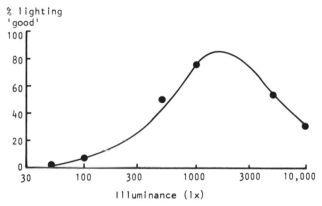

Fig. 5.7. Mean detection speeds for locating a specified number from amongst others at different illuminances, and the percentage who consider the lighting good at each illuminance. (After [49].)

Rather more credible is the conviction that lighting which creates discomfort will lead to a reduction in performance, although this too has been difficult to demonstrate in the laboratory. For example, Muck and Bodmann [49] had subjects searching a display of all the integer numbers from 1 to 99 printed in black on grey and laid out on a table. The measure of performance used was the time taken to find a named number. Figure 5.7 shows the performance achieved at each illuminance used by a group of 20–30-year-olds and the corresponding percentage who considered the lighting good. The interesting point here is not the increase in performance with illuminance but rather the maintenance of the maximum level of performance even when the ratings of comfort are decreasing. This suggests that uncomfortable lighting conditions do not affect performance, at least over the range of conditions studied. A similar lack of effect of discomfort on performance has been found for other environmental factors. For example, laboratory studies have shown that overheating only affects the performance of simple tasks when the temperature is much higher than that necessary to cause discomfort [7]. However, experience suggests that the explanation for these rather unexpected results lies with motivation. In a laboratory experiment people are usually highly motivated to succeed and will ignore any discomfort. It must be doubted if the same degree of motivation occurs outside the laboratory, in which case lighting which causes discomfort may lead to a decline in performance. The important word here is 'may'. There can be no doubt that motivation affects performance and little doubt that lighting conditions can affect motivation, but so do many other factors. These other influences ensure that whilst lighting may sometimes affect performance by changing motivation, it is unlikely to have any effect that is consistent across individuals or even across time with the same individual.

5.5.5 Conclusions on Indirect Effects

The three approaches by which lighting might indirectly affect task performance are suggested more by analogy and common experience than by careful experimentation. This should not be taken to mean that the effects are not real, rather that they are unproven. In fact distraction is a near certainty, arousal by lighting is probable, and the link between lighting and performance via mood and motivation is possible. Studies of the indirect effects of lighting would be welcome. Such studies should use a non-visual task so that any effects on performance could not be attributed to a direct effect of the lighting on the visibility of the task. There already exist several experiments which claim to be examining indirect effects of lighting, but because they have used visual tasks, the extent to which any effect is indirect is uncertain [50, 51]. However, before any experiments on the indirect effect of lighting on performance are done it is worth considering the ease of application of any results obtained. Knowledge of the identity and relevant levels of the features of lighting that can cause distraction, under a range of identifiable conditions and for a range of tasks, would be useful since such information could be readily applied in practice. The possibility of application for arousal is rather more remote. The problem with arousal has already been mentioned. It is that different tasks have different arousal levels to reach the optimum level of performance. To add further to the complications there is the fact that different people will require different amounts of environmental stimulation to reach the same total arousal level. In addition there is the unpredictability of where any increased arousal will be directed. Given this uncertainty it seems highly improbable that arousal could be used in any precise way to vary performance. A better approach in practice would be to ensure the lighting is interesting but not distracting. There are plenty of other ways of changing arousal which would be much more effective and controllable than using lighting.

As for mood, if some consistent effects of lighting were found, then application would be easy. Indeed lighting for mood is already applied in the theatre but its use for everyday lighting is unexplored. This is understandable because it is a big problem to study. To investigate it requires examination of several fields of study, psychology, sociology, physics and physiology; but it is an area of work that could be rewarding both for the lighting engineer and the worker.

5.6 INDIVIDUAL DIFFERENCES

5.6.1 Introduction
One aspect of the relationship between lighting and work which always causes arguments is that of individual differences. When viewing the same scene, different people are looking through different eyes. The information collected from the scene by each individual will be influenced by the optical properties of his eyes and the neural organisation of his retina and visual cortex. The way that information is interpreted will be affected by his experience and knowledge. In these circumstances it is hardly surprising that there are large differences in the levels of performance that can be achieved on the same task under the same lighting conditions by different people. This variability in turn leads to firmly held differences of opinion, based on personal experience, as to what are the minimum or optimum lighting conditions necessary for a task. In some practical applications this difference of opinion does not matter because each individual can be provided with lighting under his own control. However in many circumstances this is not possible. Where lighting is intended to meet the requirements of a group of individuals, the inevitable variation in performance implies that some people will find the lighting satisfactory and

others find it unsatisfactory. In an ideal world it would be possible to estimate the proportions in each of these classes. But this is not an ideal world. Very little information is available on which to base such estimates. Virtually all the data presented previously have been concerned with the average level of performance achieved under different lighting conditions. There are few occasions where the extent of individual differences has been

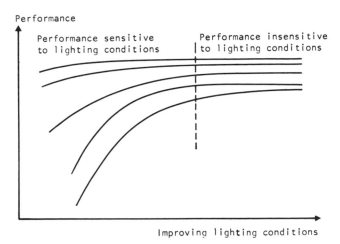

Fig. 5.8.   A schematic illustration of the overlapping of the maximum performance region for different individuals.

described let alone considered in detail, even though it is certain that such differences will be large. There are three redeeming features of this neglect, illustrated schematically in Fig. 5.8. The features are: first that the approach to maximum performance by each individual is usually gradual; second that the saturated level of performance is maintained over a wide range of lighting conditions; and third that although the magnitude of changes of performance with lighting conditions will vary between individuals the direction of change is broadly the same for everyone. The overall picture is of different people's performance reaching saturation at different lighting conditions, but the lighting conditions over which each individual shows his best level of performance overlapping to a considerable degree. This implies that there should be a range of lighting conditions over which individual differences in task performance cannot be reduced by changes in lighting. It is this insensitivity close to the maximum level of performance which allows individual differences to be largely ignored in lighting recommendations. But they are not ignored completely.

There is one aspect of individual differences which has such a large and consistent effect on the relationship between light and work that it has sometimes been considered in lighting recommendations. This aspect is age.

5.6.2 Age and Visual Performance

Figure 5.9 shows the threshold contrast achieved by different people of different ages for detection of a 4 min arc disc presented on a 20° uniform background field for 200 ms at 1·66 s intervals [52]. Three interesting points are evident from this figure. The first is the expected decrease in threshold contrast with increasing luminance. The second is the large difference between individuals in threshold contrast, even for individuals of the same age. The third is the upward trend in threshold contrast, i.e. diminishing sensitivity, with increasing age. The question of concern here is why this change with age occurs. To answer this question it is necessary to examine the details of the experiment. The actual threshold contrasts were obtained by getting the observer to adjust the luminance of the disc until he could just detect its presence. Thus the criterion for threshold contrast is left very much to the individual observer. It is possible that part of the scatter is due to differences in the criterion adopted by different individuals. While this may explain some of the overall scatter it does not explain the evident trend with age. An explanation for this trend can be derived from the changes that occur in the eye with age [53]. The changes that are likely to be relevant in visual performance are as follows: (a) the gradual retreat of the near point and a consequent reduction in the range of accommodation of the eye, (b) a considerable increase in the absorption and scattering of light in the eye itself and (c) a reduction in the pupil diameter for a given adaptation luminance. The first factor is the reason for the marked increase in spectacle wearing with age. In the experiment under consideration the subjects were carefully selected to control the effects of this variation with age. All the observers, of whatever age, and wearing spectacles if necessary, had to have a minimum visual acuity of 1·5 min arc at the observation distance. The other two factors were not controlled. Their combined effect is to reduce the amount of light actually reaching the retina of the old as compared with the young and to increase the scattering of light in the old as compared with the young. Weale [54] has estimated the retinal illumination of a 60-year-old as only one-third of that of a 20-year-old when both receive the same luminous flux at the front surface of the eye. Now consider the 60-year-old at a nominal adaptation luminance of $34 \, \text{cd m}^{-2}$. In fact, because of the reduced pupil area and the greater absorption of light in the eye, the true

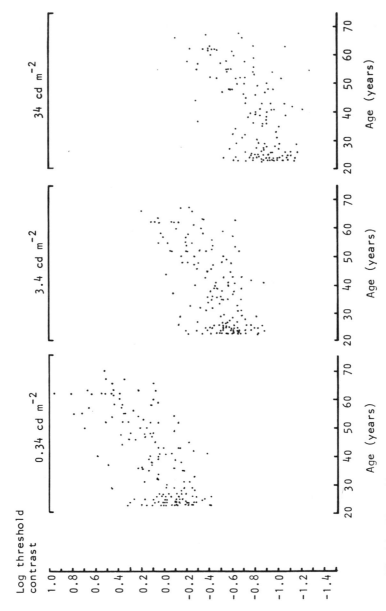

Fig. 5.9.   Threshold contrasts for individuals of different ages at three different adaptation luminances. (After [52].)

adaptation luminance for the 60-year-old should be about 10 cd m $^{-2}$. Comparison of the threshold contrasts for the 60-year-old group at a nominal 34 cd m $^{-2}$ adaptation luminance and the 20-year-old group at 3·4 cd m $^{-2}$ in Fig. 5.9 shows a much greater similarity than between the two age groups at the same nominal adaptation luminance. In other words the difference between ages in threshold contrast is considerably diminished if allowance is made for the changes that occur with ageing. This suggests that

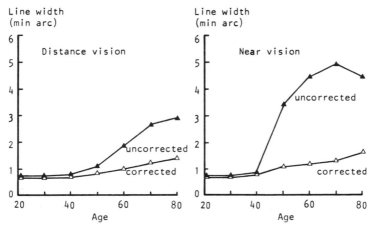

Fig. 5.10. Subtended line widths of letters that can just be read by 50 % of observers, for distance and near vision, with and without their usual spectacles, plotted against age. For distance vision the test letters were 6 m from the observer whilst for near vision they were 0·36 m from the observer. (After [55].)

reduced retinal illumination is part of the explanation for the effect of age on visual performance. However, it is not the only factor involved. The increased scattering is also likely to be important as it blurs detail. It may well be scattering rather than reduced retinal illumination which accounts for the decrease in visual acuity that occurs with age [55]. The form of the relationship is shown in Fig. 5.10 where it can be seen that even when any refractive error is corrected with spectacles, there remains a decrement in acuity which accelerates after about 40 years. Nor are contrast sensitivity and visual acuity the only aspects of the visual system performance capabilities that change with age. Sensitivity to flicker and the perception of colour also change [53].

These results suggest that the performance of tasks requiring examination of detail and/or involving low contrast and/or colour judgement should be strongly influenced by age. Figure 5.11 shows some results from a study which examined the effect of age on the performance of a range of tasks [56]. In Fig. 5.11 the mean time taken by different age groups to read through a Landolt ring chart is plotted against illuminance on the chart, for gaps of different sizes and contrasts. The interesting point is that the only significant difference between performance at the various illuminances for the young age group (16–30 years) occurred on the most

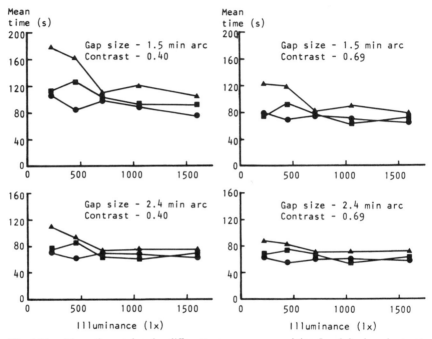

Fig. 5.11.　Mean times taken by different age groups examining Landolt ring charts at different illuminances. The age groups are: ●, 16–30 years; ■, 31–45 years; ▲, 46–60 years. (After [56].)

difficult task (i.e. gap size = 1·5 min arc, contrast = 0·40), but for the middle (30–45 years) and old (45–60 years) age groups significant differences in performance for different illuminances occurred for all the tasks. In addition, these differences were in the expected direction with increased time being taken for the lower illuminances. Further, the differences were greater for the more difficult, i.e. smaller and less contrast, Landolt rings. These results are almost certainly related to the reduced retinal illumination and increased scattering of light that occurs in the eye with increasing age. The differences between the ages in near point was controlled by visual screening to ensure that all subjects had correct refraction.

Similar results, indicating the interaction of age and illuminance on task performance have been found by other experimenters. Figure 5.12 shows the results of Muck and Bodmann [49] on a search task which required the subjects to find a specified integer number from a display of all the integer

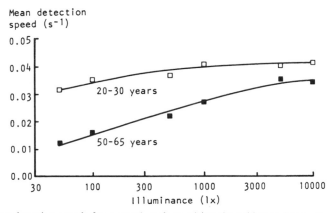

5.12.   Mean detection speeds for a search task requiring the subject to locate one integer number from 100 integer numbers, plotted against task illuminance. (After [49].)

numbers from 1 to 99. Figure 5.13 shows the results for proof reading under different illuminances obtained by Smith and Rea [57]. Both figures reveal the expected pattern of improved performance with increasing illuminance but the interesting point is that the increase in performance with illuminance is greater for the older age groups. This demonstrates the general rule that older people gain more from improvements in task illuminance than younger people.

Whilst there is no doubt that this general rule is true, the experiments described differ as to whether the older people can be brought to the same level of performance as the younger people simply by increasing illuminance. The experiments shown in Figs. 5.11 and 5.12 [56, 49] showed no significant difference between the age groups at the highest illuminances used (1600 lx and 10 000 lx respectively) but Smith and Rea do [57] (Fig. 5.13), even at 4880 lx. The probable cause of this difference is the nature of the work. Both former experiments set the observers relatively simple visual tasks. The Smith and Rea experiment involves higher cognitive processes with many more alternatives to be considered. Older people are notoriously

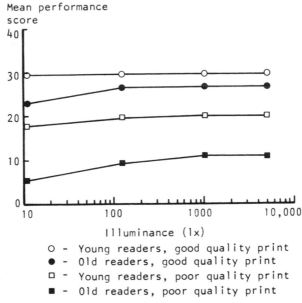

Fig. 5.13.   Mean performance scores for proof reading good and poor quality text, plotted against illuminance for young (18–22 years) and older (49–62 years) subjects. The performance score was derived from the accuracy of proof reading with a bonus for speed. The best possible performance score was 35. (After [57].)

slower at such work [58]. This will tend to reduce the absolute level of performance of the older age groups and hence lead to residual differences between the age groups. Another way to consider this is as a reduction in the visual component of the task with age. The importance of the visual component is emphasised by the fact that in tasks where the visual component is small, the effect of illuminance over a wide range is likely to be non-existent for any age group. This is not to say that there will be no differences between the ages. It is simply to point out that the illuminance will have little effect on any differences. An example of this occurred for a tracking task in which the subject had to maintain a needle on a point in the presence of electrical 'noise' which caused the needle to oscillate about the point. There were statistically significant differences between the age groups, with the oldest age group providing the worst performance, but there was no variation in performance with illuminance over the range 210–1650 lx [56].

One particular ability which is influenced by age and illuminance is the ability to discriminate colours. Figure 5.14 shows some results

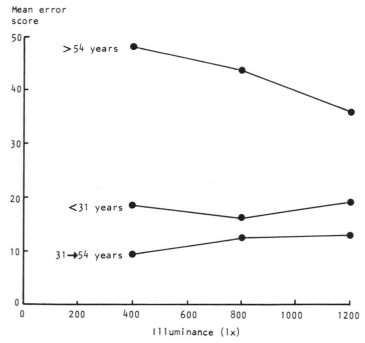

Fig. 5.14.   Mean error scores on the Farnsworth–Munsell 100 hue test for three different age
groups plotted against the illuminance on the task. (After [59].)

of an experiment in which different age groups did the Farnsworth
Munsell 100 hue test under an Artificial Daylight fluorescent lamp
providing illuminances of 400, 800 and 1200 lx [59]. Details of the
Farnsworth–Munsell test have been given in Chapter 4. Figure 5.14 shows
that the older age group made more errors in sorting the hues into the
correct order and, as might be expected from the above, showed a decrease
in errors with increasing illuminance. Even so, the lowest level of errors for
the old age group is still much greater than for the middle age and young age
groups. Similar results on the effect of age on hue discrimination have been
found by Verriest et al. [60]. This difference can be explained by another
change that occurs in the eye as we age. Increased absorption of light in the
eye has already been mentioned, what has not been disclosed is that the
absorption is not uniform across the visible spectrum but occurs
predominantly in the blue region of the spectrum. Verriest et al. [60] found
that with increasing age there was a tendency for more errors in hue
discrimination to occur in the blue–green and red regions than in other

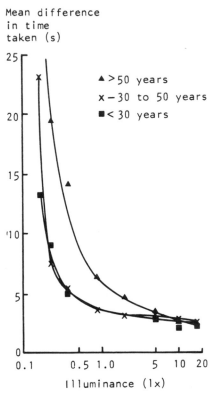

Fig. 5.15.   Mean difference in time taken by different age groups to thread a needle following
a change from 1000 lx illumination to the illuminance shown [62].

regions of the hue circle. Smith [61] has also shown that there is a marked
decline in the ability to match colours in old age.

Another aspect of practical performance for which the changes that
occur with age are important is the time taken for people to dark adapt. For
a fixed external illuminance, older people have a lower retinal illumination
to adapt to, and because of the non-linear nature of the adaptation process,
tend to take much longer to adapt than younger people. Simmons [62]
demonstrated this effect in an experiment concerned with emergency
lighting. His subjects were first adapted to an illuminance of 1000 lx. The
lighting providing this illuminance was then extinguished and replaced by
emergency lighting producing a very low illuminance. When the emergency
lighting was switched on the subject's task was to thread a needle as soon as
possible. Figure 5.15 shows the mean differences between the time taken to

thread a needle at various low illuminances and under the 1000 lx illumination, for three different age groups. The reason for subtracting the time taken to thread the needle under the 1000 lx lighting from the time taken under the emergency lighting is to correct for the variation in manual dexterity with age. The interesting points from Fig. 5.15 are the agreement between the age groups over the range 5–20 lx, where vision is predominantly photopic, and the marked increase in time taken by the over 50s at lower illuminances, where their vision will be predominantly scotopic. This result suggests that age is an important factor when considering lighting conditions which involve a sudden change in the state of adaptation, e.g. at the entrance to cinemas.

The above effect can be explained by the reduced retinal illumination allowed by the older eye but there is another feature which can be important. In addition to reducing retinal illumination, ageing produces an increased scattering of light in the eye. Since disability glare is known to be related to the occurrence of scattered light [63] it does not seem unreasonable to expect that the extent of disability glare should vary with age. Christie and Fisher [64] have studied the extent of disability glare from street lighting experienced by drivers of different ages, using a simulated street scene. As discussed previously, the magnitude of disability glare is quantified by equivalent veiling luminance. The general form of the equation for equivalent veiling luminance is

$$\text{Equivalent veiling luminance} = k \frac{(\text{illuminance at eye})^m}{(\text{deviation from line of sight})^n}$$

Christie and Fisher found that both the exponents m and n were independent of the observer's age, but that the constant was significantly related to age, the older the subject, the greater the value of k. This implies that the old do experience greater disability glare than the young for the same lighting conditions. Wolf [65] reached the same conclusion by measuring the ability of people of different ages to detect the orientation of the gap in a Landolt ring, in the presence of disability glare.

To summarise, as we get older the properties of the eye change, as do the properties of the whole nervous system. At first some of these changes can be offset by experience or by the use of optical aids such as spectacles but eventually the changes become so overpowering that they intervene in the relationship between light and work. Specifically, older people usually need higher illuminances to get their level of performance to saturation and even then they may be performing less well than younger people. Older people make more errors in colour judgement work, take longer to dark adapt and suffer more disability glare for the same physical conditions. Thus there can be little doubt that age is an important factor in producing individual differences in performance. The problem in applying this conclusion in practice is that, physiologically, people age at different rates. This means that in any group of 60-year-olds there will be a wide variation in visual abilities. Although the average level will be worse than for a group of 20-year-olds, there will be some in their 60s who still have visual abilities comparable with those in their 20s (see Fig. 5.9). Therefore the conclusions on age given above should not be applied to individuals but only to groups. When considering lighting for a specific task done by a given individual only that individual's abilities are relevant, but for lighting in a factory or office where a wide range of ages is likely to be present, some consideration of these effects of age is worth while [66].

5.6.3 Partial Sight

There is one other aspect of the differences between individuals that is worth considering, although it has no influence on general lighting recommendations. This is the existence of the partially sighted. Lighting recommendations are written with the average worker in mind. This worker may well have some degree of uncorrected refractive error and other slight vision defects but he can be assumed to have some approximation to normal vision. In contrast to this group there exists another group called the partially sighted. These people have very limited vision but lighting can be of help to them. There are various definitions of what constitutes partial sight. In the UK a register of blind or partially sighted persons is kept by the Local Authorities. Before a person can be entered on this register his vision must be examined by an ophthalmologist. The examination involves two distinct aspects of his vision. The first is foveal visual acuity, i.e. the ability to discriminate detail, for the two eyes separately. In this case visual acuity is expressed as a fraction where the numerator represents the distance between the subject and an acuity test chart, and the denominator represents the distance at which the details of the smallest letters seen by the subject subtend one minute of arc. For example, if the subject's vision is 6/60 then he can only read test letters at 6 m which subtend one minute of arc at 60 m. One minute of arc is accepted as a reasonable standard of resolution for normal vision. As a rule of thumb visual acuity of 3/60 is used as a criterion of blind registration and 6/60 for partially sighted registration. It must be emphasised that this is only a rule of thumb. Other considerations such as general health and occupation are taken into account. The other aspect of the ophthalmological examination is the extent of the field of vision. Various visual defects produce different patterns of reduced vision across the visual field. The reduction can involve a loss of vision around the periphery, in the central fovea or in more random locations (see Chapter 2). The rest of the examination involves the clinical condition of the globe of the eye as a whole and a clinical assessment of the patient, including the likely progress of any disease. Other countries have similar systems for defining the partially sighted; the ultimate authority, the World Health Organisation, accepts that a distance visual acuity of less than 6/18 [67] implies that people are visually disabled, and then grades the extent of disability in five steps.

Three points should be noted from all this. The first is that the definition of what constitutes blindness and partial sight is rather arbitrary and varies between different countries. The second is that the division between the normally sighted and the partially sighted is not sharp but rather is a point on a continuum with the partially sighted lying at one end of the continuum. The third is simply that many people registered as blind have some vision which they can use. The same applies even more strongly to the partially sighted. It is also important to realise that the partially sighted are not all limited in the same way. People with defects involving the fovea are handicapped in a different way to those with limited peripheral vision. The point that needs to be considered now is the extent of occurrence of partial sightedness.

Several attempts have been made to obtain this information in the UK [68, 69], but the most practical information arises from a survey of 15 000 households in England and Wales during 1976 77 [70]. Distance and near sight acuity tests were given to all adults who admitted on the first contact that they had any difficulty seeing to read or getting about. In addition, with the individual's permission, his medical records were consulted. From this survey an interesting pattern emerges. It suggests that out of every 100 000 home-based adults, approximately 520 are visually disabled according to the WHO definition, i.e. with the better eye and the optimal correction, visual acuity is still less than 6/18. The visually disabled are predominantly old. Of the adults found to be visually disabled, 80% were in their retirement years and almost 50% were more than 75-years old. In spite of their being classed as visually disabled, 45% of those so classified did not mention their poor vision at all. They had other more severe disabilities, usually involving arthritis or cardiovascular problems. Of the rest, 11% considered their poor vision as a secondary difficulty whilst only 44% considered it their greatest hardship. Interestingly, only 60% of the visually disabled had ever had a specialist examination of their eyes. For this 60%, medical records revealed that 25 % were suffering from cataract, 16 % from macular degeneration and 8 % from glaucoma. Other defects such as myopia, retinal detachment and optic atrophy açcounted for smaller percentages. The overall picture derived from this survey is of the extent of visual disability increasing rapidly in old age. The type of disability is variable, ranging from central opacities to reduced sensitivity in the peripheral visual field. The question that now needs to be considered is the extent to which lighting can aid the visually disabled.

Since the majority of the visually disabled tend to be past retirement age the work they undertake is largely domestic. It is the fundamentals of living; being able to move about, to prepare food, to work, to read, etc. These activities do not usually require a high level of visual capability, which is just as well. It is the effect of lighting on these activities that will be considered. The way in which lighting can be used depends on the nature of the visual disability. It is convenient to classify disabilities into two types: those involving reduced sensitivity in the fovea and those involving reduced sensitivity in the periphery.

The most common causes of visual disability of the fovea are cataract and senile macular degeneration. Both these maladies have two effects on light. They absorb a certain amount of light and they scatter a lot of what light they do pass which will tend to blur retinal images. An essential difference between them is that cataract, being located in the lens, can scatter light over a much bigger area of the retina than can the macular, which lies immediately in front of the foveal pit. The effects of both cataract and macular degeneration on foveal vision is to reduce visual acuity and contrast sensitivity. Figure 5.16 shows the change in threshold contrast with luminance for various degrees of blur, produced in this case by positive lenses being used by people with normal vision. It can be seen that the effect of blur is to increase the threshold contrast [71]. Figure 5.17 shows the change in visual acuity for different degrees of defocus produced in the same way [71]. Again, increased blur is deleterious in that it reduces visual acuity. Whilst there is no doubt both maladies produce these sort of reductions in visual capability, there are also differences in their effects. These differences can be revealed by examining the spatial modulation transfer function (Chapter 2). Hess and Woo [72] found that cataract frequently produced a decreased sensitivity to both high and low spatial frequencies. This differs from the reduced sensitivity for high spatial frequencies alone that occurs in patients with macular degeneration [73]. It also differs from the high frequency sensitivity loss produced by such disabilities as uncorrected refractive errors and corneal oedema. This information is valuable to the

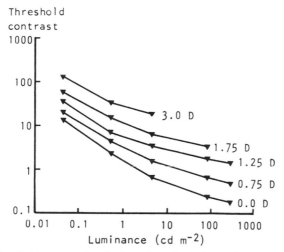

Fig. 5.16.   Threshold contrasts for the detection of a 4 min arc disc presented monocularly for 190 ms at different background luminances, by observers with normal vision defocused by positive lenses. The extent of defocus is shown in dioptres. (After [71].)

ophthalmologist in that it enables him to identify various conditions, all of which influence the performance of the visual system in the foveal region. In fact it is so valuable that a book of printed gratings of varying contrasts is now being produced for use as a rapid screening method for visual function. Using this book, Arden and Jacobson [74] have shown differences between people with normal vision and patients with different levels of intra-ocular pressure, a factor involved in the incidence of glaucoma.

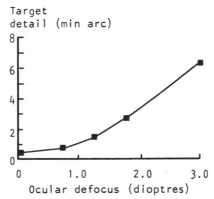

Fig. 5.17.   Size of target detail for resolution threshold for different degrees of ocular defocus. (After [71].)

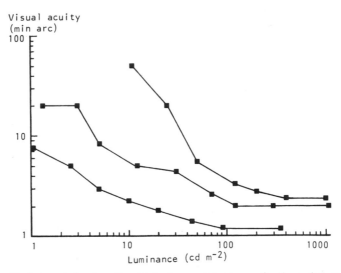

Fig. 5.18. Variation of visual acuity with background luminance for three observers with macular degeneration. The visual acuity was measured from a reading chart as the smallest letter size which allowed 70 % correct identification of the letters. (After [75].)

It should now be clear that visual disability arises from a number of different causes and that just because the area of visual function affected is in the fovea does not mean that the kind of failure is always the same. Nonetheless the common location in the fovea does produce some common features in how such disabilities affect the patient's activities. It means that anything which has to be examined closely, e.g. newspaper, faces, etc., cannot be seen clearly. Thus such people will find reading difficult but should have little trouble in finding their way about.

As for possible aids, there are a number of different approaches corresponding to different causes of the disability. For those with macular degeneration it would appear that more light is advantageous. Figure 5.18 from Sloan [75] shows the increase in visual acuity achieved by increasing the luminance of the task up to about $300\,\mathrm{cd\,m^{-2}}$ for three patients with macular degeneration. A later paper [76] confirmed this finding and examined the effect of providing very high luminances for reading. Patients suffering from diseases of the macular were tested for reading ability under normal room lighting and under a special high intensity reading lamp. It was found that with the high intensity reading lamp many of the patients were able to read continuous text without magnification or with much less magnification than was required under moderate illumination.

The mention of magnification raises the other approach to overcoming foveal visual disability, i.e. to increase the size of the retinal image. For reading this can be done either by increasing the size of the print (hence the supply of large print books available through many libraries), or by supplying an optical aid in the form of a magnifier [77]. The magnifier can be a microscope or a telescope. The microscope will have a short working distance and a wide field of view while the telescope has a long working distance and a small field of view. Whether a microscope or telescope is better for a particular application depends on the specific requirements.

It would appear from this discussion that high illuminances and/or increased retinal image size is a solution to the most commonly occurring causes of failure of foveal vision. However, there is one case involving reduced sensitivity in the fovea where a high illuminance may be less of a help. Cataract is an opacity in the lens of the eye. Increasing the illuminance on the object to be seen will lead to a reduction in pupil size thereby ensuring that more of the light that does reach the eye has to pass through the opacity in the lens. Whether an increase in illuminance is beneficial or not will depend on the balance between, the increase in acuity associated with the additional retinal illumination, and the reduction in the luminance and additional scattering of light produced by having a higher proportion of light passing through the central opacity. Lie [78] reports measurements of visual acuity over a range of distances from 0·9 to 5 m which show some benefit from increased illumination for two patients with cataract. However the same patients also demonstrated even more marked dependence on contrast. The visual acuity was much better for high contrast targets than for low contrast, whatever the illuminance.

Finally, it is worth mentioning the results of Cullinan *et al.* [79]. They found that the illuminances available in the homes of people attending a low vision clinic were very poor. By increasing the illuminances available with a reading lamp using a 60 W incandescent light source, they were able to produce an increase in visual acuity and ease of reading in the majority of people, regardless of the cause of visual disability. This suggests that even simple measures can be of some benefit to the foveal vision of the partially sighted.

The next type of disability to be considered is reduced sensitivity in the periphery. The most common cause of this defect is glaucoma. Initially, this produces a loss of vision in part of the peripheral field, but, unless treated, blindness results. The importance of the loss of peripheral vision depends upon the position and size of the field lost, at first it may be small and unnoticed but unless treated the area affected will increase in size. The value of peripheral vision to the visual system is that it provides the mechanism by which objects can be located and hence people can orientate themselves in a space. The fovea is essential for examining detail but is of little use without peripheral vision. The loss of the peripheral field can result in the individual concerned being totally visually incapacitated even though foveal visual acuity is nearly normal.

There is no simple single way of improving the vision of people with reduced sensitivity in the periphery. Peripheral vision covers a range of 170° horizontally and 140° vertically, unlike foveal vision which is concerned with the central 1° of this range. Special lighting equipment can be produced for overcoming the problems with foveal vision but for peripheral vision the solution has to cover the entire visual field. Fortunately there is one simple guideline which can be applied. It is to remember that people 'understand' a place by means of contrast between its components. On the larger scale this means that in order to aid orientation in this space there should be a clear contrast between ceiling, walls and floor. Jay [80] discusses this aspect in practical terms, suggesting that there is much to be said for white ceilings and light coloured walls with a strongly coloured frieze and dado. On a smaller scale he also suggests that more important objects in the interior should be distinguished by contrast. Door frames should be in different colours from the door, door handles should be large and coloured differently from the rest of the door. The nosings of treads on staircases should be a different colour from the rest of the step. Sicurella [81], himself partially sighted, provides a series of interesting demonstrations of the use of contrast around the house. For example, he recommends having a sheet of black paper and another of white on the kitchen wall. These can be used to provide the necessary contrast when either light or dark coloured liquids are being poured into a container.

Two other points are important. In attempting to differentiate between a number of similar articles the usual approach is to classify the articles by a number of different dimensions. The greater the number of ways in which items can be distinguished, the more likely it is that the partially sighted will be able to locate them. For example, a row of kitchen storage containers of identical shape with only a small label on them will prove difficult to distinguish, but by using containers of different sizes, shapes, colours and texture, with large labels as well, the problem for the partially sighted would be greatly eased. The other point is that plain surfaces are better as backgrounds [81]. Strongly patterned backgrounds can act as very effective camouflage.

The above very general advice has been concerned more with surfaces and objects in the room than with lighting. This is as it should be, because lighting in rooms occupied by the partially sighted needs to be uniform and without strong shadows although some degree of modelling is valuable for providing orientation. Uniformity is aided by having the light surfaces but it would also be worth while having some structure by which the surfaces could be identified. Special lighting is possible for particular work areas, as indicated earlier, but there is no special form of general room lighting for the partially sighted. Rather the aim is to provide the lighting suitable for the normally sighted but with strict regard for the quality aspects. The ability of the normally sighted people to see detail is reduced by disability glare from the luminaires and veiling reflections produced by specular surfaces. The partially sighted, especially those with cataract and/or macular degeneration, are likely to be even more sensitive to scattered light than the normally sighted and less able to cope with its effects. It is for this reason that considerable care is needed to avoid these problems in the lighting of rooms inhabited by the partially sighted. It should also be remembered that daylight can produce disability glare. Trying to watch a television which is close to a window can prove difficult for many normally sighted people but for the partially sighted it may be impossible. It is often surprising how much of an improvement can be achieved by a simple re-arrangement of furniture.

To summarise, modifications to the visual environment can do much to help those suffering from the more common visual disabilities. For those with poor peripheral vision the problem is one of orientation and location of objects and obstacles. Good contrast between important objects and the background is an invaluable aid. For those with poor foveal vision the problem is one of discrimination of detail. For them more light, improved contrast and/or some method of magnifying the retinal image are often useful approaches. The important thing to realise is that for many of the partially sighted, some care with the lighting, some understanding of contrast, and some motivation to succeed will enable many to make good use of their residual vision.

5.7 SUMMARY

This chapter examines a number of areas of uncertainty and controversy in the study of the effect of light on work. The first considered is the controversy surrounding the value of threshold performance measures for predicting the performance of real life tasks. It is argued that measurements of threshold performance can be useful indicators of the strength and direction of the effect of any lighting condition on more realistic work but the accuracy of any prediction will be strongly influenced by the conditions in which the threshold measures were obtained.

The second area examined is the implications of the Visibility Level and apparent contrast methods of quantifying ease of seeing. The important question is whether steadily increasing luminances always provide greater ease of seeing. Visibility Level suggests they do, apparent contrasts suggest they do not, as least for high contrast tasks. Which of these is correct remains to be determined.

The third area considered is that of effort and fatigue. The concern is that increased effort may be used to mask the effects of poor lighting conditions and fatigue. The problem is that there do not exist conventional objective measures of the effort and the fatigue associated with visual work. Simple physiological indices, secondary loading tasks, subjective assessments and detailed examination of the pattern of performance have all been tried, with mixed results. However, there is sufficient evidence to suggest that consistent changes in effort and fatigue do occur with different lighting conditions. This is a topic that deserves more study, but preferably with the admission that the effort being made and the fatigue felt are 'internal' to the subject and hence can most readily be obtained by enquiry of the subject.

The fourth area concerns the indirect effect of lighting on work. By indirect is meant that the lighting does not affect the visibility of the task but rather it causes distraction from the task, or changes the level of arousal or mood and motivation of the worker. Ironically there is very little evidence concerning the indirect effect of lighting on work. Therefore the discussion has to be in terms of general psychological knowledge and by analogy from other features of the physical environment. From this discussion it is concluded that lighting conditions can almost certainly affect visual performance by distraction and may affect arousal, mood and motivation. The practical problem is that such effects are unlikely to be consistent either in their magnitude or direction. Again, more work on this area would be welcome.

The final area considered is that of differences between individuals. Differences in visual task performance for the same lighting conditions are large. One of the factors that strongly influence the spread of performance is age, because the characteristics of the visual system change as people age. This explains the general worsening of visual performance with age and why older people generally benefit more from improvements in lighting conditions. A special class of individuals are the partially sighted. The nature of partial sightedness and how lighting can help such individuals is also discussed.

REFERENCES

1. Blackwell, H. R. Studies of the form of visual threshold data, *J. Opt. Soc. Amer.*, **43**, 456, 1953.
2. Blackwell, H. R. Brightness discrimination data for the specification of quantity of illumination, *Illum. Engng.*, **47**, 602, 1952.
3. Blackwell, H. R. and Smith, S. W. Additional visual performance data for use in illumination specification systems, *Illum. Engng*, **65**, 389, 1970.
4. Kulikowski, J. J. Effective contrast constancy and linearity of contrast sensation, *Vision Research*, **16**, 1419, 1976.
5. Yonemura, G. T., Benson, W. M. and Tibbott, R. L. *Levels of illumination and legibility*, National Bureau of Standards Report NBSIR-77-1307, Washington, DC, 1977.
6. Illuminating Engineering Research Institute. Annual Report Project 111-77, IERI, New York, 1978.
7. McIntyre, D. A. *Indoor Climate*, Applied Science Publishers, London, 1980.
8. Luckiesh, M. and Moss, F. K. *The Science of Seeing*, Macmillan & Co., London, 1937.
9. Luckiesh, M. and Moss, F. K. *Reading as a Visual Task*, Chapman & Hall, London, 1942.
10. McPherson, S. J. The effectiveness of lighting: its numerical assessment by methods based on blinking rates, *Trans. Illum. Engng Soc. (London)*, **8**, 48, 1943.
11. Tinker, M. A. Reliability of blinking frequency employed as a measure of readability, *J. Exp. Psychol.*, **35**, 418, 1945.
12. Bitterman, M. E. Heart rate and frequency of blinking as indices of visual efficiency, *J. Exp. Psychol.*, **35**, 279, 1945.
13. Welford, A. T. *Fundamentals of Skill*, Methuen, London, 1968.
14. Brown, I. D. and Poulton, E. C. Measuring the 'spare mental capacity' of car drivers by a subsidiary task, *Ergonomics*, **4**, 35, 1961.
15. Crosby, J. V. and Parkinson, S. R. A dual task investigation of pilots' skill level, *Ergonomics*, **22**, 130, 1979.
16. Boyce, P. R. *Illumination and the sensitivity of performance measures*, Electricity Council Research Centre Report R412, Capenhurst, UK, 1971.
17. Boyce, P. R. *Age, illuminance, visual performance and preference*, Electricity Council Research Centre Report R498, Capenhurst, UK, 1972.
18. Rolfe, J. M. The secondary task as a measure of mental load, in *Measurement of Man at Work*, eds. W. T. Singleton, J. G. Fox and D. Whitfield, Taylor & Francis, London, 1971.
19. Poulton, E. C. Observer bias, *Applied Ergonomics*, **6**, 3, 1975.
20. Boyce, P. R. *The measurement of effort in the performance of a visual task*, Electricity Council Research Centre Memorandum M370, Capenhurst, UK, 1971.
21. Helson, H. *Adaptation Level Theory*, Harper, New York, 1964.
22. Berger, C. and Mahneke, A. Fatigue in two simple visual tasks, *Amer. J. Psychol.*, **67**, 509, 1954.
23. Singleton, W. T. Deterioration of performance on a short-term perceptual-motor task, in *Symposium on Fatigue*, eds W. F. Floyd and A. T. Welford, H. K. Lewis, London, 1953.
24. Davis, D. R. *Pilot error*, Air Ministry Publication AP 3139 A, HMSO, London, 1948.
25. Bertelson, P. and Joffe, R. Blockings in prolonged serial responding, *Ergonomics*, **6**, 109, 1963.
26. Bills, A. J. Blocking: a new principle in mental fatigue, *Amer. J. Psychol.*, **43**, 230, 1931.
27. Collins, J. B. and Pruen, B. Perception time and visual fatigue, *Ergonomics*, **5**, 533, 1962.
28. Sakaguchi, T. and Nagai, H. Studies on the relations between various light sources and visual fatigue, *J. Illum. Engng Inst. Japan*, **57**, 278, 1973.
29. Simonson, E. and Brozek, J. Effects of illumination level on visual performance and fatigue, *J. Opt. Soc. Amer.*, **38**, 384, 1948.
30. Grandjean, E. and Battig, K. Das verhalten des subjectiven verschmelzungsschnellen des

auges unter verschiendenen arbeits und verspuchsbedingungen, *Helv. Physiol. Pharmacol. Acta*, **13**, 178, 1955.
31. Baschera, P. and Grandjean, E. Effect of repetitive tasks with different degrees of difficulty on critical fusion frequency (CFF) and subjective state, *Ergonomics*, **22**, 377, 1979.
32. Rey, P. The interpretation of changes in critical fusion frequency, in *Measurement of Man at Work*, eds W. T. Singleton, J. G. Fox and D. Whitfield, Taylor & Francis, London, 1971.
33. Khek, J. and Krivohlavy, J. Evaluation of the criteria to measure the suitability of visual conditions, *Proc. CIE 16th session, Washington*, 1967.
34. Lacey, J. I., Bateman, D. E. and Van Lehn, R. Autonomic response specificity, *Psychosom. Med.*, **15**, 8, 1953.
35. Woodhead, M. M. Effect of brief loud noise on decision making, *J. Acoustical Soc. Amer.*, **31**, 1329, 1959.
36. Boyce, P. R. and Brundrett, G. W. *Staff reaction to moves into integrated design offices, final report on the move to the Outland Road Office, CAO, Plymouth*, Electricity Council Research Centre Report R755, Capenhurst, UK, 1974.
37. Hokanson, J. E. *The Physiological Basis of Motivation*, J. Wiley & Son, London, 1969.
38. Schnore, M. M. Individual patterns of physiological activity as a function of task differences and degree of arousal, *J. Exp. Psychol.*, **58**, 117, 1959.
39. Fox, J. G. Background music and industrial efficiency—a review, *Applied Ergonomics*, **2**, 70, 1971.
40. Eschenbrenner, A. J. Effects of intermittent noise on the performance of a complex psychomotor task, *Human Factors*, **13**, 59, 1971.
41. Aldworth, R. C. and Bridgers, D. J. Design for variety in lighting, *Ltg Res. and Technol.*, **3**, 8, 1971.
42. Wurtman, R. J. Effects of light and visual stimuli on endocrine function, in *Neuroendocrinology*, eds L. Martini and W. F. Ganong, Academic Press, New York, 1967.
43. George, W. Z. *An investigation into a possible physiological effect of high levels of illumination*, M.Sc. Project, Loughborough University of Technology, UK, 1972.
44. Croome, D. J. *Noise, Buildings and People*, Pergamon Press, London, 1977.
45. Provins, K. A. Environmental heat, body temperature and behaviour: an hypothesis, *Australian J. of Psychology*, **18**, 118, 1966.
46. Sommer, J. and Herbst, C. H. Einfluss der beleuchtung auf die arbeit an der schreibmaschine, *Lichttechnik*, **23**, 23, 1971.
47. Spence, K. W. *Behaviour Theory and Conditioning*, Yale University Press, New Haven, 1956.
48. Kreiger, W. G. *The visual environment and task performance: a check on potential information artifacts in the TNG experimental simulation research environment*, Technical Report 45, Purdue University, USA, 1972.
49. Muck, E. and Bodmann, H. W. Die bedeutung des beleungnchtungsniveaus bei praktischer sehtatigheit, *Lichttechnik*, **13**, 502, 1961.
50. Giessmann, H. G. and Lindner, H. Der einfluss des beleuchtungsniveaus auf die konzentrationsleistung, *Arbeitsmed. Fragen in der Ophthalmologie*, **4**, 108, 1972.
51. Stenzel, A. G. and Sommer, J. Der einfluss de beleuchtungsstarke auf weitgehend sehunabhangige manuelle tatigkeiten, *Lichttechnik*, **21**, 143A, 1969.
52. Blackwell, O. M. and Blackwell, H. R. Visual performance data for 156 normal observers of various ages, *J.I.E.S.*, **1**, 3, 1971.
53. Weale, R. A. *The Ageing Eye*, H. K. Lewis, London, 1963.
54. Weale, R. A. Retinal illumination and age, *Trans. Illum. Engng Soc. (London)*, **26**, 95, 1961.
55. US Department of Health, Education and Welfare. *Binocular visual acuity of adults—US 1960–1962*, PHS Publication 1000, series 11, No. 3, Washington, DC, 1964.

56. Boyce, P. R. Age, illuminance, visual performance and preference, *Ltg Res. and Technol.*, **5**, 125, 1973.
57. Smith, S. W. and Rea, M. S. Proof reading under different levels of illumination, *J.I.E.S.*, **8**, 47, 1978.
58. Welford, A. T. Performance, biological mechanisms and age: a theoretical sketch, in *Behaviour, Ageing and the Nervous System*, eds A. T. Welford and J. E. Birren, Charles C. Thomas, Springfield, USA, 1965.
59. Boyce, P. R. and Simons, R. H. Hue discrimination and light sources. *Ltg Res. and Technol.*, **9**, 125, 1977.
60. Verriest, G., Vandevyvere, R. and Vanderdonck, R. Nouvelles recherches se rapportant a l'influence du sexe et de l'age sur la discrimination chromatique ainsi qu'a la signification pratique des resultants du test 100 hue de Farnsworth–Munsell, *Rev. D'Optique*, **41**, 499, 1962.
61. Smith, H. C. Age differences in colour discrimination, *J. Gen. Psychol.*, **29**, 191, 1943.
62. Simmons, R. C. Illuminance, diversity and disability glare in emergency lighting, *Ltg Res. and Technol.*, **7**, 125, 1975.
63. Fry, G. A. A re-evaluation of the scattering theory of glare, *Illum. Engng*, **49**, 98, 1954.
64. Christie, A. W. and Fisher, A. J. The effect of glare from street lighting lanterns on the vision of drivers of different ages, *Trans. Illum. Engng. Soc. (London)*, **31**, 93, 1966.
65. Wolf, E. Glare and age, *Archives of Ophthalmology*, **64**, 502, 1960.
66. Commission Internationale de l'Eclairage. *Guide to interior lighting*, CIE Publication 29, 1965.
67. World Health Organisation. *The prevention of blindness*, WHO Technical Report Series No. 518, Geneva, 1973.
68. Sorsby, A. *The incidence and causes of blindness in England and Wales, 1948–1962*, HMSO, London, 1966.
69. Harris, A. I. *Handicapped and impaired in Great Britain, Part 1*, HMSO, London, 1971.
70. Cullinan, T. R. *Visually disabled people in the community*, Health Services Research Unit Report No. 28, University of Kent in Canterbury, UK, 1977.
71. Johnston, A. W., Cole, B. L., Jacobs, R. J. and Gibson, A. J. Visibility of traffic control devices: catering for the real observer, *Ergonomics*, **19**, 591, 1976.
72. Hess, R. and Woo, G. Vision through cataracts, *Invest. Ophthalmol. and Visual Sci.*, **17**, 428, 1978.
73. Sjorstrand, J. and Frizen, L. Contrast sensitivity in macular disease, *Acta Ophthalmol.*, **55**, 507, 1977.
74. Arden, G. B. and Jacobson, J. J. A simple grating test for contrast sensitivity: preliminary results indicate value in screening for glaucoma, *Invest. Ophthalmol. and Visual Sci.*, **17**, 23, 1978.
75. Sloan, L. L. Variation of acuity with luminance in ocular diseases and anomalies, *Docum. Ophth.*, **20**, 384, 1969.
76. Sloan, L. L., Habel, A. and Feiock, K. High illumination as an auxiliary reading aid in diseases of the macula, *Am. J. Ophth.*, **76**, 745, 1973.
77. Fonda, G. E. Ways to improve vision in partially sighted persons, *Geriatrics*, **30**, 49, 1975.
78. Lie, I. Relation of visual acuity to illumination, contrast, and distance in the partially sighted, *American Journal of Optometry and Physiological Optics*, **54**, 528, 1977.
79. Cullinan, T. R., Silver, J. H., Gould, E. S. and Irvine, D. Visual disability and home lighting, *The Lancet*, March 24th, 642, 1979.
80. Jay, P. A. Fundamentals, in *Light for Low Vision*, ed. R. Greenhalgh, Partially Sighted Society, Hove, UK, 1980.
81. Sicurella, V. J. Colour contrast as an aid for visually impaired persons, *Visual Impairment and Blindness*, **71**, 252, 1977.

Part 3:
The Appreciation of Lighting

6

Revealing Impressions

6.1 INTRODUCTION

Lighting does more than just make things visible. Lighting in a space inevitably contributes to people's impressions of that space. The impression may be good or bad, appropriate or inappropriate, firm or vague but it will exist. There are positive and negative aspects to the influence of lighting on impression. The negative aspect is dominant when there are complaints of discomfort. The positive aspects are apparent when feelings of pleasure are experienced. But discomfort and pleasure are not the only possible impressions lighting can give. It is much more complicated than that. Lighting can give an interior a character. This character may be dramatic, inviting, depressing, boring, relaxing, interesting, functional, etc. It is the subtlety of the various impressions that can be evoked and the practicality of doing so that makes lighting such an important means of manipulation. Further, no specific visual task is required for lighting to create an impression. Lighting can be effective in influencing the impression given by the spaces used for a wide range of human activities no matter what the extent of the visual system's interest in those activities.

This chapter deals with the features of lighting that influence the impression given by a space, the methods by which they have been identified and the limitations on the applicability of the knowledge obtained.

6.2 CORRELATION STUDIES

Until recently virtually all the studies of the impressions created by lighting have been conceived at a very simple level. The concern has usually been either to establish how some variation in lighting appears to people or to validate some proposed new measure of lighting. Almost all these studies

255

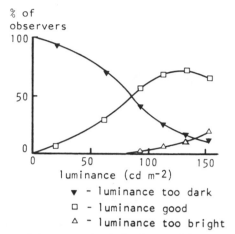

Fig. 6.1.   The percentage of observers rating the luminance of their desks as too dark, good, or too bright. (After [1].)

have used a correlation method. The basis of this method is to relate some subjective judgement of the lighting to a physical descriptor of the lighting. Typical of this approach is an extensive study by Balder [1]. Using a large room with medium reflectance surfaces he arranged for groups of people to experience a wide range of different combinations of the luminances of desks, walls and ceiling. For each combination of these luminances the observers were asked to place the luminance of each surface into one of three categories: good, too dark or too bright. The luminances on the desks were assessed first, followed by the luminances of the walls and the ceiling. The observer's experience of each condition was limited to a few minutes and the lighting was provided from a large box-like structure suspended above each desk. Figure 6.1 shows the percentage of observers who placed each luminance condition on the desk in each category. Figure 6.2 displays the most preferred luminances of the side wall and ceiling plotted against the luminance on the desk.

A rather different study, but one which also follows the correlation approach, is the examination of the factors relating to discomfort glare in interior lighting [2]. Here, the observers were asked to view a model of a schoolroom in which the luminance of the light sources was fixed but the observer could control the luminance of the rest of the interior, i.e. the background luminance. Each observer had to adjust the background luminance until his sensation of discomfort glare caused by the lighting was at one of four criterion levels: just imperceptible glare, just acceptable glare,

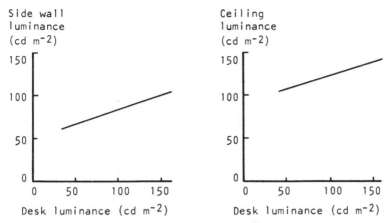

Fig. 6.2. The most preferred luminance of the side wall and ceiling plotted against the luminance of the desk top. (After [1].)

just uncomfortable glare, and just intolerable glare. The physical luminance conditions which related to each of these settings were then used to derive a general equation the aim of which was to predict the level of discomfort caused by any particular physical condition. This equation has been used to establish the Glare Index system currently used in lighting practice [3].

Both the studies described above have been concerned with how different lighting conditions appear to people. A study that demonstrates the other use of the correlation method, namely to validate a proposed lighting measure, is one by Cuttle *et al.* [4]. These workers were concerned to evaluate a number of proposed measures of the distribution of light to be used in predicting the effect of lighting on the appearance of three-dimensional objects. They asked observers to match the appearance of a model interior under different forms of lighting to that of a similar model under fixed lighting. The relationships between the various measures, namely the horizontal illuminance, the mean spherical illuminance† and the mean cylindrical illuminance‡ and the matches made were then examined. The measure which showed the most consistent relationship, the mean spherical illuminance, was considered the best predictor.

A number of criticisms can be made of these studies, but before doing so

† The mean spherical illuminance is the average illuminance over the surface of a small sphere at a point.
‡ The mean cylindrical illuminance is the average illuminance over the curved surface of a small vertical cylinder at a point.

it should be made clear that they are not alone. They have been selected for comment because they are widely quoted and/or have had considerable influence on the development of lighting criteria. It would have been possible to choose other studies on similar subjects all of which suffer from the same faults. The criticism can be divided into two classes. First there are the criticisms of the particular methods used, but which accept the correlation approach as a valid one. Second there are those who doubt the value of the correlation approach as such. Prominent among the first class are criticisms of the lack of specific context of the conditions used and the unrepresentative nature of the observers. For example, Balder's room looked like nothing so much as an experimental one, whilst the glare studies were conducted by looking at what was virtually a photograph of a schoolroom. As for the observers, the glare experiments deliberately used a set of what are called experienced observers. The nature of their experience was that they were used to making glare judgements with little variability. Whilst this may have helped in establishing a predictive equation by reducing the scatter of the results, it must leave doubt about the applicability of the equation to ordinary observers. In fact later measurements showed that the experienced observers were much more sensitive to discomfort glare than the general public [5]. Further criticisms can be made of the limited time allowed for the observers to experience the conditions and the vagueness or 'leading' nature of the instructions given to them.

Another criticism involves the usual means of obtaining the subjective judgement, a rating scale. Poulton [6] asserts that an observer's use of rating scales is strongly affected by the range of conditions experienced, either in the experiment itself or, if no range is available there, in everyday life. The effect is simply that observers tend to match the middle of the rating scale with the middle of the range of conditions experienced. Poulton shows that this is clearly so for ratings of the acceptability of noise and there seems no reason to suppose that the same effect would not occur, for example, in the judgement of the suitability of luminances.

The above are all valid criticisms of most of the studies that have been done concerning lighting but they can all be overcome by careful experimental practice. For example, it is possible, at great expense and inconvenience, to use real interiors, a representative sample of observers who are able to experience the lighting being studied for reasonable periods, and a method of obtaining subjective judgements which diminishes the range effect. However, it must be doubted if this would be worthwhile because there are more fundamental objections to the correlation method

[7]. The objections are those of relevance and independence. The point is that obtaining a correlation between a physical measure and a subjective judgement is no reason to suppose that the psychological attribute represented by the judgement is relevant to people's everyday experience, and further, that even if it is, when a number of such relationships are established there is no reason to suppose that they represent independent dimensions. However, the ultimate criticism of the correlation method is that it invariably starts from the lighting end of the problem rather than the people end. Questions, framed in terms of lighting, are asked of people. This may not matter if the intention is to validate a proposed measure of lighting but it is of little use if the aim is to discover how lighting affects people's impressions of a space. Lighting may be largely irrelevant. To identify the place of lighting, more general questions have to be asked and ideally people have to be given an opportunity to speak for themselves.

All these criticisms would seem sufficient to undermine any faith in the lighting recommendations derived by the correlation method but such loss of faith would be unwise. Some of these criteria have been confirmed in field experiments [8, 9] and, in any case, all have been used in practice for many years. If they had been seriously wrong, in the sense of producing discomfort, then this would quickly have become evident. In fact very few complaints have been heard. The most likely explanation for this apparent lack of complaint is that the lighting conditions necessary simply to avoid discomfort are not very stringent. Given the adaptability of the visual system this should not be surprising. But avoiding discomfort, although valuable, is an essentially negative function and everyday experience shows that lighting can play a much more positive role. In order to give guidance about this role a much clearer understanding of the ways in which different lighting conditions influence impression is required. The correlation approach is not suitable for achieving this understanding. It is valuable for validating proposed lighting measures but it can produce only a very limited understanding of the ways in which lighting influences people's perceptions of an interior. To move from the present state of avoiding discomfort to a positive role for lighting a more fundamental approach is needed; one that starts by considering people rather than equipment.

6.3 MULTI-DIMENSIONAL PROCEDURES

One way to achieve a greater understanding of the effects of different lighting conditions is illustrated by the work of Kimmel and Blasdel [10]. In

260   HUMAN FACTORS IN LIGHTING

a study of the impression created by different forms of lighting in library reading rooms they had 112 observers rate the luminous environment in the libraries on several seven-point rating scales. The rating scales used were concerned with the features of the lighting considered in the design, e.g. the amount of light, glare from the lights and the colour of the light. From this description it would appear that the method is little different from the correlation method. However, it is different because no physical

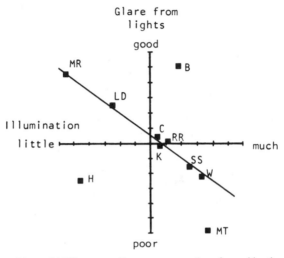

Fig. 6.3.   The positions of 10 library reading rooms on a plane formed by the dimensions of illumination and glare from lights. (After [10].)

measurements were taken so the subjective ratings were not correlated against physical measures. Rather the ratings were used to form a matrix of differences of ratings between the libraries on the same scale for the same observers. From the difference score matrices formed in this way, the minimum number of independent dimensions necessary to explain the patterns of differences was identified by a form of multi-dimensional scaling (MDS) analysis [7]. Figure 6.3 shows the sort of information that can be achieved by this method. It displays the location of the various libraries on a plane defined by two dimensions identified as independent: illumination and glare from lights. The most obvious point to note is that there is a consistent trend for poorer glare assessments to be associated with higher perceived illumination, although there are three installations which are exceptions to this trend. Installation B is a high-level luminous ceiling and has less glare than expected. Installation H is a high-source-luminance downlighting installation and has more glare than expected. Installation MT should have produced much less glare, at least according to the luminance of the luminaires, but in fact it was the worst. The discrepancies from the general trend can be explained if discomfort glare involves an element of distraction away from the area of interest towards striking and irrelevant high-luminance stimuli. If this is so the very uniformity of the luminous ceiling (B) could be expected to cause less distraction and hence less glare, whereas the non-uniformity of the downlighting installation (H) would add to the distraction. The installation MT could have been more distracting than the luminances suggest because the luminaires were arranged in a very striking pattern on the ceiling. Lynes [11] has demonstrated that the equation governing the conspicuity of signals has a similar form to that relating the sensation of discomfort glare to the physical luminances, which suggests that distraction may be involved in discomfort glare. It is valuable to have this theoretical relationship suggested by field data.

The method used in this study of libraries has much to recommend it. It offers the possibility of identifying the physical factors which are important to people's perception of an interior from simple ratings of a range of real interiors. The method certainly puts people first. The relevance of the various physical factors has to be inferred from the pattern of relationships revealed by the subjective ratings. However it does not overcome all the problems of the correlation approach. Although the MDS analysis ensures that the dimensions used are independent, the questions are related to the prevailing criteria of lighting. Thus the answers are still concerned with lighting rather than impression, and there is little opportunity for the observer to speak for himself.

A further step on the path to overcoming these defects is demonstrated in a study by Flynn et al. [12]. These workers had arranged for 96 observers to experience a conference room lit in six different ways. The observers' responses were collected on rating scales but this time the questions were concerned with the observers' impressions of the room. Factor analysis was then applied to the data collected [13, 14]. Factor analysis is a statistical technique by which an enormous quantity of such data can be ordered, so that the minimum number of underlying independent dimensions on which the observer is basing his responses to the various stimuli, in this case lighting conditions, are revealed. The factor analysis also gives a measure of the extent to which each individual rating scale is related to the independent dimensions. Figure 6.4 shows the six installations used by Flynn et al.

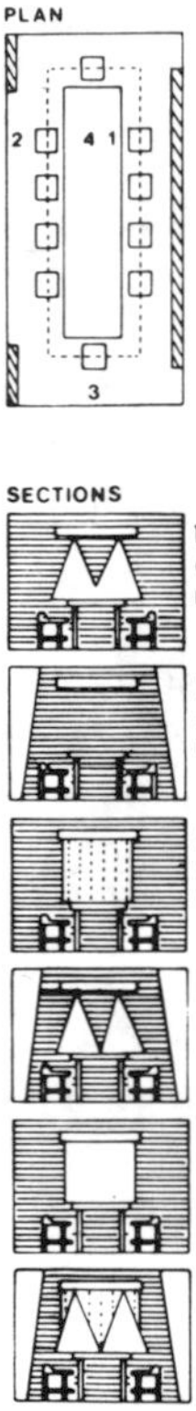

	LUMINANCE (cd/m²) SEE PLAN FOR POSITION				ILLUMINANCE (LUX)
	1 TABLE	2 WALL	3 WALL	4 CEILING	TABLE
1 OVERHEAD DOWNLIGHTING LOW INTENSITY	1·2	0·3	0·0	1·7	110
2 PERIPHERAL WALL-LIGHTING. ALL WALLS	7·9	19·9	32·5	8·9	110
3 OVERHEAD DIFFUSE, LOW INTENSITY	2·9	3·7	2·6	80·5	110
4 COMBINATION: OVERHEAD DOWNLIGHTING(1) + END WALLS	1·4	1·0	26·3	1·7	110
5 OVERHEAD DIFFUSE, HIGH INTENSITY	21·9	119·9	26·7	1404·7	1100
6 COMBINATION: OVERHEAD DOWNLIGHTING(1) + PERIPHERAL(2) + OVERHEAD DIFFUSE(3)	11·6	33·5	38·3	82·2	320

Fig. 6.4. A plan of the conference room, six schematic illustrations of the six lighting installations used, and some physical measurements of the lighting conditions produced. (After [12].)

Figure 6.5 shows the five independent dimensions on which the impressions of the room under the six lighting installations are based. Also given are the individual rating scales that are most strongly related to each dimension. The five dimensions found are named evaluative, perceptual clarity, spatial complexity, spaciousness, and formality. It should be noted that these names for each dimension are inspired guesses, produced by an inspection of the rating scales most strongly related to the dimensions.

Figure 6.5 also shows the mean rating for each installation on each rating scale and demonstrates the richness of the information available. As an example, consider the pleasant/unpleasant scale, which is the scale most strongly related to the evaluative dimension. The two installations which

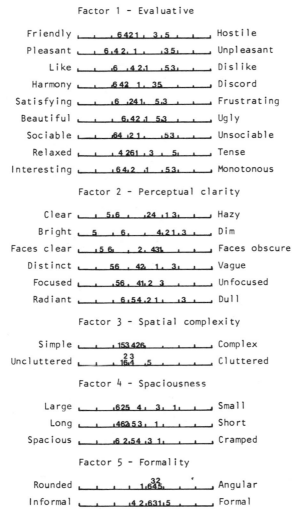

Fig. 6.5.   The five dimensions identified from the rating of the conference room lit by the six lighting installations shown in Fig. 6.4. The rating scales most strongly related to each dimension are shown under each factor title. The mean rating for each installation on each scale is given by the appropriate installation number, i.e. installation 6 has the mean rating closest to the pleasant end of the pleasant/unpleasant scale. (After [12].)

are regarded as most pleasant have a common feature as do the two installations that are considered most unpleasant. The two most pleasant installations (6 and 4) provide a combination of lighting on the table and lighting on the walls. The two most unpleasant installations (3 and 5) provide diffuse lighting only on the table, one with an illuminance of 110 lx, the other with 1100 lx. This last result is interesting in view of the tendency to use a correlation method to relate pleasantness to illuminance alone. Although these results have been discussed in terms of the pleasant/unpleasant ratings, an inspection of Fig. 6.5 reveals a consistent pattern for the six lighting installations on all the other scales related to the evaluative dimension. Thus installations 6 and 4 consistently attract favourable assessments and installations 3 and 5 perjorative ones.

The second dimension, perceptual clarity, is related to the illuminance on the table. As might be expected, the higher illuminance installations are considered to give a greater impression of clarity. Further, the separation on the clear/hazy rating scale, which is the one most strongly linked to this dimension, is much less between 320 lx (installation 6) and 1100 lx (installation 5) than between 110 lx (installations 1, 2, 3 and 4) and 320 lx (installation 6). This is just what would be expected from many of the correlation studies, but it is important to remember that the factor analysis reveals that the observers are treating pleasantness and clarity as independent dimensions.

The third and fifth dimensions, spatial complexity and formality, do not strongly separate the lighting installations; but there is a clear difference in the fourth dimension, spaciousness, although the relevant features of the lighting installation are varied. The installations which provide light on the table alone, particularly those with a low illuminance, e.g. installations 1 and 3, give the room an impression of being small and cramped. The installations in which both the wall and table or even just the wall alone were lit gave an impression to the room of being large and spacious.

Such information can be used to answer questions relevant to design. For example, suppose the designer has a choice of either lighting the conference table to a high illuminance or having a low illuminance on the table but some light on the walls. From the data discussed above, it is apparent that the former approach will give the impression of a room which is clear but rather unpleasant, whilst the latter will be assessed as less clear but more pleasant and spacious. The results will not tell the designer what to do but they do give him more information on which to base his decision.

It can be concluded that this rating scale/factor analysis procedure produces much useful information. It also overcomes some of the criticisms of the correlation method. It ensures that the dimensions identified are independent and by using rating scales related to the impression of the room rather than just the lighting, ensures that the results reflect the importance of the lighting variables to the particular context. However, the rating scales are still chosen by experimenters and not by the observers, even when they are taken from lists of rating scales thought suitable for assessing the environment [15]. A cynic would describe this as putting words into the mouths of the observers.

Fortunately there is an alternative method which avoids this problem. This involves asking the observers to assess the differences (or similarities) between all pairs of installations from a range of installations, and then applying a multi-dimensional scaling (MDS) statistical analysis to the differences. No rating scales are involved in this method and the observers use whatever basis they wish to assess the differences. Flynn *et al.* [12] adopted this method to examine the same six lighting installations in the same conference room they had used in their rating scale/factor analysis study, but with a different set of 46 observers. The MDS analysis estimates the number of independent dimensions that are needed to describe the differences between the installations. These *n* dimensions are used to form an *n*-dimensional space in which the individual installations can be located. Flynn *et al.* found that for the six installations in the conference room, a three-dimensional space was the minimum necessary to explain the data. Figure 6.6 shows the three dimensions and the location of the individual installations. The three dimensions are named peripheral/overhead, uniform/non-uniform and bright/dim. As in the rating scale/factor analysis procedure these names are inferred. They are derived from the positions of the different installations relative to each other and to the dimensions. As the dimensions are identified from the installation characteristics there is no possibility of directly evaluating the meaning of the dimensions. The best that can be done is to tell which installations are considered more or less different from each other and the independent dimensions on which they differ. This may, at first sight, seem a rather small achievement but a comparison of the results obtained by the two procedures, rating scale/factor analysis and difference/MDS, on the same installations shows how the two methods can be complementary. From the rating scale/factor analysis, the three dimensions which most clearly separate the installations are perceived clarity, evaluative and spaciousness. Choosing a rating scale strongly related to each of these dimensions allows stepwise multiple regression equations to be calculated for the mean ratings on each scale and on the three MDS dimensions. The perceptual clarity dimension gave a 0·99

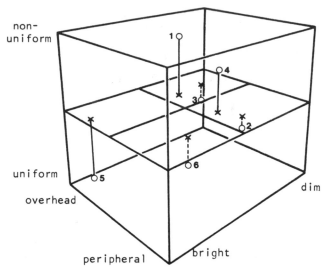

Fig. 6.6.　The positions of the six lighting installations in the conference room shown in Fig. 6.4 in the three-dimensional space derived from multi-dimensional scaling analysis. (After [12].)

correlation on the bright/dim MDS dimension; the evaluative dimension gave a correlation of 0·83 on the overhead/peripheral MDS dimension and the spaciousness dimension gave a correlation of 0·69 with the uniform/non-uniform MDS dimension alone. These correlation coefficients demonstrate that there is good but not perfect agreement between the dimensions obtained by the two methods and the placing of the various installations on those dimensions. This agreement allows the MDS dimension which the observers use to distinguish between the installations to be related to the observers' rated impressions of the interior. For example, it is clear that the extent to which the room is considered pleasant is mainly related to the existence of peripheral lighting, although as the introduction of the uniform/non-uniform MDS dimension increases the correlation with rated pleasantness to 0·92, having a non-uniform distribution of light is also of value. This conclusion could have been reached from Figs 6.4 and 6.5, but these were obtained by asking the observers specific questions. It is nice to have the conclusions confirmed by a procedure that allows the observer to use whatever basis he wishes in order to make his judgements.

It can be concluded that both the rating scale/factor analysis and the difference/MDS analysis procedures are reasonable approaches to examining the impressions created by lighting, and, if used together, can complement each other. However, before considerable effort is devoted to their use, it is worth while considering the extent to which the dimensions obtained, and the positions of specific installations on those dimensions, are stable. Flynn *et al.* [16] studied this question in two stages. First they examined the dimensions of impression obtained in rooms which were different but which had the same context for the observers. Specifically, they obtained ratings on a number of scales for three different conference rooms experienced by different groups of observers. One of the rooms was that described previously (Fig. 6.4), the other two being a similar size room but irregular in shape, and a large rectangular room. Each room was lit in five different ways using fluorescent and incandescent light sources and overhead and peripheral lighting. Factor analysis of the ratings revealed four main factors which were named perceptual clarity, spaciousness, evaluative and spatial complexity. The results are shown in Fig. 6.7 as reported by Flynn *et al.* [16]. The mean ratings for each installation in each room on six different rating scales are given. Four of the scales are obviously related to the four dimensions: perceptual clarity, spaciousness, evaluative and spatial complexity. The other two scales, private/public and relaxed/tense, were chosen for their interest. The similarity in the profiles shown for the five installations in the three rooms on each scale in Fig. 6.7 indicates that the shape, size and furnishing of the rooms are not as important as might be expected, provided of course, they are typical of the context. The basic factor structure for the three rooms taken together is similar to that obtained previously for the one room. Further, comparisons of Figs 6.7 and 6.5 show that the positions of the individual installations on the dimensions are similar in both of the experiments involving conference rooms [12, 16]. It can be concluded that there is evidence for some stability in the dimensions on which people assess the lighting of conference rooms and the way in which particular lighting conditions are related to those dimensions.

   The question that needs to be considered now is how stable these dimensions are when transferred to other contexts, e.g. a shop, factory or restaurant. This was the second aspect of stability of impression studied by Flynn *et al.* [16]. They used an auditorium which was assessed by two separate groups of people. The first group saw seven lighting conditions and recorded their impressions on 13 rating scales. The second group saw 13 lighting conditions and used a questionnaire with 15 rating scales. The results for the five installations seen by both groups are recorded on five representative rating scales: clear/hazy, public/private, unpleasant/pleasant,

Lighting installation

A — Overhead, diffuse, medium intensity

B — Peripheral only

C — Overhead, diffuse, medium intensity + peripheral

D — Downlighting, low intensity

E — Downlighting, low intensity + one wall

Room description

——— Medium-sized, irregular

------ Large, rectangular

— — — Medium-sized, rectangular

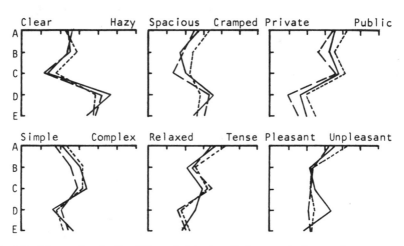

Fig. 6.7. Mean ratings for five different lighting installation types in three conference rooms on six rating scales. The mean ratings for each room on each scale are connected. (After [16].)

spacious/cramped and simple/complex. The two groups of people showed considerable consistency in their placings of the installations on each scale. Also the features of the lighting that are important for each scale are as before: high illuminances are related to greater clarity; low illuminances from incandescent sources and peripheral lighting are related to an impression of privacy; peripheral lighting and low illuminances produce a pleasant impression; the use of high illuminances and peripheral lighting is related to spaciousness; and, finally, general lighting with wall

Table 6.1: Installations used by Hawkes *et al.* **[17]**

Installation number	Installation
1	Regular array of ceiling recessed fluorescent units with opal diffusers
2	Incandescent downlights in regular array plus fluorescent wall washing of two end walls
3	Regular array of ceiling recessed fluorescent units with bats-wing prismatic panels
4	Fluorescent wall washing of two side walls
5	Fluorescent desk lights at either side of each desk
6	Incandescent spot lights at end of room and on desk
7	Incandescent downlights in a regular array
8	Incandescent spot lighting of side walls plus fluorescent wall washing of right-hand wall
9	Regular array of ceiling recessed fluorescent units with prismatic panels
10	Fluorescent desk lights at either side of desk plus fluorescent wall washing of left-hand wall
11	Regular array of incandescent downlights plus incandescent spot lighting of two side walls
12	Regular array of ceiling recessed fluorescent units with specular louvres plus fluorescent wall washing on right-hand wall
13	Fluorescent wall washing of right-hand wall
14	Regular array of ceiling recessed fluorescent units with specular louvres plus incandescent spot lighting of two side walls
15	Incandescent spot lighting of all walls and desks
16	Regular array of ceiling recessed fluorescent units with specular louvres
17	Fluorescent wall washing of all four walls
18	Fluorescent desk lights at either side of each desk plus incandescent spot lighting of two side walls

washing leads to an impression of complexity. Bearing in mind that these findings and those shown in Figs 6.7 and 6.5 were obtained with different people, in rooms with different contexts, furnished in different ways, the similarity of the results is comforting. They certainly suggest some stability in the assessment of lighting.

This conclusion is important, so it is worth while seeing if it can be supported by the work of other people. Hawkes *et al.* [17] collected ratings related to people's impressions of a small rectangular office lit in different ways. The illuminance on the desk was always approximately 500 lx but the distribution of light in the rest of the room varied widely. Eighteen different lighting installations were used made up of different combinations of conventional uniform lighting, wall washing, downlighting, local desk lighting and spot lighting (Table 6.1). Factor analysis of the extensive data collected, excluding the evaluative scales, revealed two independent dimensions. One was the brightness dimension, so named because it had

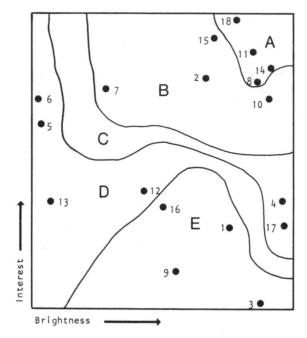

Fig. 6.8.   A map showing the location of the types of office lighting listed in Table 6.1 on the two dimensions, interest and brightness, identified by factor analysis. Superimposed on this map are iso-preference contours based on preference ratings of the same installations. These contours define areas of equal preference from area A (most preferred) to area E (least preferred). (After [17].)

such rating scales as dim/bright, strong/weak and clear/hazy strongly related to it, although it also had the cramped/spaciousness scale associated with it. The other dimension was simply named interest, because the rating scales strongly related to this dimension were simple/complex, mysterious/obvious, uninteresting/interesting and commonplace/special. The positions of the different installations on the plane formed by these two dimensions are shown in Fig. 6.8. From this figure it can be seen that the brightness dimension is clearly related to the amount of light in the room, whilst the interest dimension separates installations into those which use only diffuse sources (e.g. fluorescent lamps), which were considered uninteresting, and those which used focused sources (e.g. spot lights), which were considered interesting.

Also shown in Fig. 6.8 are iso-preference contours indicating those areas on the plane which are preferred to different extents. The top right-hand corner contains installations that are most preferred. Such installations are both bright and interesting. It should be realised that these two dimensions are independent. Making an uninteresting situation brighter does not make it better, neither does making a dim situation more interesting. Both dimensions have to be right. The interesting point for practice, available from an examination of Table 6.1 and Fig. 6.8, is that the most preferred installations have certain common features as do the least preferred installations. All the most preferred installations contain some element of variety produced by spot lighting or wall lighting. All the least preferred installations consist only of recessed fluorescent luminaires, providing uniform diffuse lighting. This division between different installation types is to be expected from the results of Flynn *et al.* [12] (Fig. 6.5).

In spite of the similarity, it must be admitted that there are some differences between the dimensional structures of this experiment and those described earlier [12, 16]. This is not too unlikely given that (a) different rating scales were used in the different experiments, (b) Hawkes *et al.* [17] deliberately removed all the scales likely to be loaded on an evaluative dimension from their factor analysis, and (c) Hawkes *et al.* [17] asked their observers to assess the lighting of the room whilst Flynn *et al.* [12] asked theirs to assess the room as such. Even so both sets of results have a bright/dim (perceptual clarity) dimension. In addition, although Flynn *et al.* [12] do not identify an independent interest dimension they did have an interesting/monotonous scale related to the evaluative dimension, and the location of different installations types along it is similar to that found by Hawkes *et al.* [17]. Overall it can be concluded that there is some evidence to support the view that different people do assess lighting in different rooms on a consistent basis.

So far consideration of the stability of impression has been restricted to experiments which have used the rating scale/factor analysis procedure. Flynn *et al.* [16] used the auditorium mentioned earlier with six installations similar to those shown in Fig. 6.4 and one high intensity downlight installation, for an experiment using the difference/MDS analysis procedure. The results for the conference room suggest that the last installation would be considered as bright, overhead and non-uniform. MDS analysis revealed that the observers used the same three dimensions as in the conference room for making judgements of differences between the installations in the auditorium. Further, the six repeated installations were located in similar places in the three-dimensional space created by the three dimensions, and the high intensity downlight installation was placed as predicted. This result offers some support for the stability of the dimensions obtained using the difference/MDS analysis procedure. Unfortunately other studies do not. Hawkes [18] used the difference/MDS analysis procedure to examine impressions of lecture theatres. He found four dimensions were necessary to explain the results but was not able to name the dimensions from the location of the particular theatres in the four-dimensional space. Stone *et al.* [19] also used the difference/MDS analysis procedure on lecture theatres, but they needed only three dimensions to explain the pattern of differences. Even Flynn *et al.* [16] have had some difficulty in reproducing their three-dimensional structure in the auditorium when other types of lighting installations are used. It must be concluded that much more information is needed before the way the dimensions identified by the difference/MDS analysis procedure vary with context can be understood.

So far this discussion has been centred on the stability of the impressions evoked by lighting in different context when seen by different people. It is now necessary to consider a different orientation, namely whether different lighting conditions can lead to the same impression. The answer must be positive. Flynn and Spencer [20] asked people to rate their impressions of a large room with no specific context other than that it was a room to be assessed. The room was lit by four different light sources in various combinations of general and wall lighting. The general lighting was produced from a luminous ceiling. The illuminances were very similar in all eight conditions used. From the ratings obtained, factor analysis revealed four familiar dimensions: evaluative; perceptual clarity, to which the scales bright/dim, hazy/clear, etc. were related; spaciousness; and something that the authors called complexity/novelty. The first three of these dimensions are very much like the ones identified in Flynn's earlier experiment. The interesting point is not that these dimensions were obtained, since by now they might be expected, but rather that in this case it is the light source colours which arrange the interiors along the dimensions. Specifically, installations with the Cool White fluorescent lamps were rated as being clear and distinct but the high pressure sodium discharge lamp installations were considered as hazy and vague. Another interesting result was that interiors lit by fluorescent lamps were rated as more pleasant than those lit by high pressure mercury discharge lamps, and that they were considered even more pleasant when they provided both general lighting and wall lighting rather than just general lighting alone. In fact installations which had wall lighting alone produced impressions of greater clarity, spaciousness and pleasantness than those which did not. Again this is consistent with the results discussed earlier.

The overall conclusion from this review of multi-dimensional methods must be that they represent a very attractive way of examining the impressions created by lighting. Certainly they are a considerable improvement over the correlation approach. Both rating scale/factor analysis and difference/MDS analysis procedures overcome the problems of independence. The two procedures can be complementary and both can produce information that is rich and valuable. Further, the results obtained

Table 6.2: Lighting cues for subjective impressions [21]

Subjective impression	Lighting cues
Impression of perceptual clarity	Bright, uniform lighting; some peripheral emphasis, such as with high reflectance walls or wall lighting
Impression of spaciousness	Uniform, peripheral (wall) lighting; brightness is a reinforcing factor, but not a decisive one
Impression of relaxation	Non-uniform lighting; peripheral (wall) emphasis, rather than overhead lighting
Impression of privacy	Non-uniform lighting; tendency toward low light intensities in the immediate locale of the user, with higher brightnesses remote from the user; peripheral (wall) emphasis is a reinforcing factor, but not a decisive one
Impression of pleasantness	Non-uniform lighting; peripheral (wall) emphasis

by these methods suggest that there is some stability across people and context in the dimensions on which the impression created by lighting is judged, and in the way in which particular features of the lighting influence that impression. This supports Flynn's concept [21] that lighting provides a number of cues which people use to interpret or 'make sense' of a space and that these cues are at least partly independent of the room that is being experienced. However, the evidence available is rather limited. It is restricted to a few contexts and lighting conditions and has mainly been obtained by the rating scale/factor analysis procedure. It seems likely that Flynn's basic point is correct but until a much wider set of interiors and lighting conditions has been examined the generality of the cues must be open to question. Table 6.2 gives a summary from Flynn [21] of the lighting cues he has identified and the impressions they produce. The value of something like this to the lighting designer, and hence the desirability of testing it, is obvious.

Even though the simple guidelines offered in Table 6.2 are useful they do call for a skilled engineer to use them if a successful installation is to be produced. To ease this situation some link between the subjective judgements of lighting and the physical variables that the lighting engineer is experienced in handling is desirable. Hawkes [18] in his lecture theatre study achieved this. By taking multiple regressions between the MDS dimensions and a mixture of physical variables he was able to correlate each of his dimensions with some combination of the physical quantities. Stone *et al.* [19] achieved the same thing but with different physical variables. There is thus some prospect of providing a complete route, from people's impressions of an interior, to the physical conditions influencing them.

6.4 BEHAVIOUR

The multi-dimensional procedures described in the previous section are the most advanced approach to recording and analysing what people say about an interior. However they relate to only one aspect of the appreciation of lighting. The impression created by lighting can also influence what people do. This does not refer to the performance of a task but rather to the way in which the space is used. To study this question the experimenter uses simple observation and recording, usually covert observation and recording. In many ways the procedures used are analogous to the methods of the zoologist observing the behaviour of an animal in its habitat. In this case the animal is man and the habitat is the luminous environment. There are other similarities. Advanced statistical analysis is rarely used. Analysis is often intuitive. Strictly, behavioural studies can be regarded as a retreat to a coarser method of obtaining data compared with the question-asking approaches discussed above, but it is several steps forward in realism. Patterson and Passini [22] have summarised the advantages and disadvantages of behavioural studies and questionnaire studies for examining the physical environment.

Behavioural studies related to lighting form a very *ad hoc* collection. Taylor and Sucov [23] explored the effect of lighting on the choice of which passageway people would use. They found that the more brightly lit passage was the one that was used by most people. Another example of a behavioural study which shows the power of lighting to direct activity is that of La Guisa and Perney [24]. They observed the duration of the attention schoolchildren paid to a display card, when it was of the same luminance as the rest of the room, and when its luminance was much higher. They found that the duration of attention was much greater for the highlighted display. Sanders *et al.* [25] measured the effect of illuminance and light distribution

SEAT SELECTION PATTERN

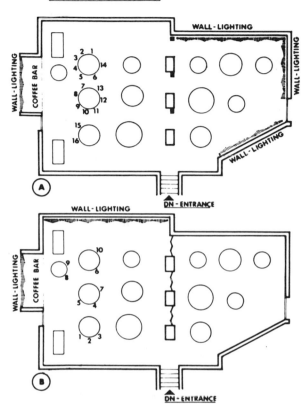

Fig. 6.9.   Selected seating for two lighting installations in a coffee bar. The numbers give the
position and order of seating chosen. (After [12].)

on the noise produced by groups of people talking whilst waiting in an
assembly room. They found that a low, uneven illuminance pattern was
associated with less noise than a higher, more even illuminance pattern. It
would be interesting to know if this result was repeated in a restaurant.
Flynn *et al.* [12] had a coffee bar lit in two different ways. The seating
arrangement chosen by the people was noted (Fig. 6.9). For the first lighting
system (upper diagram) the subjects chose to look towards an area of the
room lit by wall lighting in a way that had been rated as interesting and
pleasant. Although they chose to look in this direction they actually sat in
an unlit area. Possibly being closer to the bar was considered more
important than the benefits of being able to see what they were drinking

more clearly. The lower diagram shows the change in seating pattern that occurred when the lighting was changed. In this second installation the apparent room size was halved and only one wall was lit. Again the subjects chose to sit in an unlit area of the bar but they changed direction so that they could look towards the lit area. Obviously lighting does influence behaviour but it is only one factor amongst many.

A final example which demonstrates the understanding that can be achieved by simple observation is one by Flynn et al. [12]. They recorded the observers' behaviour and their remarks to each other on entering the conference room lit as described in Fig. 6.4. Diffuse overhead lighting producing a low illuminance (installation 3) was greeted with such remarks as 'reminds me of an elevator' and thumbs-down gestures. Installation 5 which had a high illuminance created by diffuse overhead lighting caused rubbing of the eyes and remarks like 'we have ways of making you talk'. Both these installations were considered unpleasant on the evaluative dimension in the factor analysis. However installation 5 was also rated as giving good perceptual clarity and a moderate degree of spaciousness. The evaluative dimension is apparent in these comments but the others are not.

It should be clear not only that lighting can affect behaviour but also that behavioural studies have their limitations. The information they produce is coarse and the relevant variables usually have to be inferred. Nonetheless, behaviour is sometimes a pertinent aspect of the effect of lighting, so it should not be neglected.

6.5　LIMITATIONS

All these various methods of examining people's appreciation of lighting have their advantages and disadvantages, both theoretical and practical. The underlying problem with the correlation approach is that it is not suitable for investigating the impression the lighting of an interior makes on people. It does not give people an opportunity to speak for themselves, rather it requests their opinions about a change in a particular physical variable. In a few of the more obvious cases, such as the effect of illuminance on the perceived brightness of an interior, this may give useful information but in general the results are of dubious value. The reason is that there is usually no evidence that the subjective judgements made in a correlation experiment are relevant in everyday life, or that the relationships established between different pairs of physical values and subjective judgements really refer to independent psychological dimensions.

The multi-dimensional procedures are a much more informative approach. They can give people an opportunity to speak for themselves. Also, both factor analysis and MDS analysis overcome the independence problem. However, it is by no means certain that the point about relevance is always met. The only way to deal with relevance is to ensure that the judgements are made in an identifiable context by people who have experience of that context, i.e. offices should be assessed by office workers. If this is done the observers can bring their experience to bear on what are the relevant factors about such a space. Given that this is done, the multi-dimensional procedures can produce a wealth of information of a much finer nature than is available from the correlation method.

Earlier, evidence has been given that both the multi-dimensional procedures considered lead to similar structures of impression of the same place, in which case the question arises as to which is the better. The answer is neither. The two procedures are complementary. The difference/MDS analysis procedure has the attraction of leaving observers to make their judgements of difference on any basis they wish; but it is not as immediately understandable as the rating scale/factor analysis procedure, because the dimensions have to be inferred from the locations of different installations and their physical differences. Neither does this difference/MDS analysis procedure provide a set of rating scales which can be examined individually and related to the independent dimensions identified. It would seem that the difference/MDS analysis procedure is both less open to experimenter bias and more difficult to comprehend. It should be noted that a hybrid multi-dimensional method exists in which an MDS analysis is applied to differences between environments expressed on rating scales [10]. In this form the difference/MDS analysis procedure has many of the advantages and disadvantages associated with the rating scale/factor analysis procedure, so it is not really an advance.

The basic practical problem with all the multi-dimensional procedures is that of range. For a start there is the range of conditions to be examined. If a variable is not presented to the observer then no dimension related to that variable can be found. For example, none of the studies discussed earlier, not even that of Flynn and Spencer [20], which used different light source colours, provided any dimension of colour. Yet, in general, the colours of the room surfaces seem likely to be important in determining the impression created in an interior. The answer for the Flynn and Spencer study is that the room deliberately had only low chroma colours. Thus there was little difference in the colours of the interior for the different light sources. The general point is that in evaluating a multi-dimensional study it is always necessary to consider the range of conditions examined: the wider the range the more valid the results will be.

Range is also important for the rating scales used in the rating scale/factor analysis procedure. If rating scales which might be expected to relate to a particular dimension are not used, e.g. scales relating to colourfulness in the above example, then no such dimension will be found. This gives one reason why the manner of selecting rating scales for a factor analysis study is important. The other is simply that the scales are being chosen by the experimenter and not by the observer. However, there is some advice available on suitable rating scales for assessing the physical environment [15, 26]. If this advice is followed the selection of rating scales should not present too great a problem in practice.

The final aspect of range that is important is the interaction between the range of conditions and the rating scales, in other words the range effect demonstrated by Poulton [6]. The foundation of this effect is that, when asked to make a subjective judgement, an observer will always use something as a reference. It may be the mid-point of the range of conditions to which he has been exposed in the experiment or in the absence of such exposure, a relevant previous experience. In a sense, such an effect should not be unexpected, in that no human activity takes place in a psychological vacuum, even an experiment has the context of an experiment. For subjective judgements of impression by multi-dimensional procedures or by the correlation approach, it seems reasonable to suppose that the anchoring of the judgements to previous experience would be valuable when faced with a range of different lighting conditions. To achieve such anchoring it would be necessary to firmly establish a context for the interior to be judged and to have it judged by a sample of people representative of those who would use such a space.

All the above have been practical problems involving experimental technique but the multi-dimensional procedures also pose problems involving expenditure of effort. The point is that although the multi-dimensional procedures are much superior to the correlation approach in the quality of information they produce, they are equally superior in the resources they demand. Ideally, a large number of real installations of a similar context are needed for assessment by a representative sample of people. For the difference/MDS analysis procedure each observer has to see all possible pairs of installations, whilst in the rating scale/factor analysis procedure all observers should see all installations. The logistics of such experiments would be very complex and costly. An obvious way to reduce this expenditure is to substitute colour slides or models for the real interiors, which can then be shown to a large number of observers at once. Shafer and Richards [27] have tried this approach for outdoor viewing, and Hershberger and Cass [28] have used it for housing and commercial buildings. Comparisons of people's ratings of the real scene and colour slides of the scene show strong similarities, as do the underlying factor structures. This suggests that slides which depict most of the variation of the environment can be used as substitutes for the real thing. Flynn *et al.* [16] have examined this possibility for interior lighting. Slides were made of the six installations in the conference room (Fig. 6.4). The dimensions identified from the ratings made by people in the room and people viewing photographs of the room were very similar. Further, the overall pattern of mean ratings for the six installations was similar for both methods of observation. Their conclusion was that carefully taken slides can give a reasonable overall picture of the dimensions involved but may lead to some erroneous conclusions about particular installations on particular scales. Lau [29] reached a similar conclusion in a study of the assessment of lighting in models and full-scale simulations of the same room. Clearly slides and models are not a perfect substitute for a real interior presented to a representative group in a realistic manner; but it could be argued that with care in checking some conditions against real interiors it would be reasonable to use them, when circumstances demand it.

None of these practical problems surround the observation of behaviour. Practically this is a simple matter of observation and recording. The only problems likely to arise are ethical, particularly if covert observation is involved. If the scruples aroused by covert observation can be overcome then care will allow the collection of some interesting data about this rather different aspect of people's appreciation of lighting—and it is different. Wicker [30] has shown that the link between expressed opinion and overt behaviour is rather tenuous, so that the one cannot necessarily be predicted from the other. Thus a complete examination of people's appreciation of lighting should involve both opinions and behaviour. Whilst opinions have been widely studied in the past, behaviour is only just starting to be considered [31]. Therefore it remains to be seen if some general pattern of behaviour can be related to different forms of lighting.

6.6 CONCLUSIONS

The main conclusion of this consideration of the methods of examining the appreciation of lighting is that the correlation method, from which most of the present lighting criteria have been derived, is unlikely to be of much use in the future. Its main use will be to validate any proposed physical measure of lighting conditions against people's responses. For an understanding of how lighting affects impressions of an interior, the multi-dimensional procedures are far superior. This conclusion might be taken to imply some criticism of the current lighting criteria. This is not so. The current lighting criteria have undergone tests more severe than any laboratory experiment could provide, namely survival in the real world. More precisely the current lighting criteria have been found in practice to be satisfactory in that they do not cause discomfort. This is a valuable step forward in knowledge but it is essentially negative knowledge. It is to be hoped that in the future more information will be gained on the positive attributes of lighting and particularly how lighting may be used to form the impression given by an interior.

However, a knowledge of the methodology of multi-dimensional procedures alone [32] will not be enough to ensure a successful study of the appreciation of lighting. This area of study, like so many others, is unlikely to be fruitful unless those working in it show some imagination and initiative. It is very easy to apply the multi-dimensional procedures automatically, particularly in this age of the computer, but if dull data are fed in dull results will come out. In the selection of the range of interiors to be examined, scales to be used, and the type of lighting studied, imagination is needed. The interpretation of the relationships between the various installations on the dimensions ultimately identified also calls for insight and understanding. The same applies to observation of behaviour. Without an alert observer, and one who understands the people being observed and the relationships between them, little of value will be obtained.

Finally it is worth while pointing out a feature of all that has gone before. It has only been concerned with lighting. This may seem reasonable enough in a book concerned with lighting but it should be realised that for the user of lighting it is but one aspect of the environment around him. It would be of considerable interest to have some understanding of the relative importance of lighting in different contexts (for what is known see Chapter 8). Such understanding might be achieved by using some of the techniques available in the field of general environmental evaluation [33]. These techniques include the trade-off approach in which people are asked to allocate limited resources to various environmental features [34], the use of model kits to determine preferred layouts [35], and the idea of games with different people taking different roles [36]. If these techniques were used, together with some specific studies such as those described earlier, then a more complete understanding of the many facets of lighting as they influence people might be achieved. Unfortunately all this is in the future, at the moment the available information on the appreciation of lighting consists of a number of studies, mainly by the correlation approach, of people's assessments of different lighting conditions. The results for the main contexts studied will be examined in the next chapter.

6.7 SUMMARY

Lighting can do more than just make things visible. It can impart a particular impression to an interior and influence people's behaviour. This chapter presents a survey of the methods that have been used to examine these effects of lighting and the type of data they have produced. The most widely used method in the past, and one that has been used to derive most of the existing lighting criteria, has been the correlation approach. The basis of this method is to relate some subjective judgement of the lighting to a physical descriptor of the lighting. It produces only a limited amount of information about the effect of one specific variable and there are sometimes doubts about its relevance and independence. Even so, the existing lighting criteria have been proved in practice. At the very least they eliminate discomfort. To move from this negative state to one which supports the positive role of lighting in manipulating impressions a different approach is required. A suitable one is a multi-dimensional procedure, e.g. rating scale/factor analysis or difference rating/multi-dimensional scaling analysis. These procedures can allow the observer to speak for himself and overcome the problems of independence if not always relevance. Evidence is presented of the richness of the information produced by these multi-dimensional procedures and the stability of the dimensions on which lighting is judged. The results obtained give some support to the idea of lighting as a means of producing cues by which people interpret a space. It is concluded that the multi-dimensional procedures are a good way of examining the appreciation of lighting. The practical problems they present to the experimenter are considered and some solutions suggested.

The final area examined is that of behaviour. Attitudes and behaviour appear to be rather tenuously linked but for a full understanding of the appreciation of lighting both need to be considered. The presently available data on how lighting affects behaviour are a very *ad hoc* collection but there is no doubt that lighting can influence the way in which people use a space.

REFERENCES

1. Balder, J. J. Erwunschte leuchtdichten in buroraumen, *Lichttechnik*, **9**, 455, 1957.
2. Hopkinson, R. G. *Architectural Physics: Lighting*, HMSO, London, 1963.
3. Illuminating Engineering Society. *Evaluation of discomfort glare—the IES Glare Index System for artificial lighting installations*, Technical Report No. 10, London, 1967.
4. Cuttle, C. C., Valentine, W. B., Lynes, J. A. and Burt, W. Beyond the working plane, *Proc. CIE 16th session, Washington*, 1967.
5. Hopkinson, R. G. and Petherbridge, P. Two supplementary studies on glare, *Trans. Illum. Engng Soc.* (*London*), **19**, 220, 1954.
6. Poulton, E. C. Quantitative subjective assessments are almost always biased, sometimes completely misleading, *Brit. J. Psychol.,* **68**, 409, 1977.
7. Hawkes, R. J. Multi-dimensional scaling: a method for environmental studies, *Building*, 19th June, 69, 1970.
8. Luminance Study Panel of the IES Technical Committee. The development of the IES Glare Index System, *Trans. Illum. Engng Soc.* (*London*), **27**, 9, 1962.
9. Hewitt, H., Kay, J., Longmore, J. and Rowlands, E. Designing for quality in lighting, *Trans. Illum. Engng. Soc.* (*London*), **32**, 63, 1967.
10. Kimmel, P. S. and Blasdel, H. E. Multi-dimensional scaling of the luminous environment, *J.I.E.S.*, **2**, 113, 1973.
11. Lynes, J. A. Discomfort glare and visual distraction, *Ltg Res. and Technol.*, **9**, 51, 1977.
12. Flynn, J. E., Spencer, T. J., Martyniuck, O. and Hendrick, C. Interim study of procedures for investigating the effect of light on impression and behaviour, *J.I.E.S.*, **3**, 87, 1973.
13. Canter, D. A. *Psychology for Architects*, Applied Science Publishers, London, 1974.
14. Harman, H. M. *Modern Factor Analysis*, University of Chicago Press, Chicago, 1970.
15. Kuller, R. *A semantic model for describing the perceived environment*, National Swedish Building Research D12, Stockholm, 1972.
16. Flynn, J. E., Spencer, T. J., Martyniuck, O. and Hendrick, C. *The effect of light on human judgement and behaviour*, IERI Project 92, Interim Report to the Illuminating Engineering Research Institute, New York, 1975.
17. Hawkes, R. J., Loe, D. L. and Rowlands, E. A note towards the understanding of lighting quality, *J.I.E.S.*, **8**, 111, 1979.
18. Hawkes, R. J. *A study of lighting quality*, Report to the Illuminating Engineering Research Institute, New York, 1970.
19. Stone, P. T., Parsons, K. C. and Harker, S. D. P. Subjective judgement of lighting in lecture rooms, *Ltg Res. and Technol.*, **7**, 259, 1975.
20. Flynn, J. E. and Spencer, T. J. The effect of light source colour on user impression and satisfaction, *J.I.E.S.*, **6**, 167, 1977.
21. Flynn, J. E. A study of subjective responses to low energy and non-uniform lighting systems, *Lighting Design and Application*, 7(2), 6, 1977.
22. Patterson, A. H. and Passini, R. The evaluation of physical settings: to measure attitudes, behaviour or both, in, *EDRA-5 Man–Environment Interactions: Evaluations and Applications*, ed. D. Carson, Dowden, Hutchinson and Ross, Inc., Stroudsburg, USA, 1975.
23. Taylor, L. H. and Sucov, E. W. The movement of people towards lights, *J.I.E.S.*, **3**, 237, 1974.
24. La Guisa, F. and Perney, L. R. Further studies on the effects of brightness variations in attention span in a learning environment, *J.I.E.S.*, **3**, 249, 1974.
25. Sanders, M., Gustanki, J. and Lawton, M. Effect of ambient illumination on noise level of groups, *J. Appl. Psychol.*, **4**, 527, 1974.
26. Vielhauer, J. A. *The development of a semantic scale for the description of the physical environment*, Ph.D. Thesis, Louisiana State University, USA, 1965.

27. Shafer, E. L. and Richards, T. A. A comparison of viewer reactions to outdoor scenes and photographs of those scenes, in *Psychology and the Built Environment*, eds D. Canter and T. Lee, Architectural Press, London, 1974.

28. Hershberger, R. and Cass, R. C. Predicting user responses to buildings, in *EDRA-5 Man–Environment Interactions: Evaluations and Applications*, ed. D. Carson, Dowden, Hutchinson and Ross, Inc., Stroudsburg, USA, 1975.

29. Lau, J. J. H. Use of scale models for appraising lighting quality, *Ltg Res. and Technol.*, **4**, 254, 1972.

30. Wicker, A. W. Attitudes vs actions: the relationship of verbal and overt behavioural responses to attitude objects, *Journal of Social Issues*, **25**, 41, 1969.

31. Hunt, D. R. G. The use of artificial lighting in relation to daylight levels and occupancy, *Building and Environment*, **14**, 21, 1979.

32. Flynn, J. E., Hendrick, C., Spencer, J. and Martyniuck, O. A guide to the methodology procedures for measuring subjective impressions of lighting, *J.I.E.S.*, **8**, 95, 1979.

33. Canter, D. A. Priorities in building evaluation: some methodological considerations, *J. Arch. Res.*, **6**, 38, 1977.

34. Banerjee, T. K., Baer, W. C. and Robinson, I. M. A trade-off approach for eliciting environmental preferences, in, *EDRA-5, Man–Environment Interactions: Evaluations and Applications*, ed. D. Carson, Dowden, Hutchinson and Ross, Inc., Stroudsburg, USA, 1975.

35. Rivlin, L. G., Rothenburg, M., Justa, F., Wallis, A. and Wheeler, F. G. Children's conceptions of open classrooms through use of scale models, in, *EDRA-5, Man–Environment Interactions: Evaluations and Applications*, ed. D. Carson, Dowden, Hutchinson and Ross, Inc., Stroudsburg, USA, 1975.

36. Mackie, D. Gaming as a research tool in architectural psychology, in, *Architectural Psychology*, ed. R. Kuller, Dowden, Hutchinson and Ross, Inc., Stroudsburg, USA, 1973.

7

Avoiding Discomfort

7.1 INTRODUCTION

The preceding chapter contains a discussion of the methods whereby the features of lighting that influence people's responses to an interior can be identified. Some of the approaches discussed there give great hope of a better understanding of the influence of lighting in the future. Unfortunately at present there are only a few studies based on these approaches available. This is a pity because the world cannot wait for a better understanding. Lighting installations are being designed now, so use has to be made of the understanding available now. The aim of this chapter is to describe, critically, the current understanding of the influential features of lighting manifest in the form of lighting criteria.

What understanding is available has almost all been produced by the correlation approach, i.e. a physical characteristic of the lighting has been varied and the subjective responses produced have been noted. Then the level of the physical variable necessary to reach a desired level of response has been used to derive a criterion for lighting. The problems of independence and relevance associated with this approach have been discussed in Chapter 6 but there is another limitation. This approach inevitably produces coarse-grained information, although not necessarily incorrect information. The data obtained are likely to be sufficient to define lighting conditions capable of producing definite discomfort but insufficient to identify lighting conditions which are pleasant in different ways.

These are the philosophical problems of the correlation approach but a study of the literature shows there also exist a number of more practical problems. Many correlation studies have used some form of rating scale to record the subjective judgements, often with scant regard for the semantic details of such scales. Subjects have been 'led' towards a particular conclusion by instructions which specify what are 'good' and 'bad' conditions. Some studies have been undertaken in abstract rather than in realistic surroundings. For example, work on discomfort glare has sometimes required the observer to set the luminance of a small disc of light in a uniform luminance field of view to the boundary between comfort and discomfort. This may produce repeatable results but what they mean for real interiors is open to question. People have considerable experience of judging environments but that experience is based on judging the complete environment, i.e. an environment furnished with people and objects which have some function. The assessment of whether a given physical condition is suitable for an interior will depend on that experience and the 'meaning' of the interior. In other words, the judgement of suitable and pleasant lighting conditions will depend on the context. It is for this reason that the present discussion is divided into different contexts.

7.2 HOUSES

In terms of area and numbers, one of the most common types of building is a house of some form; yet very little research has been done on the lighting of houses. What has been done has concerned the provision of daylight and sunlight. This is as it should be, because nearly all houses are designed to utilise natural light. Artificial lighting in housing is usually treated as a means of extending the day in a decorative manner. The emphasis given to natural lighting in housing has led to recommendations being made [1, 2] of daylight factors† and hours of exposure to sunlight (Table 7.1). It is not at all clear how these recommendations were derived but they undoubtedly reflect people's desire for natural lighting. In view of this desire it is rather surprising that so little information is available on the preferences for different quantities. In one of the more exact pieces of work, a survey of 47 houses was undertaken in which measurements were made of daylight factors in living rooms, kitchens and bedrooms. These were related to the householder's opinions as to whether these rooms were light enough [3]. No clear relationship between complaints of insufficient daylight and whether or not the recommended daylight factors had been met was found. However, nearly all the living rooms and main bedrooms did meet the

---

† The daylight factor at a point in an interior is the ratio of the illuminance produced at that point by daylight (excluding sunlight) from a sky of known or assumed luminance distribution to the illuminance on a horizontal plane due to an unobstructed hemisphere of this sky.

Table 7.1: Recommended daylight factors and sunlight exposures for housing [1]

Condition	Location	Recommendation
Daylight	Kitchens	A 2% daylight factor over 50% of the floor area within a minimum of 4·7 square metres and with the sink, cooker and preparation table well within this zone†
	Living rooms	A minimum 1% daylight factor to cover at least 7·0 square metres. Penetration of the 1% daylight factor zone should extend to at least half the depth of the room facing the window†
	Bedrooms	A minimum 0·5% daylight factor should cover at least 5·6 square metres with the penetration of this zone being not less than three-quarters of the depth of the room facing the window†
Sunlight	Living rooms and where practicable kitchens and bedrooms	The entry of sunlight to be possible for at least one hour at some time of the day during not less than the 10 months of the year from February to November. It is preferable for sunlight to enter living rooms in the afternoon and kitchens and bedrooms in the morning. Sunlight should not be considered to enter a room if the horizontal angle between the sun's rays and the plane of the window is less than 22·5°

† For implementation, the following surface reflectances are assumed: walls = 0·40, floor = 0·15, ceiling = 0·70.

daylight factor standards but half the kitchens did not. At least some of the reduction in daylight in the kitchens could be attributed to the occupants' choice of decorations and curtains.

This survey was part of a much larger survey involving about 1300 households on four estates, all constructed in the 1960s. In general, the number of complaints about lack of daylight was very low, especially when compared with other aspects of housing layout. Most of the complaints about daylight involved bathrooms, three-quarters of which were windowless.

Whilst there is no doubt that daylight in houses is widely desired, the value of sunlight is not so obvious. Sunlight is light direct from the sun which has not been scattered in the atmosphere. Sunlight can quickly cause overheating and rapid fading of fabrics. Daylight is much less drastic in its effects. Daylight, because it is largely diffuse radiation, is available for all orientations of windows; but a north-facing window in the northern hemisphere will receive only a very limited amount of sunlight at high angles of incidence. This effect of orientation is shown in the results of the survey of about 1300 households already described in relation to daylight [3]. The householders were also asked if they got enough sunlight in their living rooms. About one-third of those whose living room windows faced the sector between north-west and east-north-east agreed that they received insufficient sunlight. For other orientations of windows the percentage who thought they had too little sunlight was less than 5%. Although sunlight was desired it was not closely related to a criterion that one hour of sunlight be possible on the 22nd November. Nearly two-thirds of the householders who said they had enough sunlight had living rooms which would not meet this criterion. However, practically all those who were dissatisfied with the sunlight availability had living rooms which fell below the criterion. This suggests that while some sunlight is desired the amount is far from critical.

It should be realised that amount is a very primitive basis on which to evaluate the effects of sunlight. Bitter and van Ierland [4] carried out a survey of the opinions of 1000 Dutch housewives on the value of sunlight in their homes. They concluded that sunlight was valued not simply because it provided light but also because it provided warmth and what was called 'atmosphere'. A similar survey of people in the UK reached the same conclusions [5]. Specifically, four positive and three negative aspects of sunlight were eventually obtained by interviewers, after some initial difficulty in persuading people that the desirability of sunlight could be open to question! The four positive attributes were the light, the warmth, the improved appearance it gave to interiors, and what were called the therapeutic effects. This last category was used to classify such responses as 'sunlight makes you feel better'. The three negative attributes were the fading of furniture and soft furnishings, the thermal discomfort and the visual discomfort. To put these positive and negative attributes in perspective it is only necessary to mention that 93% of the responses were for positive attributes whilst only 4% were for negative attributes. Basically people were aware of the negative aspects of sunlight but considered they could be easily overcome with blinds or ventilation. The positive attractions of sunlight were overwhelming.

Bitter and van Ierland [4] elaborated their findings of a desire for sunlight into the time and place sunlight was wanted. Not surprisingly they found a desire for sunlight in rooms at the times they were occupied, that is, kitchens and bedrooms in the morning, living rooms in the afternoon. An unexpected result was the priority the housewives gave to sunlight entering their home, over a pleasant view. In the study by Ne'eman *et al.* [5] the reverse priority was evident. When forced to make a choice between a home with sunshine indoors but with a poor view, or a home with a pleasant view but no sunshine indoors, 58% of the people opted for the view whilst only 28% chose the sunshine. The rationalisation of this was that you can look at a nice view all the year round but the sun could be enjoyed much less frequently. The results of this survey support those of an earlier study [3] which found the view was of more importance than sunlight.

Providing daylight in housing means having a window. But windows do much more than admit daylight. They also provide a view out. For this function there are no simple recommendations because the influence of a window is not a simple matter. An appreciation of a window's effects involves considering several different aspects, e.g. light, view, noise, heat, ventilation, etc., and achieving the correct balance for the circumstances. Window design is also subject to the whims of fashion. Not surprisingly studies of people's responses to the daylighting of their houses have found little link with the simple parameter of window size. Rather Markus *et al.* [6] found that the general level of satisfaction with the house was strongly linked with the end result of the size and positioning of the windows, i.e. the daylight provided and the view. What constitutes a good view depends on the context. There is a definite preference for views with natural elements, i.e. trees, grass, etc. and some activity [5, 7]. The element of activity is particularly desired by the old or housebound. In a review of this area Clamp [8] concluded that the old valued a view with an element of activity for its interest, but for other groups the dominant factor was that the view should reflect their status in society. This may explain the common preference for a 'natural' view. Natural views are considered to have a high status.

To summarise, the dominant aspects of research on lighting in houses has involved daylight and sunlight. There is no doubt that in temperate climates there is a strong desire for daylight and sunlight admission, any problem of thermal or visual discomfort being easily overcome by ventilation or blinds. This conclusion seems unlikely to be applicable to hotter climates, especially as regards sunlight. Certainly vernacular window design in tropical countries aims to limit daylight and sunlight rather than to admit it. The desire for daylight and sunlight in housing in temperate climates has led to recommendations for the provision of daylight and sunlight being made (Table 7.1). Whilst these recommendations are precise, field results suggest that meeting them exactly is not critical. This should not be too surprising given the simple nature of the recommendations and the complex and multiple effects of windows.

As for the preferred characteristics of artificial lighting in houses, virtually no research has been done, except on a few specific problems [9]. This is partly because artificial lighting in the home is considered largely as an element of decoration into which some aspects of function intrude.

Conventional lighting research procedures are not well suited to examine decorative effects of lighting although some of the methods discussed in Chapter 6 are. Another factor influencing the lack of research is the point that, unlike daylighting, the decisions on artificial lighting in houses are taken by the individual householder. Thus each householder can choose the lighting he thinks most suitable, so there is little place for general recommendations based on research. In spite of the absence of information, illuminance recommendations for houses have been made [10]. However, the few field measurements that are available [11] suggest that the illuminances in use are much lower than those recommended. It seems unlikely that this situation will change much in the future unless research based on the multi-dimensional approach discussed in Chapter 6 reveals some consistent cues for generating impressions by lighting. Until such information is available, 'recipe books' demonstrating what is possible with artificial lighting in the home can be used and provide the level of advice needed [12, 13].

7.3   OFFICES

The vast majority of studies of the effect of lighting conditions on people's impressions of an interior have been done in the context of offices. There are two reasons for this. The first is the ease of application of results. As regards both the work done and the equipment used, one office is very much like another. Therefore by examining one office the experimenter can produce results which are relevant to a large number of buildings. The second is that offices are where the money is, and that always creates interest. People expect a good working environment in an office, so the owners of such buildings expect to pay for good lighting.

7.3.1 Daylight, Sunlight and View
Electric lighting can be considered at a late stage in the design of a building (although it should not be) but daylighting cannot. If the lighting of an office is to be principally by daylight then the whole shape and structure of the building is affected. This may cause the architect problems but they are worth overcoming because office workers express a strong preference for daylight as a method of lighting offices. Markus [14] in a study of the attitudes to daylight, sunlight and view of over 400 office workers on nine floors of a 12-storey, open plan building found that 95% of the people preferred to work by daylight. There was a marked variation in satisfaction

with the daylight available at the working position, those nearer windows being much more satisfied than those further away, particularly on the lower floors. This is to be expected, in that it simply reflects the distribution of daylight across the office and a reduction in its availability, caused by external obstructions, for the lower floors. Another, not unreasonable, finding was that the people closest to the north-west and north-east walls were less satisfied with the daylight available than those near the south-west and south-east walls. All these results emphasise the importance people place on daylight, an emphasis found in other surveys in offices of different sizes [15, 16, 17]. However this should not be taken to mean that unlimited daylight is desirable. Langdon [18] in a survey of offices in London in the 1960s found that the 100 % glazing then in fashion was producing extensive overheating problems, particularly where windows could not be opened to increase ventilation because of the consequent noise and draught problems. The most common solution to overheating was to use blinds, which rather defeated the purpose of the window in admitting daylight and in providing a view. More recently what are called solar control glasses have been produced which work by either reflecting or absorbing a large proportion of the solar radiation. Unfortunately they also reflect or absorb a large portion of the visible radiation. Thus although they preserve the view through the window they do reduce its ability to admit daylight. In some cases because the glass is tinted they also change the spectral distribution of the light admitted. However Cooper *et al.* [19], in a survey of 11 office buildings found no serious problems with the use of solar control glass. Adaptation took care of both reduced daylight and the colour change. It can be concluded that most people in offices would like to work by daylight provided the overheating problems associated with large amounts of glazing could be overcome, which they can, at a price.

Yet the fact remains that many office buildings are constructed with very modest amounts of glazing. The reason for this is simply that the decision to use predominantly daylighting imposes severe limitations on the form of the building. Unless rooflights are used or excessive inter-floor height is available, none of the rooms can be very deep and a narrow plan form results. The basic problem is that of daylight penetration. Figure 7.1 shows the change in daylight factor from side windows with distance from the windows. If the 2 % daylight factor recommended for office work [1] is required at the back of a normal-height office lit by side windows, then the width of the building is limited to about 8 m (two rooms and a central corridor). For many sites such a restriction does not allow the most economic use of the land nor is it a very efficient shape in terms of minimising heat loss. It is factors such as these, all of which have financial implications, which explain the deviations from the clearly expressed wish of most office workers to work by daylight.

With such weighty factors being considered it is hardly surprising that the question of sunlight has been pushed into the background. Markus [14] found that 86% of the people in the 12-storey office building he studied preferred sunlight in the office all the year round but there was a tendency

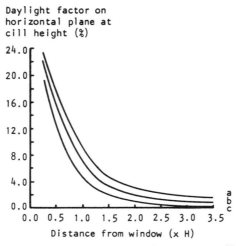

Fig. 7.1. The variation in daylight factor on a horizontal plane at cill height with distance from the window. The distance from the window is expressed in multiples of H, where H is the height of the window head above the cill. Curves: a, window width = 100% of room width; b, window width = 60% of room width; c, window width = 30% of room width. (After IES Technical Report 4, *Daytime lighting in buildings*, 1972.)

for those who sat nearest to the windows to be less enthusiastic about it than those who worked further away. Ne'eman *et al.* [5] claimed a similar result. Over the four different office buildings they studied there was a strong emphasis on the desirability of sunlight and its 'therapeutic' effects but there was also a clear underlying complaint of thermal and visual discomfort. The strength of the complaints depended on the facilities in the office for controlling the thermal and visual environment. Where blinds were available and/or windows could be opened without creating draughts or admitting noise then little problem was experienced. When these facilities were not available then the level of complaint tended to increase. The effectiveness of the heating or air conditioning systems was also important. In some poorly heated offices the thermal aspects of sunshine were beneficial but where the heating or air conditioning worked well in overcast conditions it often failed to cope with sunlight.

The factor which explains the reduction in enthusiasm for sunlight in offices as compared with houses is the degree of individual control available. In a house the occupant can do what he likes about the physical environment. In an office the worker is usually sat at a fixed location and is severely restricted as to what modifications he can make to his environment. The office worker is often dependent on facilities beyond his immediate control to alleviate the effects of sunlight. If these facilities are effective then sunlight is considered a pleasure but if they are not it can easily become a curse. Thus sunlight, like daylight, has much wider implications than just its immediately obvious effects. The designer of office lighting will need to balance all these factors before deciding on his window design.

This balancing process indicates that there is no possibility of specifying a generally applicable optimum window size based on daylight and sunlight provision. However it has been possible to specify a minimum window size based on the other main function of windows, providing a view. Ne'eman and Hopkinson [20] examined this question using a model open-plan office through which subjects could look out at real views. The window in the model office was adjustable in width and observers were asked to set the width to give the minimum acceptable window size. It was found that the observers could do this consistently except when the window was uniformly bright and featureless. For more usual views the minimum acceptable window size was determined by the amount of visual information provided by the view; near views requiring larger window sizes than distant views. The judgement of minimum acceptable window size was not much influenced by the amount of daylight or sunlight that was admitted to the model, the interior illuminance or by the position of viewing the window. About 25 % of window wall area was the minimum acceptable window size for 50 % of the observers but this had to be increased to about 32 % if 85 % of the people were to be satisfied.

Keighly [21] has also studied this problem. Using a model of an open-plan office, 40 observers were shown a range of windows which differed in size (11–65 % of the window wall) and in the number and layout of apertures. Through these windows a number of different views were presented on film. The observers did not adjust the window size but simply rated their satisfaction with the windows on a five-category rating scale with the categories labelled: 1 = entirely satisfactory, 2 = fairly satisfactory, 3 = neutral, 4 = rather unsatisfactory, 5 = very unsatisfactory. Figure 7.2 shows the mean ratings of satisfaction for single windows of different sizes,

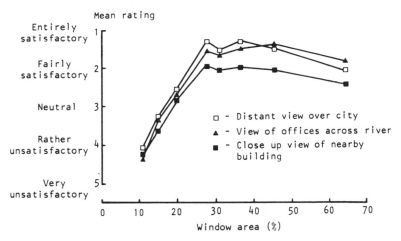

Fig. 7.2.   Mean ratings of satisfaction with windows for different window areas and views. The window area is expressed as a percentage of the window wall area. (After [21].)

for the three views used. It can be seen that glazed areas of less than 15% window wall area are considered disagreeable but above 30% almost complete satisfaction occurs, although there is a hint of a reduction in satisfaction for large window sizes. These results are in reasonable agreement with those of Ne'eman and Hopkinson, although the importance of the type of view used is much less, possibly because of the static and limited nature of the views used in Keighly's experiment. As for the effect of dividing a given glazed area into different numbers of apertures this produced some very marked preferences. People disliked the use of several windows of different sizes, a large number of regularly arranged very narrow windows, and wide mullions. The common characteristic of these aspects of window design is that they break up the perception of the view. In general people preferred windows to be large, regularly arranged and horizontal. This last aspect raises the question of window orientation and position.

Keighly [22] has examined this specific question. Using the same model of an open-plan office 30 observers were asked to adjust a variable rectangular window with a constant area of 20 % of the window wall until it was in the preferred position and of the preferred shape. The most frequently preferred condition was a central horizontal aperture, with the elevation being determined by the skyline of the view until the skyline approached the level of the ceiling. The horizontality of the preferred window shape may have been due to the largely horizontal natures of the views used in this experiment. Certainly the horizontal nature is in opposition to Markus's [14] support for predominantly vertical windows. A relevant study here is that of Ludlow [23]. Using a similar apparatus to Keighly he asked 20 observers to adjust a window aperture to the preferred size and shape for the view presented. He concluded that the specific view did have a large effect on the preferred size and shape. These results can be understood if it is assumed that above a minimum size, windows are considered essentially as a frame to a view. If this is accepted, then arguments about fundamental preferences for vertical or horizontal windows become sterile.

Ludlow [23] also found that the preferred window area was between 50 and 80 % of the window wall. These values are much higher than those considered entirely satisfactory by Keighly's observers [21]. This difference may have been simply between what is satisfactory and what is preferred but it may have been due to differences in the views or models. In either case it should be realised these window areas have been determined in a model so they are based on visual factors only; the thermal implications of the windows have not been considered.

The basis of the usual simple recommendation for temperate climates that window sizes should be somewhere between 20 and 40 % of the window wall area [10] should now be apparent. Below 20%, dissatisfaction with the windows is likely to arise, particularly if they are concentrated into one or two large areas so that many people do not have any view at all. Above a 40 % level, satisfaction with window area will be high but unless special measures are taken, such as a solar control glass, the incidence of thermal and visual discomfort is likely to increase. At 40 % glazing, the level of satisfaction will still be high but the admission of daylight and sunlight will not affect the other physical conditions unduly. In considering these results it should be borne in mind that they have almost all been found for models of large open-plan offices assessed by observers who are used to temperate climates. Experience suggests that the differences between the models and real offices and differences in office sizes are likely to be of minor consequence, but there can be no doubt that differences in climate are very important. In tropical climates the dangers of overheating place much more stringent restrictions on the admission of daylight and sunlight and hence on window design. The available information has little to say about window design in tropical climates. Vernacular architecture is the best guide.

7.3.2 Illuminance
Even if an office is designed to be predominantly daylit it is unlikely that the

staff go home at dusk in winter. Every office has an electric lighting installation, the vast majority of which are designed without any consideration of daylight availability. This is not for want of an alternative. About 20 years ago an approach called PSALI (Permanent Supplementary Artificial Lighting Installation) was proposed [24] in which the electric lighting was so designed that when used with daylight there was a rough balance of brightness between all areas of the room. This approach has been much discussed but little used. More recently, interest in matching electric lighting to daylight availability has revived because of its possible energy saving implications [25]. The proposed means of matching is by switching or dimming the electric lighting but the latter method, which is the one preferred, is expensive. The result is that virtually all electric lighting in offices is still designed without regard to daylight. It is also common practice to design it to produce a uniform illuminance across a fictional horizontal working plane. This has the practical advantage that the occupier of the office can then arrange the people and furniture in the office as he wishes in the certainty that the lighting will be sufficient everywhere. This is a very desirable feature when, as usually happens, lighting has to be designed without knowledge of the layout of workplaces. Given this widely used approach to the lighting of offices it should come as no surprise that the most common, and often the only, lighting criterion used for offices is the illuminance on the working plane. The importance given to illuminance on the working plane has led to a considerable number of studies of dubious value. Demonstrations in which people are asked to adjust the illuminance on a desk until they consider it satisfactory are a common feature of many commercial exhibitions. Tregenza et al. [15] have shown one of the basic problems with this method of adjustment. In a well controlled experiment they found that the preferred illuminance for office work was markedly dependent on the starting conditions. This fact, together with the limited time available, so that no allowance is made for the effects of adaptation, and the limited range of illuminances usually presented, is enough to place such demonstrations in the realm of sales promotion rather than science.

One of the best conducted studies of people's reactions to different illuminances is that by Saunders [26]. He had a windowless room equipped as an office and lit by conventional ceiling mounted luminaires. Subjects came into the room, were given time to look around and adapt to the conditions and were asked to sit at a desk and read from a book. The subjects were then asked to give their opinions on the suitability of the lighting for reading and the general quality of the lighting. The subjects then moved to another desk, the questions were repeated and any

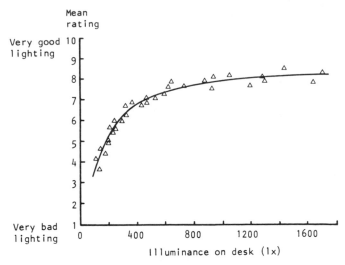

Fig. 7.3. Mean ratings of the quality of the lighting in a windowless office at different desk illuminances. (After [26].)

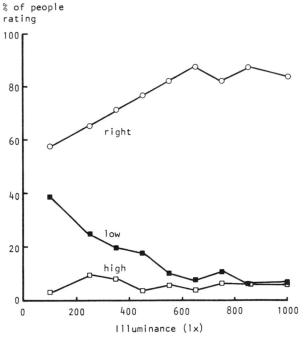

Fig. 7.4. The percentage of people rating the illuminance on their desks: ■, too low or rather low; ○, right; □, too high or rather high. (After [27].)

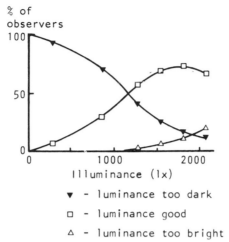

% of observers

Illuminance (lx)

▼ - luminance too dark

□ - luminance good

△ - luminance too bright

Fig. 7.5.   Percentage of observers rating the luminance of their desks too dark, good, or too bright, plotted against the illuminance on the desk. The desk top was a matt, pale green surface with a reflectance of 0·23. (After [28].)

differences between the two desks they had occupied were examined. In all 33 subjects took part in the experiment. Figure 7.3 shows some of the results obtained for the rating of quality of the lighting. It can be seen that illuminances below 200 lx were considered poor, but increased illuminances produced opinions of increased quality following a law of diminishing returns.

This sort of result is also shown by van Ierland [27]. He carried out a field survey in which real lighting installations in real offices, with all their defects, were used. The workers in the offices were asked to rate the illuminance on their desks at night on a five-category scale. Figure 7.4 shows the results obtained, this time for about 2000 subjects and a wide range of offices. It can be seen that once again there is a reasonable level of satisfaction at 500 lx which increases slightly as the illuminance moves up to 1000 lx.

Against these results can be set those of Balder [28] in which 296 subjects rated various aspects of the luminance distribution in a large room. A three-point scale was used, the three categories being 'too dark, good, and too bright'. Figure 7.5 shows the results obtained when Balder's observers were asked to consider the luminance of the desk. The percentage of people rating 'good' for an illuminance on the desk of 500 lx is about 10 % and only peaks at 75 % around 1800 lx. This is markedly different from the results of Saunders and van Ierland, a fact emphasised by Saunders' observation that, for an illuminance around 1600 lx, about half of his observers were saying that the task was too bright. At the same illuminance only 10% of Balder's observers considered the luminance of the desk top too bright.

An explanation for this apparent discrepancy probably lies in the different surfaces being considered. Balder's observers were asked to assess the desk top, which was a matt, pale green surface with a reflectance of 0·23. For such a surface an illuminance of 1800 lx will produce a luminance of 132 cd m^{-2}. Both Saunders and van Ierland asked more-general questions about the lighting on desks. It does not seem unreasonable to assume that the observers answered these questions with reference to the materials they used on the desks, i.e. to papers and books. If a reflectance of 0·75 is assumed for these materials then an illuminance of 500 lx will produce a luminance of 120 cd m^{-2}, which is not very different from the optimum luminance of desk top found by Balder. If this explanation of the apparent discrepancy between the studies is reasonable then it remains to assess which of the studies is most applicable. There can be little doubt that the van Ierland study [27] has the greatest face validity and that it supports the current illuminance recommendations of 500 lx for general offices and 750 lx for deep-plan offices [10].

But there is more to office lighting than illuminance. Figure 7.6 emphasises this fact. It shows the percentage of people in each of 20 different deep, open-plan offices who considered the lighting comfortable, plotted against mean desk illuminance [29]. The variation in comfort and the lack of consistent relationship with illuminance is obvious. This suggests the conclusion that it is not so much what you do in providing an illuminance, but the way that you do it, that determines satisfactory lighting conditions. However, it must be realised that nearly all the illuminances used were above 600 lx. From Figs 7.3 and 7.4 it can be seen that the level of satisfaction with this illuminance on the work should be high and should stay high over the illuminance range used. From the same figures (7.3 and 7.4) it would seem very likely that illuminances below 300 lx would be the cause of dissatisfaction. Really this whole topic demonstrates the difficulty of isolating one aspect of the lighting conditions for subjective assessment. The holistic approaches discussed in the preceding chapter are much superior.

Nonetheless some criterion is necessary for designers to aim at, and illuminance on a horizontal plane (horizontal illuminance) does have the practical advantage of being easy to design for and to measure. But this has not stopped the search for alternative quantities. Illuminance on a horizontal plane is particularly unsuitable for describing the effects of

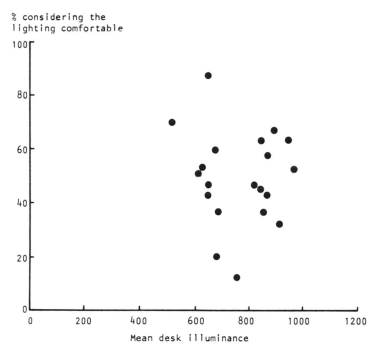

Fig. 7.6.   The percentage of people in each of 20 deep, open-plan offices who considered the lighting comfortable, plotted against mean desk illuminance in the office. (After [29].)

lighting on three-dimensional objects, such as people. Lynes *et al.* [30] have developed a measure for describing the lighting of three-dimensional objects, called 'mean spherical illuminance'. This is the average illuminance over the surface of a small sphere at a point. Others [31] have proposed a measure for a similar purpose but this time called 'mean cylindrical illuminance'. This is the average illuminance over the curved surface of a small vertical cylinder at a point. Different experiments have shown that one or other of these measures is better in specific circumstances [32, 33]. Such results are of little relevance for present day office lighting practice. The vast majority of offices are decorated in high reflectance colours which ensures a large amount of inter-reflected light. This means that the horizontal illuminance, mean spherical illuminance and mean cylindrical illuminance will all be strongly correlated. Until this practice changes, the practical advantages of horizontal illuminance ensure it will retain its pre-eminence amongst the possible illuminance measures.

One aspect of illuminance which is implicit in many designs but is only rarely considered is its uniformity. Having specified a design illuminance the actual installation will produce a range of illuminances over the interior. The problem is how large that range should be. Saunders [26], by having his subjects occupy two desks successively, was able to ask them how reasonable they thought it was to have the apparent difference in illuminance between the two desks in an office. The results obtained are

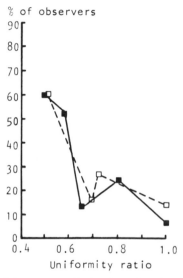

Fig. 7.7. The percentage of observers considering the uniformity of illuminance between two desks unreasonable, plotted against uniformity ratio. Uniformity ratio is the lower illuminance/higher illuminance. For the continuous line the variation in uniformity was achieved by dimming one luminaire, for the broken line different numbers of luminaires were used. (After [26].)

shown in Fig. 7.7. It can be seen that as the uniformity ratio (lower/higher illuminance) dropped below 0·7 there was a marked increase in the percentage of people considering the lighting pattern unreasonable. This confirms present practice which is to use a uniformity ratio of 0·8 (minimum illuminance/average illuminance) which is equivalent to a uniformity ratio of 0·7 on the minimum/maximum definition [10].

This conclusion applies to an office in which the design approach adopted is that of uniform lighting. Recently there has been a renewal of interest in local lighting, again because of its energy saving potential. Local lighting can be defined as a type of installation in which the lighting of the task or the desk is done by a luminaire attached to the desk and the rest of the room is lit, usually to a lower illuminance, by a uniform lighting system. Uniformity criteria developed for uniform lighting alone do not apply to this situation, but there should still be some sort of limitation on the difference between task and background illuminances. Bean [34] had subjects assessing just this problem in a small windowless office. He found that for a task illuminance of 400 lx, background illuminances in the range 200–1300 lx were considered acceptable or adequate by the majority of his subjects. In another experiment [35] background illuminances of much less than one-quarter of the task illuminances gave rise to complaints. Findings such as these support the current recommendation that the background illumin-ance in a local lighting installation should not be less than one-third of the task illuminance [10].

Whilst the above has been concerned with acceptable differences in horizontal illuminance for different types of installation, there are other differences that occur within any one installation. These are usually revealed by changes in luminance, i.e. by the combined effects of illuminance and reflectance. Touw [36] examined the question of what was the most suitable ratio for the luminances of the task and immediate surroundings. Almost 40 observers were asked to copy figures onto white paper whilst sitting at six different grey desks, each with a different reflectance. The results obtained showed that the preferred luminance ratio (desk/paper) was about 0·4 although it decreased slightly with higher illuminances. Since white paper has a reflectance of about 0·75 and the illuminance across the desk would usually be fairly uniform, this implies a desired desk reflectance of about 0·3.

This result applies to the immediate task area but what about the other surfaces of the room, especially the walls. The most comprehensive experiment on this subject is that by Balder [28] described earlier. The results obtained are shown in terms of the optimum ratio of the luminances of the facing wall and the side wall to the task luminance, in Fig. 7.8. The optimum ratio shows some decrease with increasing task luminance and hence with task illuminance. However, at the illuminances recommended for offices of 500–750 lx the optimum luminance ratio is about 0·55 for facing wall to task and 0·48 for side wall to task. Some support for these results is given by Tregenza et al. [15] who asked subjects to adjust the illumination of different surfaces in a single-person windowless office. The mean preferred ratios were 0·52 for the facing wall to task luminance ratio, and 0·53 for the side wall to task luminance ratios, but there was a wide scatter in the results. The only discrepancy is that Tregenza et al. did not find a significant variation in preferred ratios with changes in task

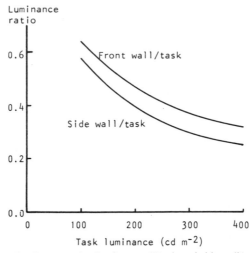

Fig. 7.8.   Optimum luminance ratios for front wall/task and side wall/task luminances, for different values of task luminance. (After [28].)

illuminance. Fortunately, the scatter in these results suggests that there is little need for a high level of precision in specifying the luminance ratios. This view is supported by the work of Barthés and Richard [37]. They have shown that the preferred luminance of a facing wall relative to a task depends not only on the task luminance but also on the colour and reflectance of the wall. Table 7.2 gives the ranges of facing wall/task luminance ratios for which a comfort score of over 80 % was obtained. It is

Table 7.2:   Range of facing wall/task luminance ratios for which comfort scores of over 80 % were obtained for walls of different colour and reflectance [37]

Facing wall colour	Facing wall reflectance	Facing wall/task luminance ratio range for a task illuminance of 300 lx (task luminance = 78 cd m^{-2})	Facing wall/task luminance ratio range for a task illuminance of 1000 lx (task luminance = 260 cd m^{-2})
Grey	0·21	0·22 → 0·72	0·09 → 0·41
Grey	0·83	0·33 → 1·08	0·23 → 1·04
Red	0·22	0·10 → 0·47	0·08 → 0·40
Yellow	0·22	0·17 → 0·76	0·14 → 0·50
Yellow	0·77	0·10 → 0·64	0·08 → 0·62
Blue–green	0·22	0·12 → 0·44	0·10 → 0·45†
Blue	0·21	0·10 → 0·69	0·12 → 0·44

† Upper limit obtained by extrapolation.

apparent that although these ranges are large, and are different for different conditions, there is a considerable degree of overlap. This confirms the loose nature of the preferred wall/task luminance ratios in offices.

A similar variability is apparent in the preferred ceiling /task luminance ratios identified by different experimenters. For example, Balder's results [28] suggest that at task luminances typical of those produced in offices a

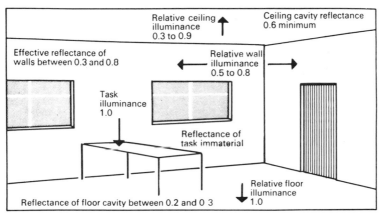

Fig. 7.9. Recommended ranges of surface reflectance and illuminance ratios relative to task illuminance [10].

ceiling/task luminance ratio in the range 0·5 to 0·8 would be considered good by more than 70 % of people. Tregenza et al. [15] found a mean preferred ceiling/task luminance ratio of 0·85 for their small, windowless office, although the range of preferred ratios was from 0·37 to 1·35. It can be concluded that the preferred ceiling/task luminance ratio is similar to the preferred wall/task luminance ratio in its dependence on the particular conditions. Nonetheless, from such data it is possible to derive criteria to give some guidance to the lighting engineer, in terms of ranges of acceptable luminance ratios. Since the ratios involve the luminance of the task, they link back to the principal lighting design criterion, the task illuminance. The luminances required for each ratio could in theory be achieved by a very large number of combinations of illuminance and reflectance. Fortunately for simplicity, physics and building practice limit what is likely to occur. The inter-reflections occurring within a conventional office mean that it is valid to specify a limited range of reflectances and illuminances which, in combination, will provide the desired conditions. Figure 7.9 shows such recommendations [10]. These recommendations, if followed, should ensure a comfortable balance between the luminances of room surfaces.

One assumption implicit in these ratio criteria is that the luminances of the surfaces are uniform. This is a reasonable assumption for walls and for indirectly lit ceilings like those used by Balder [28] and Tregenza *et al.* [15]. However, it is hardly valid for ceilings on which luminaires are mounted.

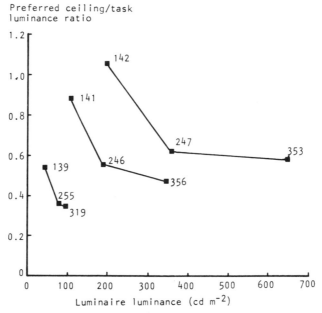

Fig. 7.10.   Preferred ceiling/task luminance ratios plotted against luminaire luminance. The number beside each point is the corresponding task luminance (cd m^{-2}). Each line links points for the same luminaire. (After [38].)

The effect of non-uniform ceiling luminance has been investigated by Collins and Plant [38] using a model windowless office. Figure 7.10 shows the preferred ceiling/task luminance ratios for three different, recessed, louvred luminaires. It is clear that the luminaire and its luminance have a marked effect on the preferred ceiling/task luminance ratio, the preferred ratio for each luminaire decreasing with increased luminaire luminance and task luminance. However, this is a rather unusual way of considering the effect of the presence of luminaires on comfort. In general the effect of luminaires on comfort has been examined for many years as part of the study of discomfort glare.

7.3.3 Discomfort Glare

When faced with a very high luminance in an otherwise low luminance field the natural responses are to blink and look away, or to shield the eyes in some way. These responses can be taken as behavioural indications that an excessive range of luminance can cause discomfort. The conditions inducing such responses are said to be producing discomfort glare in an extreme form.

This raises an immediate problem, namely how discomfort glare differs from the disability glare considered earlier. The conventional response to this problem is to say that they are two distinct forms of glare. This is partly correct but not particularly helpful. A better way of structuring the problem is to consider that they represent two different end results of variation in luminance across the visual field. When a high luminance, seen against a low luminance background, is close to the line of sight, so that light scattered in the eye produces a luminous veil across the retinal image of an object of interest and/or changes the local state of adaptation, then some reduction in object visibility will usually be evident. In these conditions disability glare is said to be occurring. Strictly, discomfort glare has nothing to do with any reduction in visibility of objects but only with any discomfort produced by the variation in luminance across the visual field. There is no unique link between these effects and the physical conditions. The same lighting conditions can produce disability and discomfort simultaneously and, to further confuse matters, different lighting conditions can cause disability and create discomfort independently. In essence, disability glare refers to the effect of a non-uniform luminance distribution on the visibility of objects; discomfort glare refers to the sense of discomfort the non-uniform luminance distribution produces. Disability glare is common on the roads but is rarely experienced in offices because the light sources are conventionally mounted well away from the usual lines of sight and the luminance differences across the space are relatively small. However, discomfort, related to the distribution of luminance across an interior, is a much more common experience. It is the factors that influence this discomfort in offices that will be considered here.

The study of discomfort glare is characterised by a long history producing little understanding [39]. This is partly because of an unwillingness by later experimenters to reconsider the fundamental concepts hidden beneath the agglomeration of past results and partly due to a failure to change unreasonable experimental techniques. Nevertheless some progress has been made. There now exist a number of equations which relate some of the characteristics of the lighting to subjective judgements of the degree of discomfort experienced. For a single glare source these equations all have the following form:

$$\text{Glare sensation} = \frac{\text{(luminance of the glare source)}^m \times \text{(angular subtense of the glare source at the eye)}^n}{\text{(luminance of the background)}^x \times \text{(deviation of the glare source from the line of sight)}^y}$$

Each component of the formula has a different exponent and these differ between the different equations. Examination of the form of the formula shows that increasing the luminance of the glare source, increasing its angular size or decreasing its deviation from the line of sight would all increase the sensation of glare experienced, but increasing the luminance of the background would decrease it. In practice it is not quite as simple as that because the different factors can rarely be varied independently. For instance by using a more powerful light source the luminance of the glare source is increased but so is the luminance of the background and the two effects may cancel.

Equations of this type form the basis of two of the three most widely used methods of discomfort glare prediction. Consider first the American visual comfort probability (VCP) method [40]. The fundamental formula used is based on the work of Guth [41]. The formula is

$$M = \frac{0 \cdot 50 L_s Q}{P F^{0 \cdot 44}} \tag{7.1}$$

where M = Glare Sensation Index; L_s = Luminance of the glare source (cd m^{-2}); $Q = 20 \cdot 4 W_s + 1 \cdot 52 W_s^{0 \cdot 2} - 0 \cdot 075$, where W_s = solid angle subtended at the eye by the glare source (steradians); P = an index of the position of the glare source with respect to the line of sight; and F = average luminance of the entire field of view including the glare source (cd m^{-2}).

This equation applies to a single glare source. For many glare sources the Glare Sensation Index for each glare source is combined with the others in the following way to form the discomfort glare rating (DGR).

$$\text{DGR} = \left(\sum_n M \right)^a \tag{7.2}$$

where $a = n^{-0 \cdot 0914}$, and n = the number of glare sources.

The DGR values for a particular lighting installation are converted to a quantity called visual comfort probability (VCP) which is simply the percentage of people who would be expected to find the cumulative glare sensation represented by the DGR acceptable. The VCP provides a convenient figure of merit with the added attraction that higher values indicate less discomfort. The question now is 'what lies behind this highly developed method, and particularly the complicated equation?' Unfortunately the answer is 'very little'. For a start few of the experiments used to develop the fundamental equation involved any attempt to suggest a real context to the observers. In the early experiments people simply fixated the pole of a uniform luminance hemisphere. A small high luminance source was introduced in a fixed position and the observer was asked to adjust either the luminance of the field or the source until they considered the situation to be at the borderline between comfort and discomfort. The high luminance source was presented intermittently on a schedule of 1 s exposure, 1 s interval, 1 s exposure, 1 s interval, 1 s exposure, followed by a 5 s rest interval. This intermittent exposure was used in an attempt to simulate a worker looking up from his work to a glare source. However, other experiments showed little difference in the setting of the boundary between comfort and discomfort for this intermittent presentation and for continuous exposure [42]. Later experiments involved simulated model and actual rooms, in which either the luminance of the luminaires was adjusted to the borderline between comfort and discomfort or the luminance of a comparison source was adjusted to give the same overall glare effect as the complete installation. In none of these experiments was it possible for the observer to speak for himself other than through the medium of an adjusted luminance. This and the lack of context, are the main failings of this experimental method.

In spite of the lack of experimental rigour behind the equation the VCP system has been used in practice and has apparently been found to be satisfactory. This says more for the precision required in practical application than it does for the correctness of the experimental approach. This view is supported by the simple guideline which is given about the level of VCP required. The Illuminating Engineering Society of North America [40] simply says that provided a VCP value of greater than 70 % occurs and the luminance uniformity and luminance distribution of the luminaire are within certain limits then the discomfort glare experienced will be acceptable.

The British Glare Index system [43] uses a similar formula, but derived by different methods. The basic formula was developed by Petherbridge and Hopkinson [44]. For a single glare source the formula is

$$\text{Glare Sensation} = G = \frac{0 \cdot 48 L_s^{1 \cdot 6} W_s^{0 \cdot 8}}{L_b P^{1 \cdot 6}} \qquad (7.3)$$

where L_s = luminance of the glare source $(cd\,m^{-2})$, W_s = solid angle subtended by the glare source at the eye (steradians), L_b = average luminance of the field of view excluding the glare source $(cd\,m^{-2})$, and P = an index of the position of the glare source with respect to the line of sight. For multiple glare sources the combined effect is given by simple addition of the individual Glare Sensation values. The Glare Sensation values are conventionally called the Glare Constants for each glare source. For convenience the total Glare Constant for any particular installation is converted into another quantity called the Glare Index by the formula

$$\text{Glare Index} = 10\log_{10}\left(\sum G\right) \qquad (7.4)$$

The experimental approach used to derive this formula was to have observers look at a photograph of a schoolroom in which the light sources, incandescent lamps in opal spheres, had been replaced by luminous discs. The luminance of the discs was set to a predetermined value by the experimenter and the observer was asked to adjust the background luminance to produce four levels of sensation: just intolerable glare, just uncomfortable glare, just acceptable glare, and just imperceptible glare. The observers used were carefully selected. Their settings of background luminance under 'standard' conditions were obtained frequently and any inconsistency resulted in elimination from the experiment. This procedure is justified on the basis of ensuring reliable judgements [45]. It can be concluded that the experimentation on which the Glare Index formula is based also has limitations, although different ones from those occurring with the VCP formula.

Nonetheless, given that the formula existed, the problem was how to set some limit to the extent of discomfort glare which should be allowed to occur. The British IES solved this problem by sending a team of observers experienced in lighting matters to a number of office and factory installations and asking them to say whether the degree of glare experienced was acceptable or unacceptable for different applications [46]. In evaluating the acceptability of the discomfort glare experienced, such factors as the function of the space lit, the amount of time people would be present, the extent to which they could look away from the glare sources, and the concentration required for the task to be done were considered. The estimates of acceptability were then related to the calculated values of Glare Index for the installations. From these data limiting values of Glare Index for different applications were derived [10]. The Glare Index system, too, has been used in practice and found to be satisfactory.

Both the VCP and the Glare Index systems are based on equations

linking the physical conditions to the glare sensation experienced. A strong
and stable link is all that is required for a predictive system. There is no need
for any causal explanation. The lack of any causal explanation is even more
apparent in the third widely used approach to discomfort glare prediction,
the European Glare Limiting Method [47]. Sollner [48, 49], who largely
developed this approach, assumed that the only feature of the luminous
environment the lighting engineer is likely to be able to control is the choice
of luminaire. With this in mind he sought a simple link between the
luminance distribution of the luminaire and the rated experience of glare. In
his experiments one-third scale models of an office were viewed by 10–15
observers. The lighting of the model was varied by using different
fluorescent luminaires, with different luminous intensity distributions, in
different numbers, in different sizes of rooms, and orientated either across
or end-on to the direction of view. In all, 750 different situations were
assessed by the observers, the glare sensation being expressed on a seven-
point rating scale, the categories being labelled: $0 =$ no glare, $1 =$
glare between non-existent and noticeable, $2 =$ glare noticeable, $3 =$ glare
between noticeable and disagreeable, $4 =$ glare disagreeable, $5 =$ glare
between disagreeable and intolerable, and $6 =$ glare intolerable. The
similarity between this scale and Petherbridge and Hopkinson's setting
points is obvious. Only four factors were found to affect the glare rating
markedly. These were (a) the luminance of the luminaire, (b) the room
length and mounting height, (c) the adaptation luminance and (d) the type
of luminaire, specifically whether it had luminous side panels. Sollner
summarised his conclusions for regular arrays of ceiling mounted
luminaires in terms of curves of luminaire luminance plotted as a function
of the emission angle, i.e. the angle between the normal to the central
luminaire and the line from the luminaire to the observer, for a specified
Glare Rating. The Glare Rating is the median rating of the glare sensation.

Figure 7.11 shows these curves for luminaires viewed crossways and end-
on. These curves apply for an installation producing 1000 lx on the
conventional horizontal working plane. For higher or lower illuminances a
correction is required which reduces the rated glare for low illuminances
and increases it for higher illuminances. The correction is given by the
expression

$$\Delta G = 1\cdot16 \log_{10} \frac{E}{1000} \qquad (7.5)$$

where $\Delta G =$ quantity to be added or subtracted from the Glare Rating
achieved at 1000 lx, and $E =$ horizontal illuminance (lx).

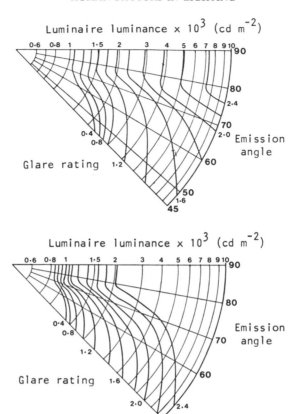

Fig. 7.11. Luminaire luminances at different emission angles for different glare ratings. The upper diagram is for viewing the luminaire end-on, the lower diagram for viewing the luminaire crossways. The emission angle is the angle between the normal to the luminaire and a line joining the luminaire and the observer. (After [52].)

Fischer [47] used these results to derive an approximate system in which the curves of luminance shown in Fig. 7.11 were redrawn on cartesian co-ordinates and split into two separate straight lines, one oblique running from 45 to 75°, the other a vertical line from 75 to 90° (Fig. 7.12). Equations specifying these lines are as follows:

(A)   for all luminaires without luminous side panels and for luminaires with luminous side panels viewed end-on

$$\log_{10} L_{75-90°} = 3{\cdot}00 + 0{\cdot}15\left(G - 1{\cdot}16\log_{10}\frac{E}{1000}\right)^2 \quad (7.6)$$

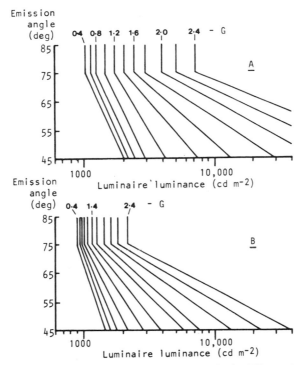

Fig. 7.12. Luminaire luminance at difference emission angles for different glare ratings (G). The data in graph A apply to all luminaires without luminous sides and luminaires with luminous sides viewed parallel to their long axis. The data in graph B apply to luminaires with luminous sides viewed at right-angles to their long axis. (After [47].)

and

$$\log_{10} L_{45^\circ} = 3{\cdot}186 + 0{\cdot}40\left(G - 1{\cdot}16\log_{10}\frac{E}{1000}\right)^2 \quad (7.7)$$

(B) For all luminaires with luminous side panels viewed crossways

$$\log_{10} L_{75-90^\circ} = 2{\cdot}93 + 0{\cdot}07\left(G - 1{\cdot}16\log_{10}\frac{E}{1000}\right)^2 \quad (7.8)$$

and

$$\log_{10} L_{45^\circ} = 3{\cdot}10 + 0{\cdot}26\left(G - 1{\cdot}16\log_{10}\frac{E}{1000}\right)^2 \quad (7.9)$$

where G = glare rating, E = illuminance on a horizontal plane (lx),

$L_{45°}$ = luminance of luminaire for an emission angle of 45°
(cd m^{-2}), and $L_{75-90°}$ = luminance of luminaire for emission
angles of 75–90° (cd m^{-2}).

Given a mean glare rating which should not be exceeded, it is possible to use these equations to derive limiting values of the luminaire luminance for a range of angles from 45 to 90°. Luminaires which have luminances less than the limiting values will produce a lower mean Glare Rating. The latter approach is the basis of the European Glare Limiting method [47] in which the luminance limits are set for luminaires for a stepped series of Glare Ratings and illuminances. This system is in use in several European countries and later experiments have extended its application to high bay installations lit by high pressure discharge lamps [50] and to luminous ceilings [51].

The quality of the experiments conducted by Sollner have placed the European Glare Limiting method on the firmest footing of the three methods considered. He used a variety of people judging an interior with an established context. The only fault is that the rating scale used by the observers to give their assessments on discomfort glare is a strange mixture of the magnitude and the acceptability of the glare experienced. In spite of the better experimental base, the European Glare Limiting method does have disadvantages in application. The main disadvantage is that it only allows for the mean Glare Rating to be changed by varying the luminaire luminance distribution. The VCP and Glare Index systems both, in principle, allow for this and for the effect of changing room reflectances and for having an irregular layout of luminaires. Thus the VCP and the Glare Index methods reflect the influence of the complete luminous environment on discomfort glare but the European Glare Limiting method only allows for differences in luminaires.

In spite of these differences in approach there is reasonable agreement between the glare sensations predicted by the various methods. Manabe [52] calculated the VCP, Glare Index and mean Glare Rating values for a range of installations and found the following correlation coefficients— VCP versus Glare Index = -0.86, VCP versus mean Glare Rating = -0.67, Glare Index versus mean Glare Rating = 0.66. Other workers [53], have shown similar correlations between the various measures of predicted glare sensation for similar ranges of installations. From such studies, conversion equations to enable a designer to move from one system to another have been derived [54]. The obvious conclusion is that all the methods point in the same direction and all will enable the lighting designer

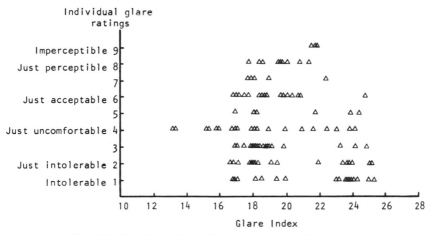

Fig. 7.13. Individual glare ratings plotted against Glare Index.

to calculate the extent to which his installations will produce discomfort glare. There is no scientific reason to suggest any of them are completely wrong but equally there is plenty of reason to suppose that none are completely right. The reason for saying this is evident from Fig. 7.13, which shows the scatter in ratings of discomfort glare made by individuals from a number of viewing positions in different installations. These ratings are plotted against Glare Index calculated for each viewing position, but the equivalent VCPs or mean Glare Ratings would produce similar scatter. With such a wide scatter in the ratings there seems little point in attempting to refine glare prediction methods to eliminate small differences in treatment of such variables as room length and illuminance [55]. To do so would be rather like worrying about the chromium plating on a car which has square wheels. The real problem of discomfort glare is to understand the reasons for these wide individual differences in reported glare sensation. To do this it is necessary to face squarely the objections to the approaches adopted so far in the study of discomfort glare.

These objections come in several forms. For a start Markus [56] doubts whether 'glare' has any inherent meaning for most people. He suggests it is an abstraction which does not correspond to any unitary experience. If this is so then asking people to rate the amount of glare they experience is likely to lead to a wide divergence of opinion as different people reach different conclusions as to what is meant by glare. Markus also emphasises the importance of context. He points out that people frequently sit for hours in front of a television set by free choice even though it should, according to the formulae, be producing intolerable glare. The point is that glare, whatever it is, does not occur in abstract. The circumstances in which the high luminance occurs, and the meaning of the source of glare can all have an important effect on whether the condition is assessed as uncomfortable. One factor likely to be important is the interest in the glare source. Usually, luminaires are, of themselves, uninteresting, so that when they attract attention by their high luminance they will probably be regarded unfavourably. This is not the case for a crystal chandelier in a great house, which is unlikely to be considered glaring even though its luminance is much greater than a conventional luminaire. Interest may also explain why there is less sensitivity to glare produced by a window, as compared with artificial lighting, until extreme conditions are reached [57]. Windows usually provide a view which can be interesting and therefore of some value, in addition to their contribution to the interior lighting. One other aspect of context that has been ignored in virtually all the studies of discomfort glare is that of the work to be done in the interior. Guth [42] found that there was a different boundary between comfort and discomfort for the same lighting conditions depending on whether a task was present or absent. Yet in most of the studies discussed there was no task, save only the assessment of glare.

It can be concluded that there is *prima facie* evidence of the need for a fundamental re-appraisal of discomfort glare. Little is known about the physiological and psychological basis of the discomfort experienced [58, 59], although distraction has been suggested as an important factor [60]. The present situation is that there are a number of different national systems for controlling discomfort glare, each of which has its own advantages and disadvantages. There are reasonable correlations for the predicted degree of discomfort glare from all these formulae but only weak correlations between the predicted discomfort glare and the ratings of individual observers. There is little point in attempting to refine the different methods further. There is unlikely to be a consistent significant difference between them, given the large individual differences in discomfort glare assessment made by people in different interiors. The more pressing need is to attempt to understand the nature of the discomfort produced by wide variations in luminance across an interior. In general, the existence of a precise mathematical formula should not be taken to prove that what it quantifies is equally precise. This is very much the position with respect to discomfort glare.

7.3.4 Veiling Reflections

A special case of a non-uniform luminance distribution with a potential for

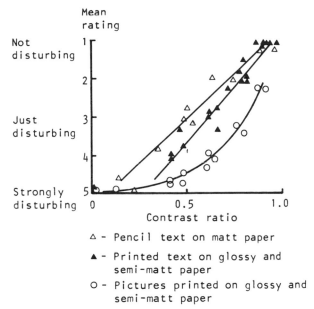

Fig. 7.14. Mean ratings of the disturbance felt with veiling reflections for pencil writing, printed text and pictures, plotted against the ratio of the contrast of the material under the lighting of interest to the contrast under a luminous ceiling. This contrast ratio is an approximation to Contrast Rendering Factor. (After [61].)

causing discomfort is the occurrence of veiling reflections. The distinction between the conditions causing veiling reflections and discomfort glare is simply the location of the glare source. Discomfort glare is related to non-uniform luminances in those parts of the visual field which do not contain the task of interest. Veiling reflections are related to non-uniform luminances in the immediate area of the task of interest. The nature of veiling reflections and their effect on performance have already been discussed and demonstrated (Chapter 3). The Contrast Rendering Factor (CRF) can be used as a convenient measure of the effect of veiling reflections on task visibility. The question of interest here is the extent to which different levels of CRF are considered uncomfortable. De Boer [61] reports some ratings of the degree of disturbance with reflections for different amounts of veiling reflection. Figure 7.14 shows the mean ratings for a close approximation to Contrast Rendering Factor, namely the ratio of the luminance contrast produced under the lighting of interest to the luminance contrast produced under a uniform luminous ceiling. The subjects were asked to read two journals; one printed on a moderately glossy paper the other on semi-matt paper, and a sheet of text written in pencil on matt paper. Figure 7.14 reveals that there is a clear relationship between comfort and veiling reflections with the 'just disturbing' category having a 'CRF value' of about 0·6. Bjorset and Frederiksen [62] measured the acceptability of the contrast reduction produced by veiling reflections for a wide range of handwritten and printed materials. Figure 7.15 shows the percentage reduction in contrast which 90 % of their subjects found acceptable, plotted

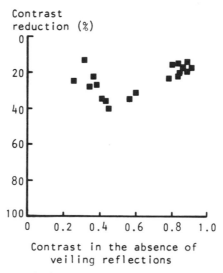

Fig. 7.15.   The contrast reduction considered acceptable by 90 % of observers plotted against the contrast of the materials when no veiling reflections occurred. (After [62].)

against the contrast of the material when no veiling reflections occurred. It is apparent that 90 % of the subjects accepted a contrast reduction of about 25 %, regardless of whether the materials had a high or low contrast in the absence of veiling reflections. This result suggests that weak veiling reflections are not critical for comfort, a conclusion supported by Reitmaier [63], at least for pencil, pen and printed text on matt and moderately glossy papers. However, he also showed that for very glossy papers people were more sensitive to veiling reflections, a CRF of 0·7 being associated with a poor opinion of the quality of the visual conditions. De Boer [61] reports a similar result for the assessment of pictures printed on glossy and semi-matt paper (Fig. 7.14).

There can be little doubt that veiling reflections can cause discomfort. The problem is to define a suitable measure of their effect and then to establish appropriate values of it. For printed text and handwritten material on matt, semi-matt and moderately glossy paper CRF appears to be a suitable measure and a value of about 0·7 would seem appropriate. However, for text on very glossy paper CRF is not a very good predictor of the discomfort caused by veiling reflections. Reitmaier [63] has shown that for pencil writing on very glossy paper it is possible for veiling reflections to be associated with CRF values greater than unity. Although this means the writing is easily visible, the visual conditions are still considered poor. Thus for very glossy papers a different criterion for limiting veiling reflections is required. Reitmaier [63] suggests a measure based on the proportion of the illuminance on the material which comes from the direction which is at the mirror angle to the line of sight. Typical maximum values for this proportion are 5% for pencil writing on very glossy paper and 25% for print on glossy paper. It remains to be seen how effective these proposed measures are in practice.

7.3.5 Light Source Colour Properties

The appearance of a uniformly lit office will be greatly influenced by its colour scheme. As indicated in Chapters 1 and 2 the perceived colour of a surface is largely determined by the light incident on it from the light source, the spectral reflectance on the surface and the state of adaptation of the observer. In practice the most important of these three is the spectral reflectance of the surface. Different colour pots of paint can produce large differences in the appearance of a room, which the light sources usually chosen for offices modify only slightly. The dominant figure in determining the colour appearance of an office is the interior designer, the lighting engineer playing a minor role. However, the properties of the lighting cannot be entirely ignored. The choice of different light sources does affect the appearance of surface colours which is why guides to lighting practice bother to give advice on appropriate light sources for different applications. The advice usually consists of the range of correlated colour temperatures and the range of Colour Rendering Indices considered suitable [10]. Correlated colour temperature is a convenient measure of the colour appearance of the light source whilst the Colour Rendering Index (CRI) indicates the effect of the light source on the appearance of surface colours relative to some standard (see Chapter 1). These two criteria are necessary because the effects they quantify can vary independently. Both high and low Colour Rendering Indices can occur with both high and low correlated colour temperatures.

The existence of these recommendations might lead one to suppose that there was some proven basis for the allocation of different light sources to different applications. There is not. The choice is usually a matter of expectation and economics. Thus incandescent lamps are extensively used in housing because they are small and have a low capital cost, not because they have a low colour temperature and a high CRI. Offices are almost always lit by tubular fluorescent lamps which have a higher correlated colour temperature and a moderate CRI. Fluorescent lamps are used because, although they have a higher capital cost, their running costs are much less than are those of incandescent lamps. In industry it is common to use high pressure sodium discharge lamps which have a poorer CRI, but a higher luminous efficacy than fluorescent lamps. These three light sources have very different colour properties and are recommended for different applications, but the recommendation is made on the basis of all their properties and not just those concerning colour.

The recommendations are, as might be expected, anything but precise. Generally a number of light sources are suggested as suitable for an application, the choice then being left to the designer. In reaching his decision he is aware that a light source with a high CRI will make colours and people look more as they do under some form of daylight or incandescent light; but he will also be aware that there is normally a price to be paid for using light sources with high CRI values, either in terms of running cost or in capital cost. In this situation it is not so surprising that a consensus should develop amongst designers as to what is the optimum choice of light source for different applications. For offices the present consensus is to use a White tubular fluorescent lamp, which has a correlated colour temperature of 3500 K and a Colour Rendering Index of about 56.

Cockram *et al.* [64], in a study of people's opinions of the suitability of different fluorescent light sources for offices at night and when lit in combination with daylight, produced some interesting and relevant results. They had eight similar offices in the same building, lit by four different fluorescent lamp types in the same type of luminaire: a ceiling mounted, louvred, trough. Because the same number of lamps was used in each case, installations with different lamp types produced different illuminances. Forty office workers viewed four of the rooms, each with a different light source, at night and during the day. They were asked to place the four rooms in order of preference for both conditions. The characteristics of the four light sources and the overall order of preference were as shown in Table 7.3. These results go some way to justify the consensus view, although the daylight lamp is the most frequently preferred for rooms with clear glazing during daytime. It is interesting to note that the panel of

Table 7.3: Rank ordering of preferred light sources in real offices assessed by office workers and experts [64]

Lamp type			Rank order (1 = most preferred)					
			Office workers			Experts		
Lamp name	Correlated Colour Temperature	CIE Colour Rendering Group†	Daytime, clear glazing	Daytime, tinted glazing	Night-time, tinted glazing	Daytime, clear glazing	Daytime, tinted glazing	Night-time, tinted glazing
Colour Matching	6 500	1	4	4	4	3	2	4
Deluxe Daylight	4 000	1	2	3	3	1	3	1
Daylight	4 300	3	1	2	2	2	1	2
White	3 500	3	3	1	1	4	4	3

† Group 1, CRI $\geq$ 85; Group 2, $70 \leq$ CRI < 85; Group 3, CRI < 70.

experts viewing the same rooms reached a rather different order of preference (see Table 7.3). Further, it is worth remarking that for the people working in the offices, the night-time order of preference followed the order of decreasing illuminance produced by the lamps. During daytime the illuminance was variable. From these results it is not at all clear whether the Colour Matching fluorescent lamp which has the highest CRI and is the least preferred by the office staff, both by night and day, achieves this unenviable position because of the low illuminance it produces or because of the very high colour temperature it has. The latter seems more likely if the comments associated with the Colour Matching lamp such as 'the room looks like a refrigerator' are accepted at face value. Against this may be put the observation of Hopkinson et al. [65] who found that for supplementing daylight a Colour Matching fluorescent lamp was satisfactory in more circumstances than any other. Which of these studies is correct is open to argument but one thing the conflict does suggest is that, within the range of fluorescent lamps examined, there is no strong consistent preference.

Although the reasons for the choice of a particular light source for offices have not been studied in detail, the interaction of light source colour properties with illuminance has. The most frequently mentioned of these studies is that of Kruithof [66] who reported a link between illuminance and preferred colour temperature. His results showed that at low illuminances, say less than 300 lx, a low colour temperature was preferred but as the illuminance increased so did the preferred colour temperature. Unfortunately Kruithof used only two subjects so his results can hardly be said to have general applicability. In addition, different light sources were used to give the different colour temperatures, so light distribution and colour rendering are confounded with colour temperature. This confusion of variables and the use of small numbers of observers is typical of many of the experiments that seek to examine the effect of different colour temperatures. With the available work having such defects there is little that can be said about the interaction between illuminance and preferred colour temperature. The most positive information is contained in an experiment by Bodmann et al. [67]. They had 80 groups of five people assess the general 'atmosphere' of a conference room. The illuminances produced under the different light sources were the same and the light sources considered all had similar high Colour Rendering Indices. For illuminances less than 700 lx, which covers the vast majority of offices, a Warm White fluorescent lamp (2950 K) gave a not unpleasant atmosphere, a White fluorescent lamp (4200 K) gave a rather dim appearance, whilst the Daylight fluorescent lamp (4000 K) gave a cool impression. For illuminances from 700–3000 lx both Warm White and White fluorescent lamps produced a pleasant atmosphere whilst the Daylight lamp gave a neutral appearance. These results are in the expected direction but, quite rightly, they do not give an aura of precision. Rather they suggest that the effect of colour temperature is not of critical importance. This conclusion is reflected in the coarse classification: cool, intermediate or warm, of the recommendations of suitable colour temperatures for different applications [10].

The interaction of the colour rendering properties of a light source and the prevailing illuminance has been studied rather more successfully. Aston and Bellchambers [68], using a box filled with different coloured objects, showed that for equal satisfaction with appearance there was a trade off between lamp type and illuminance. Specifically, the light sources with the higher Colour Rendering Indices gave a satisfactory appearance at a lower illuminance than those with poor Colour Rendering Indices. Bellchambers and Godby [69] extended this work to a simulated domestic room and reached the same conclusion. These results have been examined by using two identical, one-twelfth scale, model offices side-by-side [70]. Each observer sat with his head inside the model so that by turning his head he could see both offices. The two offices were lit in the same way by two different tubular fluorescent lamp types. The experimenter set the illuminance in one office and the observer was asked to adjust the illuminance of the other until it looked equally satisfactory in appearance. The matches were made for a range of colourfulness of interiors and for two different starting illuminances. The results obtained showed that fluorescent lamps with good colour rendering properties did indeed produce equal satisfaction with appearance at a lower illuminance than those with poor colour rendering properties. Further, this conclusion applied to offices over a wide range of interior colourfulness and for illuminances in the range 350–600 lx. Table 7.4 gives the mean ratios of illuminance between the two offices lit by different fluorescent lamp types for equal satisfaction with appearance in the different conditions. The relationship between the illuminance ratio and the difference in Colour Rendering Index of the two light sources is obvious. It is also comforting to note that, when the two light sources in the two offices were the same, illuminance ratios close to unity were obtained.

An explanation has been proposed for this rather strange phenomenon [70]. The basis of the explanation is that the observer is making his judgement of satisfaction with appearance on the relationship between the various surface colours, and particularly their saturation. Good colour rendering lamps generally produce greater saturation in surface colours

Table 7.4: Means and standard deviations of ratios of the illuminances in the two offices for equal satisfaction [70]

Interior colourfulness		Lamp colour combinations†							
		Natural/White		'Kolor-rite'/White		'Kolor-rite'/Natural		Natural/Natural	
		350 lx	600 lx	350 lx	600 lx	350 lx	600 lx	350 lx	600 lx
Low	x̄	0·73	0·79	0·77	0·79	1·05	1·07	0·94	0·97
	Standard deviation	0·11	0·18	0·05	0·15	0·17	0·15	0·10	0·06
Medium	x̄	0·76	0·78	0·75	0·68	1·09	1·03	0·98	0·95
	Standard deviation	0·13	0·16	0·14	0·11	0·09	0·08	0·08	0·08
High	x̄	0·68	0·77	0·77	0·85	1·01	1·03	0·96	1·00
	Standard deviation	0·08	0·10	0·10	0·11	0·19	0·09	0·11	0·08

† Colour Rendering Indices are: 'Kolor-rite' = 92, Natural = 85, White = 56.

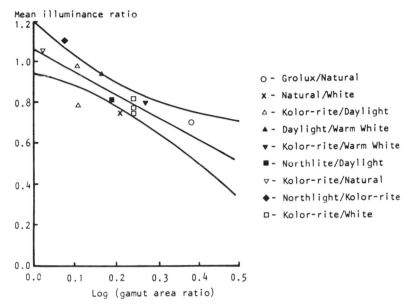

Mean illuminance ratio

o - Grolux/Natural
x - Natural/White
△ - Kolor-rite/Daylight
▲ - Daylight/Warm White
▼ - Kolor-rite/Warm White
■ - Northlite/Daylight
▽ - Kolor-rite/Natural
◆ - Northlight/Kolor-rite
□ - Kolor-rite/White

Log (gamut area ratio)

Fig. 7.16.   The linear regression line and the 95 % confidence limits through the available data on mean illuminance ratios for matched interiors lit by fluorescent lamps with different colour properties, plotted against the logarithm of the ratio of the gamut areas of the lamps. (After [70].)

than lamps with poor colour rendering properties. An alternative way to increase saturation is to increase illuminance [71]. Thus, when asked to match two interiors, one lit by a good colour rendering lamp the other by a poor colour rendering lamp, by changing the illuminance the tendency is to do this by increasing the illuminance for the poor colour rendering lamp. Now if this explanation is correct it suggests that a measure like gamut area might be better for describing the trade-off between lamp type and illuminance because it is a direct measure of the relationship between a standard range of colours produced by a given light source (see Chapter 1). This contrasts with the Colour Rendering Index which is really a measure of the accuracy of reproduction of colours relative to some standard. Further experiments [70] revealed that the effect was more closely related to the gamut area than to CRI. Figure 7.16 summarises the available results on this topic, in the form of the ratios of the illuminances matched for equal satisfaction for pairs of fluorescent lamps, plotted against the logarithm of the ratios of their gamut areas.

Whilst there is no doubt about the reality of this effect there is doubt about what people understand by it. Bellchambers and his colleagues [68, 69] coined the term 'visual clarity' to describe the balance between lamp colour properties and illuminance. This naming has led to some confusion, in that people have tended to treat visual clarity as a synonym for visibility and even visual acuity. In turn this had led them to suggest that good colour rendering should affect task performance. If the task involves colour judgements of some sort then the influence of lamp colour properties could be important, but for the more usual achromatic tasks any effect is at most small but more likely non-existent (see Chapter 3). The 'visual clarity' experiments described above have been concerned exclusively with appearance. It is misleading to attempt to extend the conclusions reached to task performance, especially when there are perfectly adequate experiments which examine the effect of light source colour properties on performance directly.

To summarise, the choice of light source in offices is largely a matter of convention. There is little reliable research information on which to base a choice, the information that is available being of very variable quality. The only well established effect is that of 'visual clarity'.

7.3.5 Conclusions for Offices

Office lighting has been studied for many years and conditions for comfort have largely been established. There is no doubt that daylight is the preferred method of office lighting but because of the limitations this approach places on building form and the financial consequences, this preference is largely ignored. Sunlight is somewhat less popular in offices than homes because people usually have less control of their environment with the result that the likelihood of thermal and visual discomfort caused by sunlight is increased. These two observations indicate why many offices are designed to use mainly electric lighting.

The criteria for comfortable electric lighting are well established if imprecise. Unfortunately much of the work on the comfort of office lighting has been of dubious value. Nonetheless there is little doubt that the basic framework within which the lighting designer should operate has been marked out. Criteria of illuminance, illuminance uniformity, surface reflectances, illuminance ratios and discomfort Glare Indices are all available [10], and if followed will ensure that office lighting is not uncomfortable. It may not be very inspiring but that is a different matter. Guidance is also offered on suitable light sources for offices [10] but this is even less critical, being mainly a matter of convention and hence expectation. The only phenomenon that can cause discomfort in offices, and which is not covered by a criterion, is veiling reflections. Different criteria have been proposed for different materials but it remains to be seen how useful they are in practice. Finally, a word of warning is appropriate. A recurring theme throughout this discussion of office lighting has been the influence of expectation on people's assessments of what constitute uncomfortable conditions. Virtually all the research discussed has been done in western countries. It seems reasonable to expect, and there is evidence to suggest [72], that lighting conditions considered comfortable would be different in countries which have different expectations.

7.4 FACTORIES

Although a considerable amount of lighting is used in factories, virtually no experimental work has been done on the comfort aspects of factory lighting. There are two reasons for this. The first is the difficulty of generalising results. Unlike offices, factories make different products with different equipment and hence have different lighting problems. Therefore any study of comfortable lighting conditions is likely to be specific to the factory studied. The second reason is that the emphasis in factory lighting is on the effect it has on work. Considerations of visual comfort are given much less prominence than is the case for office lighting. This emphasis is apparent in the recommendations made for lighting factories [10]. Whilst the illuminance recommendations are linked to the task difficulty, the criteria which relate to comfort, e.g. Glare Indices, are generally much less stringent than is the case for offices. Further, a wider range of light sources is considered acceptable, including some of the discharge light sources which have poor colour rendering properties but high luminous efficacies.

There is no logical reason why this difference in criteria should occur. People who work in factories are not fundamentally different from those who work in offices. Once again it is a matter of expectations. Many aspects of the physical environment are worse in factories than in offices and lighting is just one of them.

To summarise, the differences between the criteria for factory lighting and office lighting do not arise from knowledge, because very little is known about people's responses to factory lighting. Many of the current lighting criteria for factories are derived from work done in offices but relaxed to allow for the decreased emphasis on comfort. Really the whole field of factory lighting calls for study. Factories often deal with tasks which are much smaller and of lower contrast than occur in the office, they are usually

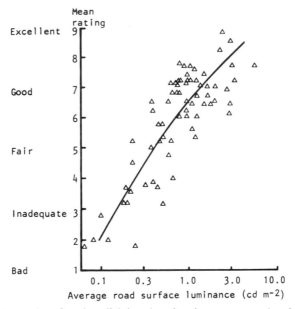

Fig. 7.17.   Mean rating of roadway lighting plotted against average road surface luminance for 70 roads. (After [73].)

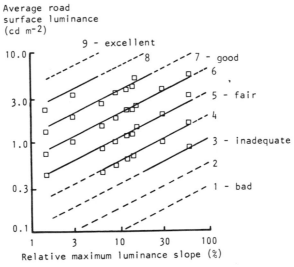

Fig. 7.18.   Lines of constant rating of roadway lighting for roads with different combinations of average road surface luminance and luminance uniformity (relative maximum luminance slope). (After [74].)

three-dimensional, often specular and the lighting sometimes suffers from a degree of obstruction. These conditions suggest that factory lighting could well benefit from different techniques and measures than those used in offices.

7.5 ROADS

Roadway lighting is primarily functional but in order for that function to be fulfilled it is necessary for the driver to be largely free from discomfort. The effects of different features of roadway lighting on accident occurrence and the visibility of objects have been discussed in Chapter 4. Here the concern is primarily the extent to which lighting conditions create discomfort. Unfortunately, very little work has been done on discomfort alone. Subjective appraisals of roadway lighting are usually made in terms of overall quality, which implies that both visibility and comfort are being judged. Nonetheless, lighting appraised in this way is unlikely to be satisfactory unless comfort is achieved. One example of a study that uses such an approach is that by De Boer *et al.* [73]. They had 16 experienced lighting engineers categorise the lighting of 70 urban streets, 46 of them in dry weather. Their impression was expressed by assigning the road to one of nine classes ranging from bad to excellent. Figure 7.17 shows the average rating for each road plotted against average road surface luminance. The interesting point to note is that the category of good road lighting occurs at an average road surface luminance of $1 \cdot 5 \, \mathrm{cd \, m^{-2}}$. Reference to Chapter 4 will show that it is at about this luminance that the curves of night/day accident ratios levels off. Such an agreement between different measures is encouraging.

But average road surface luminance is not the only factor influencing the assessment of quality in roadway lighting. De Boer and Knudsen [74] showed that the general impression of roadway lighting depended on the combined effects of average luminance and uniformity of road surface luminance. Using a model road in which different forms of luminance non-uniformity could be created, they asked 43 observers to appraise a variety of installations on the bad to excellent scale used previously. It is important to note that for these assessments the luminance of the light sources was kept low to minimise any discomfort glare. Statistical analysis of the data collected showed that the more uniform the luminance pattern and/or the higher the average road surface luminance, the better was the overall impression. Figure 7.18 summarises these results as lines of constant rating plotted against average road surface luminance and against a measure of luminance uniformity. The measure of uniformity is interesting. Study of the results showed that a simple maximum/minimum luminance ratio was unsuitable as a measure of uniformity. The important feature of the uniformity was the rate of change of luminance. The measure used in Fig. 7.18 is the percentage relative maximum luminance slope, which is defined as the maximum luminance variation found over any one metre transverse or three metre longitudinal distance expressed as a percentage of the average road surface luminance. The different emphasis given to transverse and longitudinal uniformity of luminance arises because perspective diminishes the effect of longitudinal variations but does not affect transverse variations. Other authors have confirmed the importance of longitudinal uniformity [75] to the extent that it is now used as a criterion in the CIE roadway lighting recommendations [76].

It is now possible to reconsider the previous results obtained with experienced observers [74]. There the category 'good' was reached at an average road surface luminance of $1 \cdot 5 \, \text{cd m}^{-2}$. From Fig. 7.18 it can be seen that this only applies for very uniform luminance distributions. For more non-uniform distributions, i.e. higher values of the percentage relative maximum luminance slope, a higher average luminance is required for the general impression to be 'good'.

This introduces the problem of wet roads. Wet roads almost always have a more uneven luminance pattern than dry roads. The importance of this factor is demonstrated in some results obtained by Cornwell [77]. He had people assess the visibility produced by 38 different roadway lighting installations on a similar nine-category 'bad to excellent' scale. For dry roads, the physical conditions considered most important for good visibility were road surface luminance and the extent of visual guidance given by the installation. For wet roads, luminance uniformity was considered to be the major factor affecting visibility. Criteria for luminance uniformity for comfort when viewing wet roads need to be established.

So far this examination of luminance uniformity has been concerned with the luminance of the road surface, but another potential source of discomfort is the luminous intensity distribution of the street lighting lanterns. In general, discomfort caused by street lantern luminance will be experienced under much less extreme conditions than are required to produce any change in the detection of obstacles [78]. This suggests that if glare criteria for roadway lighting are based on discomfort, the conditions provided will not adversely affect driver performance.

Studies of discomfort glare from roadway lighting initially followed a similar path to those for interior lighting, using many of the same techniques [45, 78] and are therefore open to the same criticisms. However, later studies overcame this problem. A model was constructed in which observers were given the impression that they were moving along the road at anything from 35–85 km h^{-1}. The average road surface luminance ($\bar{L}$), the luminous intensity of the lanterns at 80° from the downward vertical (I_{80}), and the relative maximum luminance slope S_{max} were varied, whilst the other factors such as the mounting height of the lanterns, the number of lanterns visible, the flashed area of the lanterns, their luminous intensity distribution and the road dimensions were kept constant [79]. A group of 30 observers said to be representative of normal road users saw a range of conditions created by different values of $\bar{L}$, I_{80} and S_{max}, and rated the glare on a nine-point scale with the categories labelled: 1 = unbearable glare, 3 = disturbing glare, 5 = just admissible glare, 7 = satisfactory glare, 9 = unnoticeable glare. Regression analysis of the data collected produced a summary equation of the form

$$G = 14 \cdot 56 - 3 \cdot 31 \log_{10} I_{80} + 0 \cdot 97 \log_{10} \bar{L} \qquad (7.10)$$

where G = the glare rating and is called the Glare Mark, I_{80} = luminous intensity of the lantern at 80° from the downward vertical (cd), and $\bar{L}$ = mean road surface luminance (cd m^{-2}). As might be expected, as the average road surface luminance increases the perceived discomfort glare decreases. The reverse happens for increases in lantern luminous intensity (I_{80}). The information contained in this equation is useful but strictly applies only to a limited range of conditions. It can be predicted that the most important of the factors fixed in the experiment would be the mounting height, flashed area and luminous intensity distribution of the lanterns. These factors were all found to modify the basic equation together with the number of lanterns visible. The final equation predicting discomfort glare sensation was

$$G = 13 \cdot 82 - 3 \cdot 31 \log_{10} I_{80} + 0 \cdot 97 \log_{10} \bar{L} + 4 \cdot 41 \log_{10} h$$
$$+ \log_{10} \left(\frac{I_{80}}{I_{88}} - 0 \cdot 9 \right) + \log_{10} F - 1 \cdot 46 \log_{10} p \qquad (7.11)$$

where G, I_{80} and $\bar{L}$ are as before; h = the difference between the height of the lantern and the height of the observer's eye (m); I_{80}/I_{88} = the ratio of the luminous intensities of the lantern at 80 and 88° from the downward vertical (this is used as a measure of the luminous intensity distribution of the luminaire); F = the flashed area of the lantern (m^2); p = the number of luminaires per kilometre of road.

The results that led to this equation were achieved in a model. Fortunately predictions made on the basis of the above equation have been verified on real roads. Figure 7.19 shows the agreement between the rated values of the glare sensation and the calculated values of Glare Mark for 19

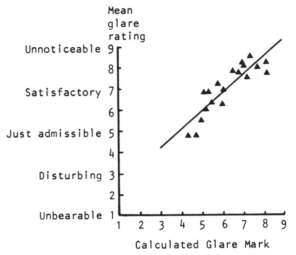

Fig. 7.19.   Mean glare rating for 19 different streets plotted against calculated Glare Mark.
(After [79].)

streets appraised by a group of lighting experts [79]. De Boer and Schreuder [79] have also found agreement between the calculated Glare Mark and the assessment of glare made by 10 observers whilst driving along a number of different streets. Cornwell [77] reports a similar result for appraisals of discomfort glare from 38 wet and dry roads. However, he also shows that the correlations between the subjective assessments of discomfort glare and the measures of the physical conditions are significantly improved if only the flashed area of the lantern and the I_{80}/I_{88} ratio are used as the latter. This suggests that the Glare Mark formula is not the ultimate predictor of discomfort glare from roadway lighting. Nonetheless there is considerable justification for its use (in a slightly modified form) as a roadway lighting criterion [75, 76]. The experiments on which it is based have been carefully conducted in models simulating roads, and the results largely confirmed on real roads.

One factor which might affect comfort under roadway lighting but which

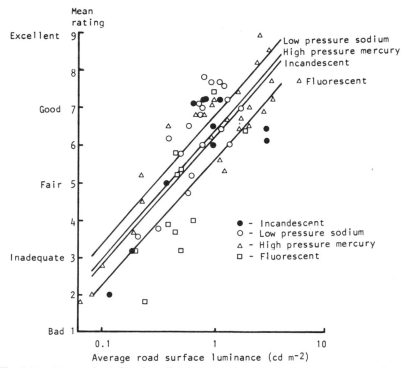

Fig. 7.20. Mean ratings of roadway lighting plotted against average road surface luminance for 70 roads classified according to the light source used. The slopes of the linear regression lines for each light source separately were not significantly different. Therefore the trend for each light source is shown by a straight line going through the overall mean for the light source but with the slope obtained from the linear regression line for all the data. (After [73].)

has not been considered in detail is that of different light source colour properties. Light sources used for roadway lighting range from incandescent lamps, which approach a full radiator spectral emission, to low pressure sodium discharge lamps, which produce a few powerful spectral lines and little else (see Chapter 1). Figure 7.20 shows the results of assessments of the general impression of lighting in 70 roads [73]. The overall effect has already been discussed in terms of preferred average road surface luminance, but the point of interest here is the difference between the various light sources. There is no strong distinction between the installations with the various lamp types. This should not be too surprising because of all the other differences between the installations that might be present. What these results do suggest is that any effect of different light source colour on general impression is far from overwhelming. This conclusion is reinforced by the failure to find any effect of light source colour in the experiment on the assessment of non-uniformity of luminance [74].

The one area where light source colour has been found to be significant is in the assessment of discomfort glare [80]. Subjects alternately viewed two identical photographs of a lit street. The photographs were lit from the front but the street lanterns in the scene had been cut out and replaced by luminous discs lit from behind by different light sources. This arrangement enabled the observers to vary the road surface luminance and the lantern luminance independently. The procedure used was to set the road surface luminance to the same value in both photographs, then to set the lantern luminance of one photograph to a fixed value and to vary the lantern luminance in the other photograph to a higher or lower value. The observer was then asked simply to say whether photograph A produced more, the same, or less glare than photograph B. Thirty subjects made the comparison for different light sources for various values of road surface luminance over a range of $0.5-5.0$ cd m^{-2}. The interesting result was that the degree of discomfort glare experienced was equal when the luminance of low pressure sodium discharge light sources was 1.4 times that of high pressure mercury discharge light sources. Whether this difference is due to the failure of the CIE relative luminous efficiency function (V_λ) to match the actual spectral response of the visual system at low luminances is a matter of theoretical but not practical importance. Roadway lighting is designed in photopic units. Attempts have been made to confirm these differences between sodium and mercury light sources in real roads, but with little success because the comparisons have been made between roads which have other differences besides that of light source colour [80]. Nonetheless it has been suggested that the Glare Mark derived from eqn (7.11) should be increased by 0.4 for low pressure sodium discharge light sources and decreased by 0.1 for high pressure mercury discharge lamps. Overall it would appear that the colour of the light source may have a small effect on the assessment of discomfort glare from roadway lighting.

To summarise, the factors which influence comfort in roadway lighting have been extensively if indirectly studied. Criteria for average road surface luminance, uniformity of road surface luminance, and discomfort glare all exist and are reasonably well founded. If these recommendations are followed then comfortable roadway lighting will result. The one area which remains to be studied is the effect of wet road surfaces. Wet roads alter the relative contribution of different features of roadway lighting to visibility, principally by increasing the importance of luminance uniformity [77, 81].

Until the effects of these changes have been taken into account the study of comfort whilst driving and the development of roadway lighting criteria will not be complete.

7.6 WINDOWLESS BUILDINGS

There is one aspect of comfort and lighting which should be considered here, although it is not associated with one specific context. This aspect is the absence of windows. Over recent years there has been a considerable swell of interest in eliminating windows from conventional buildings such as schools, offices and hospitals. At first this may seem rather strange because windows fulfill a wide range of functions and, as has been shown earlier, are widely desired in buildings. But the fact is windows cost more than a plain wall, and concern about energy conservation in recent years has tended to emphasise the window as the weak link in the insulation of buildings. These facts, together with the trend for deep building, have led to the interest in windowless buildings. Even so this interest would have fallen on stony ground if it were not for the fact that the main physical functions of windows, the provision of light and ventilation, can be conveniently provided by artifical means. As the physical functions can be covered, the main change when windows are eliminated is in their psychological effects. In an attempt to indicate the importance of this loss a number of studies have been made of people's opinions of windowless buildings.

One of the most comprehensive of these was an attitude survey done by Ruys [82] on five different buildings in Seattle, USA, each containing a number of windowless offices. Although the buildings did have windows, the structures were deep and the individuals questioned worked in offices which were themselves windowless. Most of these offices were small, ranging from $3 \cdot 2$ m^2 to $14 \cdot 8$ m^2, 60% were single person offices and only 5% were occupied by more than three people. There was little complaint about the artificial lighting provided, 88% of the 139 subjects believing the illumination to be adequate. However 87% also expressed dissatisfaction with the lack of windows. When asked to explain the reason for their dissatisfaction the subjects replied that the lack of windows meant a lack of daylight, poor ventilation and an inability to know about the weather or to have a view. The lack of windows also gave some a feeling of being cooped up and led to feelings of depression and tension. These feelings were very stable. The presence of bright lights, warm colours and access to a nearby window did not change the dissatisfaction felt at the lack of a window in the room. It should be noted that the subjects in this survey were people who occupied small offices, frequently alone, in which they did paper work. There was little opportunity for movement in the office and/or social interaction with other people. The cell-like impression of the offices can well be imagined.

Sommer [83] reports several complaints about offices underground. The major complaints were about the lack of variety in the physical environment, the stale air and the general unnaturalness of being underground all day. This is a more unusual situation than the windowless offices in deep, windowed buildings. In this case these are not complaints about lack of windows because underground there is no possibility of having windows. What the experience of underground offices tells us is the reaction of people to a completely artificial luminous environment. Obviously it is considered unnatural but it is also assessed as lacking variety, a feature that a window is well able to provide.

Another, rather unusual situation, where windows are sometimes absent is in hospitals. Modern hospitals are occasionally built with wards around a windowless core in which are housed operating theatres, laboratories and nurses' stations. A special case arises with an intensive care ward because of the need to be close to the nursing station. Wilson [84] examined 50 patients, in two intensive care wards, in two comparable hospitals, in Arkansas, USA. One ward was windowless, the other had windows. Both groups of patients were similar in terms of age, treatment and physical condition. Wilson found that 40 % of the patients in the windowless ward had developed post-operative delirium as against 18 % in the windowed ward. Further, there was an increased incidence of post-surgical depression in those patients in the windowless ward who did not develop post-operative delirium.

It may seem a long way from the small office described earlier to an intensive care ward but there are similarities. In both situations the movement of the individual is limited, as is contact with other individuals. The activities allowable may be boring or simply non-existent. In such situations it would seem that an absence of windows produces dissatisfaction with the environment and feelings of depression.

A rather different situation occurs in factories, which are often windowless. Here the windowless space is usually large and occupied by many people. Unfortunately there has been virtually no reliable research on people's responses to windowless factories, even though such factories are probably the most common form of windowless building. Most of the information about windowless factories is anecdotal. Interviews with personnel managers in several factories revealed few employee complaints about the lack of windows [85]. Whether this was because the lack of windows was not a cause of concern, or because formal complaints were troublesome and had little likelihood of success, or because there were more important things to complain about, remains an open question. Plant [86] reports that Russian and Czech experience of windowless factories suggested that absenteeism is higher in such installations and the occupants are more likely to complain of headaches and sickness. However, Weist [87] claims that there was little difference in sickness for a windowless textile factory when compared with other factories with windows. Thus the picture for factories is confused.

A type of building in which large spaces occupied by many people occur, and where people's reactions have been carefully studied, is schools. Demos *et al.* [88] examined the scholastic performance and attitudes of 10-year-old pupils in a windowless classroom and in a classroom with windows. They found no significant difference in achievement, personality tests or health records for the two classes. They did find that children in the windowless classroom viewed school more favourably although the teachers thought the same children were simultaneously more timid and more likely to complain. The children preferred the windowless classroom in the first year but disliked it intensely in the second year. Such variability is one of the problems of subjective assessments.

Larson [89] studied primary schoolchildren (five to eight years old) over a three-year period. In the first and third years the windows were present in the school but in the second year the windows were removed (leading to what was described as 'paneless education' in one newspaper). Removal of the windows did not change scholastic performance but there was a significant tendency for the younger children to be absent more frequently from the windowless classrooms. This is interesting in view of the reported higher absenteeism in windowless factories. Larson also had the teachers interview each child separately about the classrooms. Teachers reported that the children did not mention windows spontaneously but when they were mentioned some 90 % of the five-year olds and 50 % of the eight-year olds expressed a desire for windows. Interestingly enough the teachers, in spite of an initial antipathy to the windowless school, finished the year strongly in favour. They found that the absence of windows made more display space available and eliminated a potent source of distraction. Chambers [90] found similar effects in a survey of attitudes to windowless classrooms at a university in Tennessee, USA and schools in New Mexico, USA. Both teachers and students said they appreciated the lack of outside distraction and the good thermal and lighting conditions in the windowless rooms. The importance given to physical conditions is understandable given the generally sunny climate of the area and the consequent potential for solar overheating. Tikkanen [91] obtained the reactions of more than 300 students in California in four schools with windows and four without. Classrooms with windows were rated as having better lighting and also as being noisier and with having more outside distractions. Pupils in the windowless school were evenly divided in their desire for windows but those in the schools with windows were strongly in favour of windows (94 % preferred windows).

There is no strong direction to all these data on attitudes to windowless classrooms. Some children in some schools appreciate the absence of windows, other children in other schools are very much in favour of them. It appears to depend on the particular circumstances. The only firm conclusion about schools is that the absence of windows neither improves nor impairs scholastic performance. It would be surprising if it did, given the host of other variables, such as quality and method of teaching, that must be important.

Two points stand out from this collection of survey results. The first is the general lack of complaint about the physical environment in the windowless buildings, Indeed in some of the school studies the thermal environment is considered to have been improved by not having windows. Even when daylight has been mentioned as desirable there have been no complaints that the artificial lighting was inadequate. The second point that is evident is that there is no single, completely general, conclusion about attitudes to windowless buildings. The information available is too sparse, too scattered and has too many intervening variables for that to be possible. What the studies considered have done is to reveal a number of factors which influence attitudes to windowless buildings. Among them are the size of the windowless space and the level of activity in it. Schools and factories, which appear to create fewer problems when they are windowless, tend to have large spaces with many people in them, thus allowing a considerable variety of social interactions to occur. The function of windows that appears to be missed when windows are absent is their ability to introduce a variety of stimulation into people's lives. If the level of variety is already high, as in schools, factories and shops, then the absence of windows may be of little importance; but in a small office, where there is little possibility of variety, the lack of a window will be the cause of complaint. However, variety may not be the only factor. Expectation may have a lot to do with attitudes. Factories and large shops are expected to be windowless for all practical purposes, offices are not. If an unusual situation is unpleasant then people will dislike it because they know a better alternative is widely available. When a windowless environment is necessary as, for example, occurs in some pharmaceutical manufacturing plant where cleanliness is very important, then the absence of windows will be accepted although reluctantly. In such circumstances equality of treatment is important. Sommer [83] found that the dislike of the windowless environment in the underground offices he studied was amplified by the fact that the executives had offices above ground with windows revealing beautiful views. It is an observable fact that very few offices are arranged in such a way that the manager has a room without a view and the office workers all have windows. The reverse situation is much more common. Perhaps this says more about attitudes to windows than all the surveys.

7.7 COMMENTS

The most obvious feature of the results discussed in this chapter has been the variation in the depth to which different contexts have been studied. Office lighting has been extensively examined. Overhead lighting for dry roads has been well covered but wet roads have not. Natural lighting in houses has been considered but electric lighting has not. Factory lighting has been ignored completely. Thus there are still plenty of areas that need to be investigated but this should not be taken to indicate that the existing lighting criteria are wrong. They may not cover all conditions in all contexts and each has its own limitations but overall there can be little doubt that they are correct although imprecise. If the available lighting criteria are followed, discomfort will be avoided. The resulting impression will not necessarily be what is wanted but it will not be uncomfortable. This situation is at least partly due to the limitations of the correlation approach adopted in the vast majority of past studies of desirable lighting conditions. No better example of the coarse nature of the information obtained by this method can be found than Fig. 7.6. This shows the variation in satisfaction with the lighting of 20 different deep-plan offices, plotted against average illuminance. The implication of this result is that it is not how much you provide, but how you provide it, that affects satisfaction with lighting. A similar type of conclusion probably applies to other criteria. The way to create a desired impression for a particular interior can best be identified by the multi-dimensional approaches discussed in the preceding chapter. Two particular problems to which the multi-dimensional methods could be applied with advantage are (a) the effect windows have on the appreciation of lighting in an interior and (b) the strange difference that occurs between task illuminances when they are provided by electric light and daylight. The effect of window size, proportion and position on the appreciation of the interior has not been studied systematically and yet it could be very important for impression. The other effect is really a discrepancy. Illuminances on desks of greater than 1000 lx produced by electric lighting tend to be associated with complaints of over-brightness. Yet illuminances on desks produced by daylight frequently exceed 1000 lx by a large margin without complaint. It would be interesting to know why.

7.8 SUMMARY

This chapter is concerned with describing, critically, the current understanding of the important features of lighting for comfort, as made manifest in the form of lighting criteria. What is regarded as suitable and comfortable lighting for a particular interior will depend on the context of the interior. For example, what is satisfactory lighting for an office may not be suitable for a restaurant. The particular contexts considered here are houses, offices, factories, roads and windowless buildings. Space has not allowed discussion of the information available on suitable lighting for such special contexts as hospitals [92] and tunnels [93].

Little is known about the comfortable electric lighting of houses, but this is not too important because there is no doubt that the preferred form of lighting for housing is daylight. Criteria for daylight and sunlight in terms of daylight factor and duration of exposure are supported to a limited extent by people's assessments of the desirability of daylight and sunlight. Natural lighting is also strongly desired in offices but is much more rarely provided in large quantities. This is because predominantly natural lighting imposes severe limitations on building form. Consequently many offices are mainly lit by electric lighting even during daytime. Criteria for electric lighting in offices in terms of illuminance, illuminance uniformity, surface luminance ratios, discomfort glare limitations and suitable light source colour properties are discussed, and a criterion for veiling reflections is suggested. All the existing criteria, except the light source colour properties, can be shown to have some justification but the overall effect is to avoid discomfort rather than to create a specified impression. In order to gain an understanding of the impression created by different lighting conditions, multi-dimensional experiments of the type discussed in Chapter 6 are required. The same applies to roadway lighting in cities, towns and villages, where appearance is important. However, the main purpose of roadway lighting is to provide good visibility of obstacles without discomfort. To ensure this occurs, roadway lighting criteria of average road surface luminance, uniformity of road surface luminance, and discomfort glare from luminaires are available and are justified by the evidence. The only problem is that the criteria may well be too lax because they have nearly all been derived for dry roads. Wet roads create much more critical conditions. Criteria suitable for these conditions remain to be developed.

Virtually nothing can be said about criteria for factory lighting. In spite of the obvious differences between factories and offices, factory lighting criteria are usually based on information obtained in offices. Factory lighting for comfort is an area that needs investigation.

Finally windowless buildings are considered. Windowless buildings are rarely inhabited with pleasure, but if there is a good reason why they should be windowless then they may be tolerated. The deprivation of being without windows will be minimised in large spaces where plenty of activity and social interaction is possible. Where the occupants are restricted in space and/or activity, windows will be sorely missed.

REFERENCES

1. British Standards Institution. BS CP 3. Chapter 1, Part 1, *Daylighting*, 1964. Chapter 1 (B), *Sunlight*, 1945.
2. The Stationery Office. *Building Standards (Scotland) (Consolidated) Regulations*, HMSO, London, 1971.
3. Department of the Environment. *The estate outside the dwelling*, Design Bulletin 25, HMSO, London, 1972.
4. Bitter, C. and van Ierland, J. F. A. A. Appreciation of sunlight in the home, in *Sunlight in Buildings*, ed. R. G. Hopkinson, Boewcentrum International, Rotterdam, 1967.
5. Ne'eman, E., Craddock, J. and Hopkinson, R. G. Sunlight requirements in buildings—1. Social survey, *Building and Environment*, **11**, 217, 1976.
6. Markus, T. A., Brierley, E. and Gray, A. *Criteria of sunshine, daylight, visual privacy and view in housing*, Building Performance Research Unit, Department of Architecture and Building Science, University of Strathclyde, Glasgow, 1972.
7. Kaplan, S. and Wendt, J. S. Preference and the visual environment: complexity and some alternatives, *Proc. EDRA 3 Conference*, **1**, Paper 6.8, University of California at Los Angeles, 1972.
8. Clamp, P. E. An approach to the visual environment, *Arch. Res. and Teach.*, **2**, 153, 1973.
9. Chroscicki, W. Full length mirror lighting, *Ljuskultur*, **1**, 28, 1973.
10. Illuminating Engineering Society. *IES Code for Interior lighting*, IES, London, 1977.

11. Levitt, J. G. Lighting for the elderly—an optician's view, *IES/CIBS National Lighting Conference, Churchill College, Cambridge, UK*, 1978.
12. Phillips, D. H. *Lighting: The Principles and Planning of Home Lighting*, MacDonald, London, 1966.
13. The Electricity Council. *Lighting your home*, Electricity Council; London, 1971.
14. Markus, T. A. The significance of sunshine and view for office workers, in *Sunlight in Buildings*, ed. R. G. Hopkinson, Boewcentrum International, Rotterdam, 1967.
15. Tregenza, P. R., Romaya, S. M., Dawes, S. P. and Heap, L. J. Consistency and variation in preferences for office lighting, *Ltg Res. and Technol.*, **6**, 205, 1974.
16. Wells, B. W. P. Subjective responses to the lighting installation in a modern office building and their design implications, *Building Sci.*, **1**, 57, 1965.
17. Manning, P. *Office design: a study of environment*, Pilkington Research Unit, Liverpool University, Liverpool, 1965.
18. Langdon, F. J. *Modern offices: a user survey*, HMSO, London, 1966.
19. Cooper, J. R., Wiltshire, T. and Hardy, A. C. Attitudes towards the use of heat rejecting/low light transmission glasses in office buildings, *Proc. CIE Conference, Windows and their function in architectural design, Instanbul*, 1973.
20. Ne'eman, E. and Hopkinson, R. G. Critical minimum acceptable window size, a study of window design and provision of view, *Ltg Res. and Technol.*, **2**, 17, 1970.
21. Keighly, E. C. Visual requirements and reduced fenestration in offices—a study of multiple apertures and window area, *Building Sci.*, **8**, 321, 1973.
22. Keighly, E. C. Visual requirements and reduced fenestration in office buildings—a study of window shape, *Building Sci.*, **8**, 311, 1973.
23. Ludlow, A. M. The functions of windows in buildings, *Ltg Res. and Technol.*, **8**, 57, 1976.
24. Hopkinson, R. G. and Longmore, J. The permanent supplementary artificial lighting of interiors, *Trans. Illum. Engng Soc. (London)*, **24**, 121, 1959.
25. Crisp, V. H. C. The light switch in buildings, *Ltg Res. and Technol.*, **10**, 69, 1978.
26. Saunders, J. E. The role of the level and diversity of horizontal illumination in an appraisal of a simple office task, *Ltg Res. and Technol.*, **1**, 37, 1969.
27. van Ierland, J. F. A. A., *Two thousand Dutch office workers evaluate lighting*, Publication 283, Research Institute for Environmental Hygiene, TNO, Delft, Netherlands, 1967.
28. Balder, J. J. Erwunschte leuchtdichten in Buroraumen, *Lichttechnik*, **9**, 455, 1957.
29. Kraemer, Sieverts and Partners. *Open Plan Offices*, McGraw-Hill, London, 1977.
30. Lynes, J. A., Burt, W., Jackson, G. K. and Cuttle, C. The flow of light in buildings, *Trans. Illum. Engng Soc. (London)*, **31**, 65, 1966.
31. Epaneshnikov, M. M., Obrosova, N. A., Sidorova, T. N. and Undasynov, G. N. New characteristics of lighting conditions for premises of public buildings and methods for their calculation, *Proc. CIE 17th session, Barcelona*, 1971.
32. Cuttle, C. C., Burt, W. and Valentine, B. Beyond the working plane, *Proc. CIE 16th session, Washington*, 1967.
33. Bean, A. R. Lighting of occupants and objects within an interior, *Ltg Res. and Technol.*, **10**, 146, 1978.
34. Bean, A. R. *An experimental study of task to background lighting*, Department of Health and Social Security, London, 1978.
35. Boyce, P. R. Users' attitudes to local lighting, *Ltg Res. and Technol.*, **11**, 158, 1979.
36. Touw, L. M. C. Preferred brightness ratio of task and its immediate surround, *Proc. CIE 12th session, Stockholm*, 1951.
37. Barthés, E. and Richard, J. Étude expérimentale du confort visuel à travers l'équilibre des luminances, *Lux*, **54**, 298, 1969.
38. Collins, J. B. and Plant, C. G. H. Preferred luminance distributions in windowless spaces, *Ltg Res. and Technol.*, **3**, 219, 1971.
39. Ostberg, O. and Stone, P. T. Methods of evaluating discomfort glare aspects of lighting, *Goteburg Psychological Reports*, **4**, 4, University of Goteburg, Sweden, 1974.

40. Illuminating Engineering Society of North America. Visual comfort ratings for interior lighting, *Illum. Engng*, **61**, 643, 1966, and *J.I.E.S.*, **2**, 328, 1973.
41. Guth, S. K. A method for the evaluation of discomfort glare, *Illum. Engng*, **58**, 351, 1963.
42. Guth, S. K. Comfortable brightness relationships for critical and casual seeing, *Illum. Engng*, **46**, 65, 1951.
43. Illuminating Engineering Society. *Evaluation of discomfort glare: the Glare Index system for artificial lighting installations*, Technical Report 10, London, 1967.
44. Petherbridge, P. and Hopkinson, R. G. Discomfort glare and the lighting of buildings, *Trans. Illum. Engng Soc. (London)*, **15**, 39, 1950.
45. Hopkinson, R. G. *Architectural Physics—Lighting*, HMSO, London, 1963.
46. Luminance Study Panel of the IES Technical Committee. The development of the IES Glare Index System, *Trans. Illum. Engng Soc. (London)*, **27**, 9, 1962.
47. Fischer, D. The European Glare Limiting method, *Ltg Res. and Technol.*, **4**, 97, 1972.
48. Bodmann, H. W. and Sollner, G. Glare evaluation by luminance control, *Light and Lighting*, **58**, 195, 1965.
49. Sollner, G. Ein einfaches system zur blendungsbewertung, *Lichttechnik*, **17**, 59a, 1965.
50. Sollner, G. Glare assessment of luminaires with lamps of a high luminance, *Lichttechnik*, **26**, 169, 1974. (BRE Library Translation 1890.)
51. Sollner, G. Glare from luminous ceilings, *Lichttechnik*, **24**, 11, 1972. (BRE Library Translation 1760.)
52. Manabe, H. *The assessment of discomfort glare in practical lighting installations*, Oteman Economic Studies (9), Oteman Gakuin University, Osaka, Japan, 1976.
53. Aleksiev, P. K. and Vasilev, N. I. Comparison of the methods for determining the discomfort glare by regression analysis, *National Lighting Conference, Varna, Bulgaria*, 1978.
54. Bellchambers, H. E., Collins, J. B. and Crisp, V. H. C. Relationship between two systems of glare limitation, *Ltg Res. and Technol.*, **7**, 106, 1975.
55. Boyce, P. R., Crisp, V. H. C., Simons, R. H. and Rowlands, E. Discomfort glare sensation and prediction, *Proc. CIE 19th session, Kyoto*, 1979.
56. Markus, T. A. The why and the how of research in real buildings, *J. Arch. Res.*, **3**, 19, 1974.
57. Hopkinson, R. G. Glare from windows, *Construction Research and Development Journal*, **2**, 98, 1970.
58. Fry, G. A. and King, V. M. The pupillary response and discomfort glare, *J.I.E.S.*, **4**, 307, 1975.
59. Stone, P. T. Discomfort glare and its significance in visual work, *Proc. International Congress on Occupational Health, Vienna, Austria*, 1966.
60. Lynes, J. A. Discomfort glare and visual distraction, *Ltg Res. and Technol.*, **9**, 51, 1977.
61. De Boer, J. B. Performance and comfort in the presence of veiling reflections, *Ltg Res. and Technol.*, **9**, 169, 1977.
62. Bjorset, H. H. and Frederiksen, E. A proposal for recommendations for the limitation of the contrast reduction in office lighting, *Proc. CIE 19th session, Kyoto*, 1979.
63. Reitmaier, J. Some effects of veiling reflections in papers, *Ltg Res. and Technol.*, **11**, 204, 1979.
64. Cockram, A. H., Collins, J. B. and Langdon, F. J. A study of user preferences for fluorescent lamp colours for daytime and night-time lighting, *Ltg Res. and Technol.*, **2**, 249, 1970.
65. Hopkinson, R. G., Medd, D. I., Longmore, J. and Gloag, H. L. Integrated daylight and artificial light in interiors, *Proc. CIE 15th session, Vienna*, 1963.
66. Kruithof, A. A. Tubular fluorescent lamps for general lighting, *Philips Technical Review*, **6**, 65, 1941.
67. Bodmann, H. W., Sollner, G. and Voit, E. Evaluation of lighting levels with various kinds of light, *Proc. CIE 15th session, Vienna*, 1963.

68. Aston, S. M. and Bellchambers, H. E. Illumination, colour rendering and visual clarity, *Ltg Res. and Technol.*, **1**, 259, 1969.
69. Bellchambers, H. E. and Godby, A. C. Illumination, colour rendering and visual clarity, *Ltg Res. and Technol.*, **4**, 104, 1972.
70. Boyce, P. R. Investigations of the subjective balance between illuminance and lamp colour properties, *Ltg Res. and Technol.*, **9**, 11, 1977.
71. Hunt, R. W. G. The perception of colour in 1° fields for different states of adaptation, *J. Opt. Soc. Amer.*, **43**, 479, 1953.
72. Maitreya, U. K. Artificial lighting—a field study, *Design*, **21**, 32, 1977.
73. De Boer, J. B., Burghout, F. and Van Heemskerck Veeckens, J. F. T. Appraisal of the quality of public lighting based on road surface luminance and glare, *Proc. CIE 14th session, Brussels*, 1959.
74. De Boer, J. B. and Knudsen, B. The pattern of road luminance in public lighting, *Proc. CIE 15th session, Vienna*, 1963.
75. Commission Internationale de L'Eclairage. *Glare and uniformity in road lighting installations*, CIE Publication 31, 1976.
76. Commission Internationale de l'Eclairage. *Recommendations for the lighting of roads for motorised traffic*, CIE Publication 12/2, 1977.
77. Cornwell, P. R. Appraisals of traffic route lighting, *Ltg Res. and Technol.*, **5**, 10, 1973.
78. De Boer, J. B. Fundamental experiments of visibility and admissible glare in road lighting, *Proc. CIE 12th session, Stockholm*, 1951.
79. De Boer, J. B. and Schreuder, D. A. Glare as a criterion for quality in street lighting, *Trans. Illum. Engng Soc. (London)*, **32**, 117, 1967.
80. De Boer, J. B. *Public lighting*, Philips Technical Library, Eindhoven, Netherlands, 1967.
81. Gordon, P. Appraisals of visibility on lighted dry and wet roads, *Ltg Res. and Technol.*, **9**, 177, 1977.
82. Ruys, T. *Windowless offices*, MA Thesis, University of Washington, USA, 1970.
83. Sommer, R. *Tight Spaces: Hard Architecture and How to Humanise It*, Prentice Hall, New Jersey, 1974.
84. Wilson, L. M. Intensive care delirium: the effect of outside deprivation in a windowless unit, *Arch. Internal. Medicine*, **130**, 225, 1972.
85. Pritchard, D. Industrial lighting in windowless factories, *Light and Lighting*, **57**, 265, 1964.
86. Plant, C. G. H. The light of day, *Light and Lighting*, **63**, 292, 1970.
87. Weist, H. J. Medical experience with industrial buildings without windows, *Das deuchte gesundheitswessen*, **13**, 684, 1958. (BRE Library Translation 1203.)
88. Demos, G. D., Davis, S. and Zuwaylif, F. F. Controlled physical classroom environments, *Building Res.*, **4**, 60, 1967.
89. Larson, C. T. *The effect of windowless classrooms on elementary schoolchildren*. Architectural Research Laboratory, Department of Architecture, University of Michigan, USA, 1975.
90. Chambers, J. A. *A study of attitudes and feelings towards windowless classrooms*, Ed.D. Dissertation, University of Tennessee, USA, 1963.
91. Tikkanen, K. T. *Significance of windows in classrooms*, M.A. Thesis, University of California at Berkeley, USA, 1970.
92. The Chartered Institution of Building Services. *CIBS lighting guide to hospitals and health care buildings*, CIBS, London, 1979.
93. Schreuder, D. A., Narisada, K., Westermann, H. O., Bijllardt, D., von den Muller, E. and Reimenschneider, W. Symposium on tunnel lighting, *Ltg Res. and Technol.*, **7**, 85, 1965.

8

Wider Horizons

8.1 INTRODUCTION

So far, this examination of the appreciation of lighting has considered the use of lighting for creating an impression, the features of lighting which can cause discomfort and how this discomfort can be avoided. The results and techniques described previously justify many of the present lighting criteria and hold out hope for a most positive role for lighting in the future. However, the whole discussion could be considered somewhat introverted, for three reasons. First, the discussion only deals with lighting. Second, it only deals with the average man's appreciation of lighting. Third, it only deals with what might be called 'reasonable lighting', i.e. lighting which may be unsuitable or even uncomfortable but is not definitely unhealthy. The aim of this chapter is to widen the discussion of the appreciation of lighting so as to overcome these objections. To achieve this aim this chapter includes a consideration of the place of lighting in the overall evaluation of buildings, some discussion of the differences between individuals that influence their appreciation of lighting, and a description of lighting conditions that are a threat to health.

8.2 LIGHTING IN ITS PLACE

The interaction between lighting conditions and people's responses to them is of considerable interest to the lighting engineer but it is only one interest among many for the architect. As the person professionally responsible for the complete building and very often its surroundings, the architect is concerned with all the aesthetic and functional aspects of the building as well as the lighting. The skill of the architect lies in combining these features into a complete environment that satisfies as many people as possible within the legal, financial and social constraints imposed on him. To do this he has to allocate priorities. As an aid to allocating priorities the study of the relative importance people give to different facets of the environment has particular value.

Canter [1], in a useful paper, discusses some of the methods whereby this question of priorities can be examined. The first method he considers is the simplest and the most direct. It is to ask people how important they consider

Table 8.1: Median rankings of characteristics of a job in an office in decreasing order of importance (1 = most important, 8 = least important)

Characteristic	Old, compartmented office		New, landscaped office	
	Median	Rank	Median	Rank
Pay	2·7	1	2·9	2
Type of work	2·9	2	2·2	1
Security and prospects	3·1	3	4·3	4
Atmosphere amongst staff	3·5	4	3·8	3
Hours and holidays	4·8	5	4·4	5
Physical environment	5·8	6	5·4	6
Office layout	6·5	7	5·9	7
Journey to work	6·8	8	6·5	8

each item on a list to be. Table 8.1 shows the results obtained by this method in two different buildings, one a run-down, town-centre, compartmented office, the other a new, landscaped office on the edge of the town [2]. The office workers, who were the same 33 people in each case, were asked to rank the eight characteristics of a job in an office in order of importance to themselves (1 = most important, 8 = least important). From Table 8.1 it can be seen that for both buildings the physical environment comes sixth out of eight in the median rankings. The most important features were the pay and the type of work, followed by the job security and prospects, social atmosphere, hours and holidays, and then the physical environment. It would appear that for these office workers there are more important things about the job than the physical environment. However, there is one important limitation to this conclusion. For the offices studied, the physical conditions, whilst not ideal, were tolerable and this may have diminished the importance given to the physical environment. It would be interesting to know if the physical conditions were similarly ranked for a factory job in which the environment imposed a definite stress, as, for example, in the rolling mill of a steelworks. It seems likely that in this situation the physical environment would increase in importance but not to the extent that it would replace pay as the prime feature. This illustrates the problem of this whole field: the physical environment will be given a different importance in different conditions. The only conclusion that can be drawn from the limited data available is rather vague. It is that the physical conditions are less important than some financial and psychological factors, and more important than other peripheral factors.

Table 8.2: Median rankings of the importance of different environmental features in an office (1 = most important, 8 = least important)

Environmental features	New, landscaped office—Bedford [2] (n = 82)		New, landscaped office—Leeds [3] (n = 185)		New, landscaped office—Aberdeen [4] (n = 169)	
	Median	Rank	Median	Rank	Median	Rank
Good heating and ventilating	2·5	1	1·7	1	2·0	1
Good lighting	2·8	2	2·3	2	2·9	2
Good furniture and equipment	3·5	3	4·4	3	3·8	3
Peace and quiet	3·8	4	4·6	4	4·5	4
Plenty of space	4·7	5	5·2	5 =	5·1	5
Privacy	6·0	6 =	5·2	5 =	5·5	6
Nearness to a window	6·0	6 =	6·5	7	6·3	7
Pleasant decoration	6·5	8	6·6	8	6·6	8

Given that the physical conditions do have some importance, the next point of issue is the ranking given to various features of the physical environment. Table 8.2 shows the median rankings of eight different physical features made by three different sets of office workers, each occupying a different, new, air-conditioned, electrically lit, deep plan office [2, 3, 4]. It can be seen that the lighting is regarded as highly important, it being consistently placed second after good heating and ventilation. These two features were on the top of the rankings even when the samples were divided by age, sex and occupation. A plausible explanation for the importance attached to lighting, heating and ventilation is the immediacy of the effects of any departures from comfortable conditions. If the heating or ventilation is poor then thermal discomfort is likely. If the lighting on the work is poor then doing the work may be difficult and uncomfortable. Compared with these effects the impact of being too far from a window for a good view out, or of unpleasant decoration, is minimal. Unpleasant decoration may be disappointing but it is not likely to be a cause of discomfort.

From these data it can be concluded that lighting is one of the more important features of the physical environment. Whilst this conclusion seems reasonable it must be remembered that it is based solely on a few samples of office workers occupying similar but different buildings. The possibility that other groups of workers would give different emphases to lighting, depending on the type of activity and their expectations, still exists. Further, the method by which the data were obtained is subject to bias, the particular bias depending very much on the list of items available for ranking and the conditions which the people are experiencing. The problem with lists of factors is simply that the ranking may change drastically for different lists. As for the effect of conditions, if one particular aspect of the environment in a building is bad, then that aspect is likely to be given greater importance than if it is good. For example, in a well lit but overcrowded office [2] 'plenty of space' was given the highest ranking and the importance of lighting was considerably reduced. Given this possibility of bias it is important to note that the results shown in Table 8.2 have largely been confirmed in an independent study by Cooper et al. [5]. They asked office workers in 11 different offices, most with tinted glass windows, to rank 12 features of the physical environment in order of importance to them. Again lighting, ventilation and temperature were highly ranked, colourful surroundings and a good view out were considered to be much less important. It can be concluded, at least for workers in modern offices, that lighting is one of the more important aspects of the physical environment. It is also implied by the relative positions of the rankings of lighting and view out, both of which may be considered as parts of the luminous environment, that it is the aspects of lighting that directly affect the working area that are the more important.

An alternative approach to studying the relative importance of the different components of the physical environment is the use of factor analysis procedures, as described in Chapter 6. The idea is to use these statistical methods to identify the dimensions on which judgements of buildings are made. Then, the amount of the total variance accounted for by each dimension can be assessed; the greater the amount of variance accounted for, the more important the dimension. An example of this approach is a study of teachers' assessments of their schools [6]. In all, 500 teachers in 33 different schools were given a questionnaire concerning 12 different aspects of their schools. A factor analysis of the replies produced the loadings of each aspect on the dimensions given in Table 8.3. The three

Table 8.3: Loadings on three dimensions and percentages of variance explained for teachers' ratings of their school buildings

Building aspect	Factor loadings		
	Dimension 1	Dimension 2	Dimension 3
Satisfaction with heating and ventilation	0·92	0·03	0·05
Satisfaction with daylighting and view	0·74	0·02	0·08
Satisfaction with classroom lighting	0·62	0·4	0·09
Satisfaction with classroom overall	0·57	−0·01	0·45
Satisfaction with space in classroom	0·49	0·11	−0·50
Centrality of classroom	0·23	0·97	0·13
Convenience of classroom position	0·25	0·79	0·13
Degree of isolation of classroom	−0·09	0·42	0·06
Distractions from the corridor	−0·03	−0·02	−0·56
Noise from outside the building	−0·06	−0·11	−0·53
Distractions from other classrooms	−0·05	−0·01	−0·44
General satisfaction with school building	0·39	0·16	0·42
Percentage of total variance explained	56	25	19

dimensions can be identified from the aspects most heavily loaded on them. The factor which accounts for most of the total variance (56%) has satisfaction with heating and ventilation, daylight and view, artificial lighting in the classroom, and space in the classroom all heavily loaded on it. The next most important dimension, accounting for 25% of the variance, is related to the location of the classroom in the school. The third factor, accounting for 19% of the variance, is strongly loaded on another aspect of the physical environment, but this time one which has an intermittent effect and is outside the classroom. This dimension is essentially distraction, particularly that due to noise. Overall, these results emphasise the importance of heating and ventilation, and lighting for teachers' assessments of their classrooms. It is also interesting to note the division in satisfaction with the classroom, which is in the first dimension, and the satisfaction with the school building, which is mainly in the third dimension. The first dimension accounts for much more of the variance than the third. This again emphasises the importance people attach to the physical conditions which immediately affect their activities.

It can be seen from the above results that both direct ranking and factor analysis of ratings of satisfaction reveal considerable importance being attached to lighting. This agreement is very comforting but it should be realised that both methods have similar drawbacks. The problems of direct ranking have already been discussed. For the factor analysis approach, the dimensions discovered and the way in which the total variance is partitioned will depend greatly on the questions asked and the range of conditions experienced. Ideally any conclusions derived from either of the methods should be confirmed across investigations and across investigators before being generalised. Unfortunately there is insufficient information to do this for the physical environment. Until such evidence is available the best that can be done is to note that the two methods, used by different people in different buildings, are consistent enough to indicate that considerable importance is being attached to lighting.

To summarise, the aim here has been to locate the place of lighting amongst all the factors that need to be considered by the architect. The limited amount of data available inevitably means that the conclusions reached must be regarded as provisional. However, there is sufficient evidence to suggest that the physical environment is not the most important thing in the building user's life, but that it does have some relevance to his happiness. Lighting is one of the major features of the physical environment. Further there is an implication in the results that greater importance is attached to those lighting components which directly impinge on the activities of the occupants. Those features of lighting which are peripheral to the occupant's activities are less important.

8.3 INDIVIDUAL DIFFERENCES

Judgements of the meaning and suitability of any environmental condition are likely to vary from one individual to another. Judgements of lighting are no exception. The magnitudes of the differences in judgements, between individuals, and for the same individual at different times, are important because if they are very great there is no stability for the impression created by a particular set of lighting conditions. Then the related lighting criteria cease to be meaningful. Fortunately there is evidence to show that individual differences in judgements of lighting rarely reach this level. Chapter 6 has demonstrated some stability in the dimensional structure on which the impressions created by lighting are judged, and in the way in which different features of lighting locate an interior in the structure. Chapter 7 has revealed a less secure picture. At the extremes of individual variability are satisfaction with illuminance for office work and preferred luminance ratios. The former shows some variation but that for the latter is much greater. In fact the variation in preferred luminance ratios is so great that it comes close to making any recommendations meaningless. Discomfort glare is an intermediate case. Differences between individuals in judgements of discomfort glare are large, partly because of the vagueness of what constitutes discomfort glare, but there is no doubt that a lighting installation with a Glare Index of 13 will be judged as much more comfortable than one with a Glare Index of 26, by the majority of people. The extent of individual differences has not been ignored by those drawing up lighting criteria. For example, although the threshold for distinguishing between two glare conditions is one Glare Index unit [7], the recommendations of limiting Glare Index in the IES Code [8] are made in steps of three Glare Index units. Smaller steps would have been lost in the scatter of individual differences.

From this it would appear that the practical problems posed by individual differences in judgements of lighting are all solved. The existence of individual differences is admitted but they have been shown to be insufficient to prevent some stability in overall impression from emerging. The extent of individual differences for specific criteria can be taken into account in the spacing of the criteria. There is only one flaw in this rosy picture. It is that many of the studies on which the criteria have been based used groups of people who were far from representative of the general population, or even of the people most likely to experience the conditions being examined. This is not to suggest that the judgements of one group are inevitably different from those of another, but simply to note that they may be. Whether such differences do occur and whether they are large enough to matter remains to be investigated. Until it is, the only practical approach is to use the existing lighting criteria, but without zeal and subject to any modifications thought necessary for different groups of people. This leaves the lighting designer with a problem, namely in what way to modify the lighting criteria for different groups of people. If he is dealing with a specific individual, then there is no problem. The wishes of the individual override the criteria recommended for groups, but if a group is involved, such as the rapidly changing population of an old people's home, what should the designer do? Fortunately, for a few, very broad, demographic variables, such as age and sex, data are available which at least indicate the direction of any modification of criteria.

The first variable to be considered is that of age. As a general rule people's vision declines with age. Less light reaches the retina and a higher proportion of what does is scattered light. Older people have lower visual acuity, are more sensitive to disability glare and less sensitive to differences in colour. All these factors suggest that older people might prefer different lighting conditions to young people. Tregenza et al. [9] found an effect of age on the preferred desk illuminance in a simulated single person office, when 32 office workers were each asked to change the desk illuminance from an initial 435 lx to their preferred condition. The results obtained are shown in Fig. 8.1. The slight upward slope of the best fitting linear regression is obvious as is the large scatter in the results. The results are interesting but limited in that the number of people above 40 years old is small. Threshold measurements of the performance of the visual system show that there is little change in performance until the 50s are reached and then vision tends

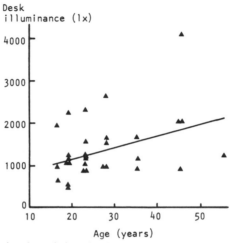

Fig. 8.1.   Linear regression through data showing the desk illuminances preferred by people of different ages [9].

to decline at an increasing rate (see Chapter 5). A limited distribution of ages may have been the reason why van Ierland [10] failed to find any significant effect of age on satisfaction with illuminance in offices, as assessed by about 2000 Dutch office workers. Any effect of age on illuminance preference is unlikely to become important until the later years.

Sex is another easily available demographic variable which has been examined but this time with little effect. van Ierland [10] again failed to find any difference between the sexes in their assessments of the illuminances on their work in the offices he studied.

Rather more relevant is the reason for people being in the space. Different people will have different judgements of the lighting conditions depending on what they are trying to do. This is nicely revealed in a survey of attitudes to sunlight in hospitals [11]. Out of 55 patients asked, 91% said they considered sunlight a pleasure, but for the 29 staff, 62% considered sunlight a nuisance. This difference is easily explained. Many of the patients were mobile and lightly dressed. They could move away from any disturbing sunlight if they wished. As for the staff, many of those in the laboratory had fixed working positions whilst the nursing staff were often engaged in quite strenuous physical work. Sunlight for the former often led to visual discomfort, and for the latter to thermal discomfort. Once again it would appear that it is how the particular lighting conditions impinge on the activities the person wishes to undertake that influences his opinions about the lighting. This implies that, where spaces are occupied by people doing many different jobs, a monolithic lighting installation may be inappropriate. Lighting should be matched to the job, so wherever areas with particular lighting needs can be identified they should be lit appropriately.

The final variable which has to be considered is that of people's expertise. A number of studies have shown that experts, in the sense of people whose business is to understand and/or design lighting or buildings, are likely to give different judgements from those who simply occupy buildings. Cockram et al. [12] examined this point in a study of people's judgements of suitability of various fluorescent light sources with different colour properties for lighting offices. Forty office staff and a number of lighting experts were asked to rank some small offices in order of preference (see Table 7.3). The rankings obtained from the two groups were very different. The authors suggest that the reason for the differences was that the office staff were influenced by the illuminance in the offices which was different for each type of light source, whereas the experts were all able to separate the colour properties of the lamps from the illuminance and hence give greater emphasis to the colour properties. In this case the difference between experts and 'naive' subjects is one of kind. Studies of sensitivity to discomfort glare have revealed differences in degree. Hopkinson [13] showed that the experienced observers he used to assess glare were much more sensitive than the general population. Specifically, the glare condition considered just acceptable by 50 % of the experienced team was judged to be just acceptable by 85 % of the sample from the general population. The implication of these results is that the designer should listen to the user with care and to the expert with some scepticism, at least when it comes to impressions of lighting in a space. Maybe once the impression has been collected, the expert will be more able to identify what needs to be done to modify the impression or improve the conditions, but it seems unlikely that he will be the best person to assess the effect.

This discussion of the demographic variables that might be important in modifying lighting criteria leaves one feeling rather uneasy. There are several indications to suggest that lighting criteria should be modified in some circumstances but insufficient information to identify precisely what those circumstances are. The most clearly determined variable is that of age, but even this seems to produce little effect until the years close to retirement are reached. The other important factor is the purpose the person has for being in the space. Lighting criteria may well require modification for different activities in the same space. Until more information is available on these and other variables there is little reason for the designer to depart greatly from the average conditions specified in lighting criteria.

8.4   LIGHT AND HEALTH

Light is a necessity for the visual system to operate but if used in the wrong way it can be injurious to health. The most common effect of light on health is the occurrence of eyestrain as a consequence of vision. However light can affect health in other ways, not related to vision. Human tissue can be damaged and a myriad of physiological processes within the human being can be influenced by light. Each of these effects of light will be considered, but the examination will be limited to the conditions likely to occur under conventional lighting.

8.4.1 Eyestrain
There is some disagreement about whether eyestrain exists at all, one side maintaining that the eye itself cannot be strained [14], whilst the other asserts that the muscles associated with the eye can, and that in any case 'eyestrain' is a useful descriptive term for a set of symptoms in much the same way as 'backache' is [15]. Really this is an argument about the usefulness of eyestrain as a concept and need not concern us here. Whether or not eyestrain exists there is no doubt about the occurrence of the symptoms which are used to identify it. These are (a) irritation of the eyes, e.g. inflammation of the eyes and lids, hot, itchy feelings about the eyes, etc.; (b) breakdown of vision, e.g. blurring of vision and double vision; and (c) the referred effects, e.g. headaches, indigestion, giddiness, etc. Most people have suffered from one or more of these symptoms at some time and know how unpleasant they can be. The symptoms are temporary but when they occur frequently then the individual can hardly be said to be enjoying the best of health.

The symptoms of eyestrain are likely to appear when the visual system is called upon to act at the limits of its capabilities for any length of time. This implies that eyestrain can occur whenever the task is visually difficult, no matter whether the difficulty arises from poor lighting, from the inherent features of the task, from inadequacies of the individual's visual system, or from some combination of these elements. Eyestrain can be caused in two ways [16]. The first is muscular strain occurring in the ocular motor system, i.e. in the muscle systems that control the fixation, accommodation, convergence and pupil size of the eyes. The second is the fatigue that occurs at higher levels of perception when there is a need to continuously interpret small, blurred and indistinct retinal images.

There is little doubt that if a task is insufficiently or improperly illuminated then eyestrain will occur. If there is insufficient light on a task involving small details it will be difficult to see, which may lead to mental fatigue. Also where there is insufficient light the natural response is to get closer to the task because this makes the detail bigger. Unfortunately it also requires greater accommodation, a smaller pupil size for each eye and more convergence between the two eyes. All these changes call for greater exertions by the muscles of the ocular motor system. If the lighting is poor in the sense that strong veiling reflections occur, the task will again be difficult to see. This time the usual response is to view the task from a different direction. In doing this, unusual directions of view and/or unusual postures may be adopted, eventually leading to strain in the extra-ocular muscles and in other muscle systems. The illuminance on the task can be too high as well as too low. If the object lit has a much higher luminance than the surround then the pupil control will be attempting to keep the pupil restricted, so again a strong muscular reaction is required. Fry and King [17] have suggested that the oscillations of the pupil control system are at least partly responsible for the discomfort experienced under some conditions of glare. However, it should be remembered that other muscles may be involved in the response to glare. Figure 8.2 shows some photographs of people trying to read a passage with a glare source nearby. The muscular contractions around the eyes are obvious and may cause discomfort after some time. Prolonged exposure to such disability glare, e.g. whilst driving, can cause mental and muscular fatigue and almost invariably results in the symptoms of eyestrain occurring. Another aspect of vision that occurs when the luminance of the task is too high is the clarity with which any discrepancies between the positions of the retinal images of the two eyes are revealed. Bedwell [18] reports that people with such fixation disparities tend to be photophobic and suffer eyestrain symptoms. Again this is to be expected. Fixation disparities can lead to mental fatigue and are often minimised by excessive use of the eye muscles to keep the two retinal images in their correct relative positions.

This observation is generally applicable. People with any form of uncorrected defect in the ocular motor system, e.g. long sight, short sight, astigmatism, squints, etc., are more prone to eyestrain than those with a well balanced ocular motor system. Even the latter may experience eyestrain if the task calls for prolonged steady fixation or for fixation in any

Fig. 8.2.   People attempting to read a passage of text placed adjacent to a glare source.

position far from the normal lines of sight or for rapid and frequent movements of the eyes.

All the above factors have involved the lighting of the task directly but there are also lighting factors which may produce the symptoms of eyestrain by distraction. An example of this is when high luminance reflections occur in a polished desk adjacent to a book one is trying to read (see Fig. 8.3). Whether the unpleasant symptoms that occur in this situation are caused by the continual correction of eye movements, and hence the greater use of eye muscles, or whether they derive from the high order sense of irritation produced by the continual distraction, or whether some combination of these two effects is involved, is unknown. What can be said definitely is that considerable discomfort will be experienced. The same applies to other forms of discomfort glare either direct from luminaires or through windows. Whenever a high luminance acts as a distraction, discomfort will be experienced and eyestrain may occur.

Flicker is another property of lighting systems which has considerable power to gather attention and hence can act as a distraction. The same remarks as were applied to reflected glare apply to flicker. But distraction is not its only possible effect. Regular light flashes, in the form of visible flicker, cause systematic variations in the electrical activity of the brain.

Fig. 8.3. A high luminance reflection in a polished desk. Such reflections can be distracting and may lead to symptoms of eyestrain occurring.

When this stimulation is strong, people who suffer from epilepsy can be sent into an attack by it [19].

To summarise, symptoms of eyestrain are produced by difficulties in interpreting poor quality retinal images, and/or by excessive use of the ocular motor system in an effort to overcome deficiencies in the visual system, or in the task, or in the lighting of the task. Too little or too much light on the task, veiling reflections, disability glare, discomfort glare (direct or reflected), and flicker, can all cause eyestrain. All this would seem to make any form of vision impossible but it should be remembered that really the visual system is extremely well suited to its function. Eyestrain is not inevitable. Carmichael and Dearborn [20] measured the eye movement patterns of people continuously reading for six hours, expecting to find signs of eyestrain. No such signs were found. Apparently the visual system is perfectly capable of prolonged activity without strain in the right conditions. Even when the conditions are not right, vision does not fail completely. Rather it protests but will rapidly recover with rest.

8.4.2 Damage to Tissue

The detrimental effects discussed above arise from attempts to use the visual system to extract information from the outside world in unsuitable conditions. The detrimental effects to be considered now can occur in the eye even when the brain is paying no attention, and on the skin even when the eyes are shut. The possibility to be considered is that of damage to the body by electromagnetic radiation as such.

8.4.2.1 Ultraviolet Radiation

For convenience the Commission Internationale de l'Eclairage has divided the ultraviolet components of the electromagnetic spectrum into three regions, UVA (400–315 nm), UVB (315–280 nm), and UVC (280–100 nm) [21]. Part of the UVA region is used to operate the visual system although the lens in the eye is a strong absorber of wavelengths shorter than 400 nm. This accounts for the conventional lower limit on the visible spectrum of about 380 nm. It has been suggested that UVA radiation is involved in the development of a certain type of cataract in the lens [22] but this possibility has not been fully investigated. A more common danger to the eye arises from the UVB and UVC wavelength regions. Radiation in these regions is absorbed by the cornea and the conjunctiva. In sufficient doses they will cause keratoconjunctivitis (or welder's flash). This is a very unpleasant but temporary condition involving severe inflammation of the cornea and conjunctiva. It is important to realise that it is the dose that counts in producing keratoconjunctivitis, i.e. the product of the irradiance and exposure time, and that reciprocity of irradiance and exposure time holds from seconds to hours. Pitts and Tredici [23] have measured the relevant action spectrum of photo-keratitis. Figure 8.4 shows the dose threshold against wavelength. The peak sensitivity occurs about 265–275 nm which is in agreement with other data [24]. Unfortunately this action spectrum does not tell the whole story. If it did it would be expected that everyone would develop keratoconjunctivitis after a few hours outdoors. In fact it is only after prolonged exposure to UVB reflected from snow that keratoconjunctivitis occurs outdoors [25]. The most probable reasons for this failure of prediction are an error in the action spectra in the region of 300–325 nm, where sensitivity is rapidly changing and measurement is difficult, and/or a breakdown in reciprocity.

The other important effect of ultraviolet radiation is on the skin. Within a few hours of exposure to wavelengths in the UVB region the skin reddens. This reddening is called erythema and extreme cases become sunburn. The classic work on erythema is that by Hausser and Vahle [26]. By exposing the skin of several individuals to ultraviolet radiation to examine the influence of wavelength, exposure time and exposure rate upon the nature, degree and course of erythema they found that some 4–8 hours after exposure, skin starts to redden. The degree of reddening continues to change with time, eventually stabilising after a few days. There is considerable variation in wavelength sensitivity for erythema but reciprocity between irradiance and exposure time has again been found to hold for times up to several hours [27]. Figure 8.4 shows the threshold dose against wavelength for *minimum*

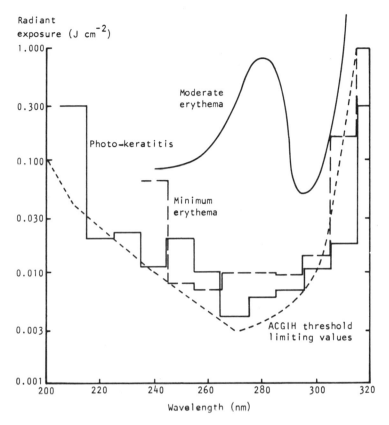

Fig. 8.4. Action spectra for photo-keratitis [23], minimum perceptible erythema [28], moderate erythema [29] and the ACGIH threshold limiting values [32].

perceptible erythema [28]. It is remarkably similar to the threshold dose for photo-keratitis. Also shown in Fig. 8.4 is the threshold dose for *moderate* erythema [29]. This curve reveals a marked increase in threshold dose overall but with strong minima occurring about 250 and 300 nm. This difference in sensitivity between minimum and moderate erythema illustrates an important point. As exposure to UVB and UVC continues, pigment migration to the surface of the skin occurs and a new darker pigment is formed. Coincident with this the outer layer of the skin thickens. All these effects screen out ultraviolet radiation and produce the socially acceptable tan.

It is just as well that this screening process occurs because chronic, i.e.

continual, exposure to high levels of ultraviolet radiation accelerates 'skin ageing' and increases the risk of developing certain types of skin cancer [30]. At present there is no agreement about the action spectra for the basal and squamous cell skin cancers that appear to be related to ultraviolet radiation and sunlight, but the ultraviolet region is thought to be involved, as there are strong correlations between the incidence of skin cancer and the local solar UVB levels [31].

It should be obvious from the above that there is a need to control people's exposure to ultraviolet radiation. This is socially very difficult out of doors, but for interior lighting, something can be done. The American Conference of Governmental Industrial Hygienists (ACGIH) have produced a recommendation [32] which limits the radiant exposure that should be experienced by people. This recommendation is supported by the UK National Radiological Protection Board [33]. The maximum permitted radiant exposure over an eight-hour period for different wavelengths is shown in Fig. 8.4. It can be seen that the ACGIH recommendation follows the photo-keratitis curve and gives a considerable safety margin over the minimum dosage for moderate erythema. This is as it should be because the eye is more sensitive to ultraviolet radiation than the skin and has no protective mechanism such as tanning.

This information is interesting, but for evaluating specific lighting equipment it is more relevant to convert the values for radiant exposure at single wavelengths to the maximum permissible time of exposure to a broad band source of known spectral irradiance. This is done by first summing the relative contribution from all wavelengths, each contribution being weighted by the relative spectral effectiveness (S_λ) (see Table 8.4). The total obtained is the effective irradiance (E_{eff}) relative to 270 nm. In other words

$$E_{eff} = \sum_{200}^{315} E_\lambda S_\lambda \Delta\lambda$$

where E_{eff} = effective irradiance relative to monochromatic wavelength 270 nm (W cm^{-2}), E_λ = spectral irradiance (W cm^{-2} nm^{-1}), $\Delta\lambda$ = bandwidth (nm), S_λ = relative spectral effectiveness (see Table 8.4). Then the maximum permissible exposure time in seconds is obtained by dividing the maximum permissible exposure at 270 nm, i.e. 0·003 J cm^{-2} by E_{eff}.

It should be noted that these limiting values only apply to the UVB and part of the UVC regions (200–315 nm), the other part of the UVC region (< 200 nm) being strongly absorbed by air. For wavelengths in the UVA

Table 8.4: Weighting functions for ACGIH recommendations [32]

Wavelength (nm)	S_λ = Relative spectral effectiveness	Wavelength (nm)	R_λ = Burn hazard function	B_λ = Blue light function
200	0·03	400	1·0	0·10
210	0·075	405	2·0	0·20
220	0·12	410	4·0	0·40
230	0·19	415	8·0	0·80
240	0·30	420	9·0	0·90
250	0·43	425	9·5	0·95
254	0·50	430	9·8	0·98
260	0·65	435	10·0	1·00
270	1·00	440	10·0	1·00
280	0·88	445	9·7	0·97
290	0·64	450	9·4	0·94
300	0·30	455	9·0	0·90
305	0·06	460	8·0	0·80
310	0·015	465	7·0	0·70
315	0·003	470	6·2	0·62
		475	5·5	0·55
		480	4·5	0·45
		485	4·0	0·40
		490	2·2	0·22
		495	1·6	0·16
		500–600	1·0	$10^{(450-\lambda)/50}$
		600–700	1·0	0·001
		700–1 060	$10 \cdot 0^{(700-\lambda)/505}$	0·001
		1 060–1 400	0·2	0·001

region (315–400 nm) the ACGIH specify that for exposure times greater than 1000 s an irradiance on the eye or skin of 1 mW cm^{-2} should not be exceeded, but for exposure times less than 1000 s a radiant exposure of up to 1 J cm^{-2} is allowed.

Finally it is worth considering under what circumstances these limits on ultraviolet radiant exposures are likely to be exceeded. Figure 8.5 shows the variation in solar irradiance on a horizontal surface for wavelengths below 313 nm throughout the day and on different clear days in Washington, DC [34]. It should be appreciated that on such clear days, around noon, exposure times of the order of minutes are all that is necessary to exceed the criteria of maximum permissible radiant exposure. In comparison the irradiances in the ultraviolet region produced by conventional light sources used in conventionally lit interiors are insignificant. Table 8.5 shows the irradiances from a number of conventional light sources measured at 1 m distance [35]. Also shown is the maximum permissible exposure time calculated by the ACGIH procedure given above. It can be seen that all

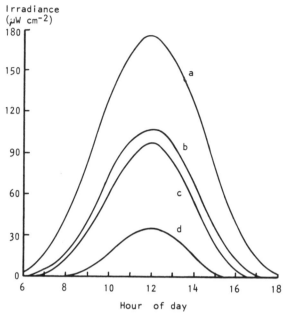

Fig. 8.5.   Irradiance from sun and sky of wavelengths less than 313 nm on a horizontal plane for four clear days in Washington, DC: a, 4th June 1943; b, 11th April 1943; c, 18th September 1943; d, 21st December 1943. (After [34].)

these times are much longer than that occurring under strong sunlight conditions, of the order of hours rather than minutes. Further, the irradiances were measured only 1 m from the bare lamp. In conventional interior lighting, such lamps are usually positioned much further from the observer, are sometimes enclosed in luminaires and are not looked at directly for long periods. All these factors will tend to diminish the radiant exposure experienced. It can be concluded that the ultraviolet radiant exposure limits are not likely to produce any problem for conventional lighting. However, there are some light sources designed specifically to produce ultraviolet radiation for various industrial, scientific and social purposes, e.g. paint curing lamps, germicidal lamps and sun tan lamps. These lamps produce considerable quantities of ultraviolet radiation, although whether the total exposure would give a large enough dose to cause problems obviously depends on the effective irradiance and the exposure duration. Whenever these light sources are to be used, protection should be considered. Guidance on the necessary administrative, engineering and personal protection measures is available [33, 36].

Table 8.5: Ultraviolet irradiances at 1 m distance and calculated maximum permissible exposure times according to ACGIH procedure for four types of discharge lamp [35]

Lamp type	Irradiances for 250–315 nm ($\mu W\,cm^{-2}$)	Calculated permissible exposure times (h)	Irradiances for 315–400 nm ($\mu W\,cm^{-2}$)	Exceeds ACGIH irradiance limit ($1000\,\mu W\,cm^{-2}$)
40 W Cool White, tubular fluorescent	0·954	21	2·9	No
400 W High pressure sodium discharge	0·164	14	18·0	No
400 W Metal halide discharge	0·480	6	126·7	No
400 W High pressure mercury discharge	0·512	3	97·6	No

8.4.2.2 Visible and Near Infrared Radiation

Radiation in the visible and near infrared region (400–1400 nm) can damage the eye. Radiation in this region is transmitted through the ocular media (see Fig. 8.6 [37]) and is absorbed in the retina. Part of the energy is absorbed by the photo-receptors initiating visual response, the rest is absorbed in the surrounding retinal pigment epithelium and choroid. The absorbed energy causes a rise in the temperature of these areas. Given enough energy the temperature can be elevated sufficiently to damage the tissue. This effect goes under the name of chorio-retinal injury. Such injuries have a long history, most deriving from looking directly at the sun. However the invention and widespread use of such powerful sources of light

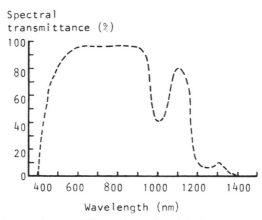

Fig. 8.6. Spectral transmittance of the ocular media of the human eye. (After [37].)

as compact arc lamps, lasers and high power flash bulbs means that nature no longer has a monopoly of high radiance sources.

The main effect of chorio-retinal injury is to destroy the part of the retina under the retinal image and hence to produce localised loss of vision. The location of the injury is important. If it occurs in the fovea then it severely interferes with vision. If it is small and occurs in the periphery it may pass unnoticed.

The probability of chorio-retinal injury by exposure to visible and near infrared radiation basically depends on the retinal radiant exposure. Whilst this is largely determined by the nature of the radiant source there are a number of intervening factors which are important. These factors are the size and quality of the retinal image, the pupil size, the spectral transmission of the ocular media and the spectral reflectance of the retina and choroid. The relevance of image size and quality is simply that tissues in the retina can much more easily conduct heat away from the point of absorption for small retinal images, say less than 50 μm diameter, than for large image sizes, e.g. 1000 μm. Therefore large retinal images are much more likely to damage the retina than will a small area of the same irradiance. All the other important factors operate by influencing the irradiance of the retina. The pupil is one of the mechanisms by which the visual system adjusts the irradiance of the retina. Thus differences between individuals in the operation of the pupil will cause differences in sensitivity to chorio-retinal injury. The same applies to the transmission of the ocular media and the spectral reflectance of the retina and choroid.

Figure 8.7 shows a compilation of retinal injury threshold data in terms of retinal radiant exposure against exposure time [38]. These data are for animals such as rabbits and monkeys. However, other work has shown reasonable agreement between the retinal radiant exposures necessary to damage the retina in these animals and in man [39]. Figure 8.7 shows that for retinal injury, radiant exposure is not the the whole story because the threshold radiant exposure varies for different exposure times. The longer the exposure time the greater is the dose required for injury, presumably because compensation by heat removal to surrounding tissue starts to operate. The curves in Fig. 8.7 can be conveniently divided into two parts, above and below an exposure time of 0·1 s. This time is of practical importance because it approximates to the time required for the operation of a simple mechanism used to protect the eye. The usual response to seeing a high brightness light is to blink and turn the head and eye away. These movements have a reaction time of 0·15–0·30 s. For exposure times below 0·15s no avoiding action is possible. Fortunately very high retinal irradiances are required to produce a damaging radiant exposure at such short times. For example, for an exposure time of 0·1 s, retinal irradiances from about 50 W cm⁻² to about 1000 W cm⁻², depending on retinal image size, are necessary for injury to occur. Only a few special light sources are capable of producing such retinal irradiances, e.g. Xenon arc searchlights and the more powerful lasers. The light sources conventionally used in

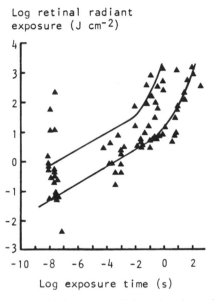

Fig. 8.7. A compilation of retinal injury threshold data plotted as retinal radiant exposure against exposure time. The upper line is for small retinal image sizes (40–70 μm); the lower line is for larger retinal image sizes (800 μm). (After [38].)

interiors produce retinal irradiances of about 10^{-4} W cm⁻². The conventional light source that comes closest to causing chorio-retinal injury in such times is the sun, with a retinal irradiance of about 10 W cm⁻².

It is now necessary to consider the situation for exposure times longer than 0·1 s. From Fig. 8.7 it can be seen that for 100 s exposures a retinal irradiance of about 10 W cm⁻² is required for chorio-retinal injury. Such irradiances are still much greater than those produced by conventional light sources but are well within the capabilities of the sun. This explains the danger of staring at the sun for any length of time.

By now it will be appreciated that establishing a criterion to avoid chorio-retinal injury calls for many factors to be considered. Nonetheless a criterion has been proposed by the ACGIH [32]. The criterion is given in the form of a limit to the weighted radiance of the radiation source. Specifically,

$$\sum_{400}^{1400} L_\lambda R_\lambda \Delta\lambda \leq \alpha^{-1} t^{-0.5}$$

where L_λ = spectral radiance of the source in bandwidth $\Delta\lambda$ (W cm^{-2} sr^{-1}), R_λ = burn hazard weighting function (see Table 8.4), t = viewing duration (s) (values of t are limited to the range 1 μs to 10 s), and α = angular subtense of the largest dimension of the source (rads).

So far the only mechanism of retinal damage by visible and near infrared radiation that has been discussed is thermal, but there is another possibility that has to be considered. This is the possibility of a phototoxic effect caused by the absorbtion of light by the photoreceptors. Following long exposure to 463 nm radiation at levels below those necessary to cause thermal damage, rhesus monkeys have suffered permanent retinal injury [40]. Further, rats and nocturnal monkeys exposed for many hours to strong broad band radiation from fluorescent lamps have also suffered retinal damage [41, 42]. Examination of the mechanisms by which such a phototoxic effect could arise suggests that the interruption of the light/dark cycle is an important factor [41]. It seems unlikely that these animal studies can be directly applied to any condition of human exposure, particularly when one considers the absence of retinal damage which outdoor workers would be expected to have experienced if the results were applicable to humans. Nonetheless, the ACGIH has proposed limits to the integrated spectral radiance of lamps in order to protect people against a possible phototoxic effect. These limits are:†

$$\sum_{400}^{1400} L_\lambda B_\lambda t \Delta\lambda \leq 100$$

for exposure times less than 10^4 s, and

$$\sum_{400}^{1400} L_\lambda B_\lambda \Delta\lambda \leq 10^{-2}$$

for exposure times greater than 10^4 s; where L_λ = spectral radiance of the

† For a source whose radiance (L) exceeds 2 mW cm^{-2} sr^{-1} in the blue region of the spectrum the permissible exposure duration in seconds is $t = 100/L$.

source in bandwidth $\Delta\lambda$ (W cm^{-2} sr^{-1}), B_λ = blue light hazard function (see Table 8.4), and t = exposure time (s).

8.4.2.3 Infrared Radiation

The CIE has treated the infrared region of the electromagnetic spectrum in the same way as the ultraviolet region, i.e. they have divided it up into three parts: conventionally IR-A (780–1400 nm), IR-B (1·4–3·0 μm) and IR-C (3·0 μm–1 mm) [21].

Figure 8.6 shows the spectral transmittance of the ocular media. It can be seen that the ocular·media absorb an increasing amount of radiant energy out to 1400 nm. For wavelengths between 1400 and 1900 nm virtually all the radiation is absorbed in the cornea and aqueous humour. Above 1900 nm the cornea is the sole absorber. The effect of the energy in the IR-A region that reaches the retina has already been considered in the discussion of chorio-retinal injury. However, the energy in the IR-A region and the IR-B and IR-C regions that is absorbed either in the ocular media or in the cornea also needs to be considered because it raises the temperature of the tissue where it is absorbed and may, by conduction, raise the temperature of other areas. Elevated temperatures in the eye are believed to play a part in the development of opacities in the lens. Fortunately extremely high corneal irradiances, of the order of 100 W cm^2 are necessary for changes in the lens to occur within the time taken to blink. Further, only 10 W cm^{-2} absorbed in the cornea will produce a powerful sensation of pain which should trigger the blink reflex. It is generally considered that the blink reflex provides protection for the eye against this effect of infrared radiation up to levels in excess of those that cause a flashburn of the skin.

This is for a very short exposure but prolonged exposure also has to be considered. Some workers exposed for 10–15 years to irradiances of 0·08–0·40 W cm^{-2} have developed cataracts [43], although whether these were caused by prolonged but small elevations of eye temperature or by occasional greater increases in temperature remains to be determined. From the practical point of view the important thing to note is that irradiances of this order will produce a marked sensation of warmth, provided more than a few square centimetres of skin are exposed. Thus when radiation heat stress is apparent the possibility of infrared radiation damage to the eye should also be considered. The basic criteria proposed for avoiding cataractogenesis is very simple [32]. The irradiance at the eye for wavelengths above 770 nm should be less than 10 mW cm^{-2}.†

† For a near infrared heat source where a strong visual stimulus is absent the near infrared radiance (770–1400 nm) should be less than 0·6/α, where α is the angular subtense of the source (in radians).

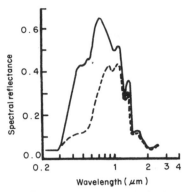

Fig. 8.8.   Spectral reflectance of white (solid line) and black (dashed line) human skin. (After [44, 45].)

As for the skin, the effect of visible and infrared radiation is simply to raise the temperature of the skin. If the temperature elevation is sufficient, then burns will be produced. It is important to realise that the focusing process of the eye makes it much more sensitive than the skin to such injury for visible radiation and IR-A infrared radiation. However, the skin and eye are equally at risk for the IR-B and IR-C radiation because the ocular media are virtually opaque for these wavelengths and the mechanism of injury is thermal. The efficiency with which a given irradiance raises the temperature of the skin depends on its area and the reflectance of the skin. Figure 8.8 shows measured skin spectral reflectances for black and white skins [44, 45]. Although they differ in the visible and the IR-A regions, both tend to low reflectances above 2000 nm. The threshold irradiance for thermal injury to the skin is greater than 1 W cm^{-2} and is highly dependent on the exposed skin area and exposure time. Such irradiances are difficult to produce over large areas. Direct solar irradiance is about 40–70 mW cm^{-2} in northern latitudes, going up to about 100 mW cm^{-2} in the tropics. Illuminances produced in interiors by conventional lighting have total irradiances of about 1 mW cm^{-2} associated with them [46]. It can be concluded that sunlight and conventional lighting in interiors are unlikely to produce any degree of thermal injury to the skin. In any case, for anything other than very short exposure times, considerations of heat stress become relevant before such damage can occur. Exposure criteria for avoiding heat stress are available [32].

8.4.2.4 Practical Exposure Control
As shown above, there exist various criteria aimed at controlling the

possible damage to tissue by ultraviolet, visible and infrared radiation. Full details of these and other criteria are available in the ACGIH recommendations [32]. These criteria are in the form of threshold limiting values, which represent conditions to which it is believed that nearly all workers may be repeatedly exposed without adverse effect. When faced with a situation where any of these criteria is exceeded, the ideal approach is to reduce the output from the radiating device to below the necessary limit. Unfortunately this is not always possible. Often the potentially damaging radiation is an inevitable product of the work being done, e.g. in furnaces, or is there to produce a particular effect on some component of the work, e.g. ultraviolet curing. In these circumstances a degree of personal protection is required. This can take the form of screening the source with suitable materials, i.e. those opaque to the damaging radiation, and/or personal protection in the form of eye filters, helmets, clothing, etc. Advice is widely available on suitable methods of protection [33, 38, 47, 48]. It is important to emphasise that conventional lighting of interiors is extremely unlikely to cause any damage to the eye or skin. However, some special light sources used for industrial or research purposes do need care. For example, lasers are often powerful enough to produce burn damage to eyes and skin within the time of the blink reflex. Lasers are outside the interests of this book but, for anyone who is concerned, there is plenty of authoritative advice available on the safety aspects of their use [32, 38, 48, 49].

8.4.3 Other Non-Visual Effects of Light

All the above has been concerned with the potentially damaging effects of excessive radiation on tissue. But too little light can also be harmful. For example, a group of normal elderly males deprived of ultraviolet light for three months developed an impairment in the ability of their intestinal mucosa to absorb calcium, whilst a control group exposed to a simulated solar spectrum did not [50]. This result can be explained by the formation of vitamin D in the skin and subcutaneous tissue that is known to occur upon exposure to ultraviolet light.

It is one example of the beneficial effects of light. Another is the bleaching of bilirubin by light. This is used as a clinical treatment but its implications are more important than that. If the plasma level of one endogenous component can be influenced by light, either natural or artificial, are there any others that are similarly affected, and if so are the effects beneficial or not?

Both the production of vitamin D and the destruction of bilirubin occur through the skin. However there is evidence that light absorbed by the retinal photoreceptors can influence physiological events other than vision. Certainly such physiological responses as body temperature and hormone secretions vary with a 24-hour rhythm related to the dark/light cycle. As if this were not enough there is also evidence that light received by the photoreceptors of mammals influences the secretion of such endocrine organs as the pituitary, the gonads and the adrenals [51]. This brief discussion should be sufficient to indicate the possibility of the non-visual effects of light. What it does not do, and cannot do because of lack of information, is to detail what features of lighting are important for these effects and to consider whether these effects are only the tip of the iceberg. The fact is we do not know. The non-visual effects of light is a territory which is only starting to be systematically explored, an event indicated by the creation of a CIE Technical Committee on this subject. At the moment the few areas that have been examined in detail have usually been those of interest to the medical profession and have resulted in clinical treatments for a variety of infections and malfunctions. Among the better known are the following.

8.4.3.1 Rickets

Rickets is a disorder of the bones in which there is imperfect calcification so that the bones are not rigid enough to support the body. The role of light in the treatment of rickets arises because ultraviolet radiation converts some of the chemicals in the skin to vitamin D which is then absorbed into the bloodstream. Vitamin D̊ is essential to healthy bone growth. The irradiances required in the treatment of rickets are much less than those required to produce erythema.

8.4.3.2 Hyperbilirubinemia

Hyperbilirubinemia, or neonatal jaundice, affects some newly born babies. The problem is that there is too much bilirubin in the blood which can lead to brain damage. Normally the bilirubin level is controlled by the liver but in new born babies the liver does not operate efficiently. In the past if the bilirubin level rose sufficiently it was necessary to change the blood by transfusion, with all its attendant risks. Light now provides an alternative. It has been established that light in the visible spectrum breaks down bilirubin and that the resultant by-products are harmlessly and easily excreted [52]. Further, the amount of light radiation required can be easily achieved with conventional light sources.

8.4.3.3 Herpes Viruses

This infection is most commonly seen as cold sores. The problem with viral infections is to destroy the virus without damaging the host. In this case irradiating the sore with light from a fluorescent lamp after the application of a photosensitisor to the affected area will destroy the infection.

8.4.3.4 Psoriasis

This is a common chronic skin disease which can vary in magnitude from the presence of a few red patches to completely covering the body with sores. Parrish et al. [53] developed a simple treatment which consists of the administration of a photosensitisor by mouth, waiting two hours and then exposing the patient to high intensity UVA radiation with a peak of about 365 nm, for about 10–15 minutes. Irradiances of about $6\,mW\,cm^{-2}$ are used. The treatment is very effective but is not a cure. The patient needs a booster treatment at regular intervals.

8.4.4 Conclusions

Although the visual system is an extremely precise yet robust mechanism it can be damaged by the very medium that allows it to operate—light. Excessive exposure to ultraviolet, visible and infrared radiation can cause health problems in the eye. The same radiation can also damage the skin. Fortunately there exist well established limits, which if followed will ensure that damage does not occur.

However, even if the radiation as such is not damaging, poor lighting can cause eyestrain. The lighting conditions which are likely to lead to eyestrain are well known and can readily be avoided.

It might appear from the above that the relationship between light and health is completely understood but this is deceptive. A great deal is known about the specific problem of how light affects the eye and skin but lurking in the background is the effect of light on more remote but important physiological activities. Light is known to be involved in the timing of metabolic and glandular responses in mammals [54]. Further, experiments on a number of different creatures has suggested that light influences all sorts of physiological responses [51]. A fairer overall conclusion would be that the direct and obvious effects of light on the human visual system and the skin are well understood but the deeper effects on general physiological responses have barely been explored yet. It may well be that when they are, a different orientation will be given to lighting equipment and design.

8.5 SUMMARY

This chapter is concerned with expanding the framework within which the appreciation of light is considered. It does this by examining three distinct areas. First the importance of lighting in the overall appraisal of a building is assessed. Second the importance of individual differences in preferred lighting criteria is examined. Third the limitations imposed on lighting by health requirements are detailed.

Lighting is part of the physical environment but the limited data available indicate that for everyday work, the physical environment is less important than financial and social factors. Nonetheless the physical environment does have some importance, and lighting, particularly those elements of lighting which directly impinge on the activities people wish to do, is considered to be one of its major components.

Differences between individuals in the appreciation of lighting are large but not sufficiently large to prevent some stability in the impressions created by lighting conditions from emerging. The magnitude of individual differences is allowed for to some extent in lighting criteria by varying the spacing of steps in each criterion. The limitation of this procedure is that the information on which the spacing is based has often been obtained from unrepresentative groups of people. The problem is to know how differences between groups are likely to affect preferred lighting conditions. Of the two demographic variables, age and sex, only age has been shown to be important. Generally old people prefer higher illuminances than young people. Other personal factors which affect the appreciation of lighting are the reasons people have for being in the space and the degree of expertise they have in lighting. What the person is trying to do in the space may influence his appreciation of the lighting. As for expertise, experts tend to be more sensitive and can separate different lighting conditions more easily than the general population.

Light is necessary for the visual system to operate but it can also be injurious to health. Excessive radiation, in the ultraviolet, visible and infrared region of the electromagnetic spectrum can cause tissue damage to eyes and skin which varies from the temporary and minor to the permanent and serious. There exist criteria which set limits to the irradiance and/or the radiant exposure received by people for all these wavelength regions. The limits are set so as to avoid adverse effects even for repeated exposure. Conventional lighting is extremely unlikely to exceed these criteria, the light source which comes closest to doing so being the sun.

Other aspects of light and health considered are the occurrence of eyestrain and the non-visual effects of light other than tissue damage. The symptoms of eyestrain are caused by attempts by the visual system to improve and identify poor quality visual information. The occurrence of eyestrain is related to deficiencies in the visual system, any inherent difficulties in the task, and the quality of the lighting, either separately or in combination. Too little or too much light, veiling reflections, disability and discomfort glare and flicker can all cause eyestrain.

The other non-visual effects of light examined are largely a step into the unknown. Light influences the activities of numerous glands in the human body and has been used as a treatment for some infections and malfunctions. What remains a mystery is whether many other non-visual effects of light occur and how important they are.

REFERENCES

1. Canter, D. A. Priorities in building evaluation: some methodological considerations, *J. Arch. Res.*, **6**, 38, 1977.
2. Boyce, P. R. *Staff reactions to moves into integrated design offices: final report on the move to the Bedford District IED office, Caxton Road, Bedford*, Electricity Council Research Centre Report R800, Capenhurst, UK, 1975.
3. Boyce, P. R. *Staff reactions to the lighting of the IED office at YEB Headquarters, Scarcroft, Leeds*, Electricity Council Research Centre Report R1164, Capenhurst, UK, 1978.
4. Boyce, P. R. *Staff reactions to the physical conditions in the Grampian Regional Council Offices, Woodhill House, Aberdeen*, Electricity Council Research Centre Memorandum M1307, Capenhurst, UK, 1980.
5. Cooper, J. R., Wiltshire, T. and Hardy, A. C. Attitudes towards the use of heat rejecting/low light transmission glasses in office buildings, *Proc. CIE Conference, windows and their function in architectural design, Istanbul*, 1973.
6. Canter, D. A. *The development of scales for the evaluation of buildings*, University of Strathclyde, Glasgow, 1971.
7. Collins, W. M. The determination of the minimum identifiable glare sensation interval using a pair comparison method, *Trans. Illum. Engng Soc. (London)*, **27**, 27, 1962.
8. Illuminating Engineering Society. *IES Code for Interior Lighting*, IES, London, 1977.
9. Tregenza, P. R., Romaya, S. M., Dawe, S. P. and Tuck, B. Consistency and variation in preferences for office lighting, *Ltg Res. and Technol.*, **6**, 205, 1974.
10. van Ierland, J. F. A. A. *Two thousand Dutch office workers evaluate lighting*, Research Institute for Environmental Hygiene TNO, Publication 283, Delft, Netherlands, 1967.
11. Ne'eman, E., Craddock, J. and Hopkinson, R. G. Sunlight requirements in buildings—1. A social survey, *Building and Environment*, **11**, 217, 1976.
12. Cockram, A. H., Collins, J. B. and Langdon, F. J. A study of user preferences for fluorescent lamp colours for daytime and night-time lighting, *Ltg Res. and Technol.*, **2**, 249, 1970.
13. Hopkinson, R. G. *Architectural Physics: Lighting*, HMSO, London, 1963.
14. Cogan, D. G. Lighting and health hazards, in *NIOSH Symposium, The occupational, safety and health effects associated with reduced levels of illumination, Cincinatti*, 1974.

15. Fry, G. A. Ocular discomfort and other symptoms of eyestrain at low levels of illumination, in *NIOSH Symposium, The occupational, safety and health effects associated with reduced levels of illumination, Cincinatti*, 1974.
16. Duke-Elder, S. and Abrams, D. Eyestrain, in *System of Ophthalmology, vol. 5*, ed. Sir Stewart Duke-Elder, Henry Kimpton, London, 1970.
17. Fry, G. A. and King, V. M. The pupillary response and discomfort glare, *J.I.E.S.*, **4**, 307, 1975.
18. Bedwell, C. H. The eye, vision and visual discomfort, *Ltg Res. and Technol.*, **4**, 151, 1972.
19. Jeavons, P. M. and Harding, G. F. A. *Photosensitive Epilepsy*, Heinneman, London, 1975.
20. Carmichael, L. and Dearborn, W. F. *Reading and Visual Fatigue*, Greenwood Press, Connecticut, 1947.
21. Commission Internationale de l'Eclairage. *International Lighting Vocabulary*, CIE Publication 17, 1970.
22. Zigman, S. Eye lens colour: formation and function, *Science*, **171**, 807, 1971.
23. Pitts, D. G. and Tredici, T. J. The effect of ultra-violet on the eye, *Amer. Ind. Hyg. Ass. J.*, **32**, 235, 1971.
24. Leach, W. M. *Biological aspects of ultraviolet radiation: a review of hazards*, US Public Health Service, Bureau of Radiological Health, BRH/DBE 70-3, Rockville, Maryland, USA, 1970.
25. Hedblom, E. E. Snowscape eye protection, *Arch. Environ. Health*, **2**, 685, 1961.
26. Hausser, K. W. and Vahle, W. Sunburning and suntanning, in *The Biologic Effects of Ultra-violet Radiation*, ed. F. Urback, Pergamon, Oxford, 1969.
27. Schmidt, K. Comparison of intermittent and continuous UV irradiation in induction of skin erythema, *Strahlentherapie*, **121**, 383, 1963.
28. Freeman, R. G., Owens, D. W., Knox, J. M. and Hudson, H. T. Relative energy requirements for an erythemal response of skin to monochromatic wavelengths of ultraviolet present in the solar spectrum, *J. Invest. Dermatol.*, **47**, 586, 1966.
29. Coblentz, W. W., Stair, R. and Hogue, J. M. The spectral erythemic reaction of the human skin to ultra-violet radiation, *Proc. Nat. Acad. Sci. (US)*, **17**, 401, 1931.
30. Freeman, R. G., Hudson, H. T. and Carnes, R. Ultra-violet wavelength factors in solar radiation and skin cancer, *Int. J. Dermatol.*, **9**, 232, 1970.
31. MacDonald, E. J. The epidemiology of skin cancer, *J. Invest. Dermatol.*, **32**, 379, 1959.
32. American Conference of Governmental Industrial Hygienists. *Threshold limit values for chemical substances and physical agents in the workroom environment*, Cincinnati, USA, 1979.
33. National Radiological Protection Board. *Protection against ultraviolet radiation in the workplace*, HMSO, London, 1977.
34. Coblentz, W. W. and Stair, R. A daily record of ultraviolet solar and sky radiation in Washington, 1941 to 1943, *Journal of Research of the National Bureau of Standards*, **33**, 21, 1944.
35. Spears, G. R. Radiant flux measurements of ultraviolet emitting light sources, *J.I.E.S.*, **4**, 36, 1974.
36. Hughes, D. *Hazards of occupational exposure to ultraviolet radiation*, Occupational Hygiene Monograph 1, University of Leeds, UK, 1978.
37. Geeraets, W. J. and Berry, E. R. Ocular spectral characteristics in relation to lasers and other light sources, *Amer. J. Ophthal.*, **66**, 15, 1968.
38. Sliney, D. H. Non-ionising radiation, in *Industrial Environmental Health*, ed. L. V. Cralley, Academic Press, London, 1972.
39. Geeraets, W. J. and Nooney, T. W. Observations following high intensity white light exposure to the retina, *Am. J. Optom. and Arch. Am. Acad. Optom.*, **50**, 405, 1973.
40. Harwerth, R. S. and Sperling, H. G. Effect of intense visible radiation on the increment threshold spectral sensitivity of the rhesus monkey eye, *Vision Research*, **15**, 1193, 1975.

41. Noell, W. K. and Albrecht, R. Irreversible effects of visible light on the retina: role of vitamin A, *Science*, **172**, 76, 1971.
42. Kuwabara, T. and Gorn, R. A. Retinal damage by visible light, *Arch. Ophthal.*, **79**, 69, 1968.
43. Matelsky, I. The non-ionising radiation, in *Industrial Hygiene Highlights*, Vol. 1, Industrial Hygiene Foundation of America, Pittsburgh, USA, 1968.
44. Jacquez, J. A., Huss, J. A., McKeehan, W., Dimitroff, J. M. and Kuppenheim, H. F. Spectral reflectance of human skin in the region 0·7–2·6 μm, *J. Appl. Physiol.*, **8**, 297, 1955.
45. Jacquez, J. A., Kuppenheim, H. F., Dimitroff, J. M., McKeehan, W. and Huss, J. A. Spectral reflectance of human skin in the region 235–700 nm, *J. Appl. Physiol.*, **8**, 212, 1955.
46. McIntyre, D. A. Radiant heat from lights, and its effect on thermal comfort, *Ltg Res. and Technol.*, **8**, 121, 1976.
47. Davis, A., Diffey, B. L. and Magnus, I. A. Personal monitoring of ultraviolet radiation, in *Light Measurement in Industry*, SIRA, London, 1978.
48. American National Standards Institute. *American national standard for the safe use of lasers*, Z136.1, New York, 1976.
49. Laser Institute of America. *Laser safety guide*, 4th edition, 1976.
50. Neer, R. M., Davis, T. R. A., Walcott, A., Koski, S., Schepis, P., Taylor, I., Thorington, L. and Wurtman, R. J. Stimulation by artificial lighting of calcium absorbtion in elderly human subjects, *Nature*, **229**, 255, 1971.
51. New York Academy of Sciences. Photo-neuro-endocrine effects of circadian systems with particular reference to the eye, *Annals of the New York Academy of Sciences*, **117**, 11, 1964.
52. Cremer, R. J., Perryman, P. W. and Richards, D. H. Influence of light on the hyperbilirubinaemia of infants, *The Lancet*, 24 May, 1094, 1958.
53. Parrish, J. A., Fitzpatrick, T. V., Tannebaum, L. and Pathak, M. A. Photo-chemotherapy of psoriasis with oral methoxsalen and longwave ultraviolet light, *New England Journal of Medicine*, **296**, 1207, 1974.
54. Wurtman, R. J. The effect of light on man and other mammals, *Annual Review of Physiology*, **37**, 467, 1975.

Part 4:
Consequences

9

Standards, Codes and Guides

9.1 INTRODUCTION

Lighting research enters the lives of most people through the medium of documents specifying standards of lighting to be achieved or offering guidance on lighting practice. Such documents, which occur in profusion in developed countries, influence the lighting of everything from roads to tennis courts, from factories to schools. It has been suggested that there is far too much advice available on lighting requirements and that this leads to an undesirable uniformity of style. Lam [1] believes that, in interiors at least, the role of lighting should be to reveal the form of the architecture and thereby to fulfill man's need for unambiguous perception. He asserts that there are many ways of doing this which are known to anyone with practical experience of lighting and that precise guidance and standards are inappropriate. He also avers that one way of not revealing form in architecture is to follow slavishly recommendations which simply specify illuminances on a mythical horizontal working plane. There is much truth in this view, as anyone who has contemplated the boring uniformity of many modern lighting schemes will realise. However, there are some functions of lighting which are too important to be left to artists/architects; and there are some applications in which revealing the architectural form is very much a secondary consideration. An example of the first case is safety lighting. Statutory requirements have existed for safety lighting in theatres and cinemas for many years [2]. This lighting has to be incorporated into the design alongside lighting which does indeed reveal the architectural form of the building. An example of the second case is industrial lighting. Anyone who thinks industrial lighting is about anything other than lighting the work at minimum cost is deceiving himself. Further, given the variety of positions, detail and form of tasks that can occur in a factory, uniform lighting across a hypothetical working plane is often the most practical approach to defining the necessary lighting conditions.

It also has to be admitted that much lighting is actually specified and/or installed by people who are not architects, artists or lighting engineers and who have little interest in lighting as such. Very often lighting is only a minor aspect of their work. These people constitute what might be called the 'watts per square metre' school of lighting design. For them, simple guidance on the standards to be adopted and suitable design methods is essential if poor lighting is to be avoided.

Thus there is a place for standards, codes and guides to lighting practice. But insisting that they are applicable in all circumstances and at all times would be as wrong as asserting that they are all irrelevant. In the field of lighting publications, as in so much lighting practice, it is very much a matter of 'horses for courses'. This chapter will examine some of these 'horses', their breeding, their faults and their virtues.

9.2 FORM

Before writing a standard, code or guide for lighting there are a number of decisions which have to be taken. These decisions will determine the form of the document and the importance that is likely to be attached to it. The first two matters for decision are shown in Table 9.1. This table classifies documents according to the criteria of legal status and precision. Documents can be statutory or non-statutory. They can also contain precise or imprecise recommendations. Each square in Table 9.1 is filled with an example of a document that fits the particular description. *The slaughterhouses (hygiene) regulations* [3] is an example of a statutory document which gives precise recommendations about the minimum illuminance that should be provided in slaughterhalls and workrooms

Table 9.1: A classification of form for lighting documents

Precision	Legal status	
	Statutory	*Non-statutory*
Precise recommendations	Statutory Instrument 2168, *The slaughterhouses (hygiene) regulations*	*IES Code for interior lighting* (standard service illuminances)
Imprecise recommendations	*Offices, shops and railway premises act*, (1963)	*IES code for interior lighting* (veiling reflections)

(215 lx) and where meat inspection is carried out (540 lx). The *Offices, shops and railway premises act* (1963) is an example of a statutory document without precise recommendations on lighting. The wording given is simply that lighting should be 'sufficient and suitable'. What is 'sufficient and suitable' is left to legal precedent to establish. Examples of both precise and imprecise, non-statutory lighting recommendations can be found in a single document, for example the 1977 *IES code for interior lighting* [4]. The recommendations given on illuminances for particular activities are extensive and exact. In comparison, the advice given about veiling reflections is imprecise. It is confined to demonstrating what veiling reflections are, and how they are produced. No limits are set for their magnitude. The advice given is that veiling reflections are a bad thing and should be avoided as far as possible.

These different levels of legal status and precision represent different ways of reconciling the consequences of the lighting being poor, with the need to make any criterion simple to understand and implement, and applicable to a variety of conditions. The more serious the consequences of failure, the more definite the recommendations need to be, i.e. statutory and more precise. Thus, some lighting standards in slaughterhouses are essential because without good lighting public health is at risk. In comparison, the consequences of failing to provide exactly the recommended illuminances for many other types of work are small. It may be that the work will be done less well or with greater difficulty but public health is not likely to be affected. However, given a large enough reduction from the recommended standards, eyestrain may occur repeatedly and safety measures may be compromised. It is for this reason that the phrase about 'sufficient and suitable' lighting occurs in the *Offices, shops and railway premises act*. It is effectively a long stop against extremely poor standards of lighting. Usually 'sufficient and suitable' lighting is taken to be anything with an illuminance of about half the values recommended as good practice [5].

The vagueness of specification represented by the phrase 'sufficient and suitable' lighting may be thought to be unsatisfactory but a lack of enthusiasm for determining lighting practice by legislation is desirable, for two reasons. First, exact specification tends to lead to a single design solution being adopted. This in turn tends to produce a boring uniformity of approach and to inhibit equipment development. Second, a statutory recommendation may have unexpected and undesirable side effects. An example of this occurred when a 2% daylight factor was specified as the minimum daylight factor that should be provided anywhere in an area normally used as teaching accommodation [6]. In order to achieve this, schools were designed with large windows and roof lights. In consequence many suffered from solar overheating. In this case the statutory element of the recommendation was excessive. There was no danger to life or limb in using less daylight. Simple advice on how daylight could be used would have been sufficient, the decision on the extent to which it should be used being left to the designer. In fact, the statutory requirement for a 2 % daylight factor distorted the design of schools for many years.

Having decided on the legal status of the document and the level of precision required, the next decision concerns the form the recommendations should take. The two extremes here are documents which specify what the lighting should do and those which specify what the lighting should be. The former codifies the ends, the latter the means. An example of a code which specifies the ends is the German DIN Standard [7]. Essentially this confines itself to specifying lighting criteria appropriate for different applications. It does not say how the criteria should be achieved. The *Code of practice for road lighting* follows the opposite approach [8]. Essentially it forms a recipe book in which the spacing of lanterns of different types on different road layouts is displayed.

The choice between these two extremes is influenced by the diversity of the situations being considered. Interior lighting covers a wide range of conditions, so interior lighting guidance is usually expressed in terms of what should be achieved. The situation for roadway lighting is rather different in that the requirements are much more consistent. Indeed, uniformity of design for different roadway types can be a useful aid to the driver in interpreting the road ahead. By specifying how the roadway lighting should be achieved, uniformity of design is assured.

Although this division between ends and means is useful for understanding, it should be realised that a complete division is the exception rather than the rule. Much design in both interiors and exteriors is done by people with limited experience of lighting. For this reason most documents [4, 9, 10] recommend conditions to be achieved and give simple design methods and lists of available equipment by which the recommendations can be met. The reader is then left to choose whether to exercise his skills or to follow the simple design methods. This compromise is a reasonable solution to what is a difficult problem.

Another point which is determined by the anticipated audience for the document is the extent of any explanation given for the recommendations. Obviously there is a greater incentive for more explanation in documents which are to be followed voluntarily, but even here there is something to be said for keeping the explanation simple. The best approach is to give the advice unambiguously with references to other documents which provide some justification for the proposals. However, the extent of justification necessary for any advice is also variable. For something, such as emergency lighting, where the consequences of any failure are likely to be grave, some guidance is essential so that the recommendations made can be based on experience and good judgement. Where · recommendations are not essential, such as for veiling reflections, then strong evidence is needed before precise recommendations are made. In a perfect world each recommendation could be completely justified, but this is not a perfect world. Recommendations are made with varying degrees of backing, from prejudice to informed opinion to clear experimental evidence. It has been one of the purposes of this book to indicate the limits of the evidence available for many lighting recommendations.

This mention of the evidence behind recommendations raises the much bigger question of how the particular criteria to be included and the levels of each criterion to be recommended are decided. The basic consideration for including a factor is that it should help to make the lighting suitable for its purpose. Different applications will have different purposes and hence give different weight to different aspects of lighting. For industrial lighting the priority is for the lighting to enable people to work quickly, easily and accurately. For the lighting of a hotel foyer, the impression created is the first priority, and assisting work performance is a poor second. Thus the recommendations for different applications can be expected to contain different criteria and different levels of the same criteria.

In general the number of criteria used in lighting recommendations has increased over the years. This is exemplified by the IES code. When this was first published in 1936 it consisted of little more than a list of illumination levels, as illuminances were then called. In 1961 a major change was made with the introduction of criteria to control discomfort glare in the form of the Glare Index system. In the latest edition [4] the criteria included for each application are the illuminances on the task, the limiting Glare Index and a range of light sources suitable for that situation. In addition criteria are given for the uniformity of illuminance across a space, the range of reflectances suitable for room surfaces and the relative illuminances on those surfaces. This increase in the number of criteria considered reflects the greater understanding of the factors that ensure comfortable lighting. The increase in number of criteria has probably gone about as far as it can with the possible exception of a criterion for veiling reflections.

In spite of this increase it remains a fact that the most fundamental

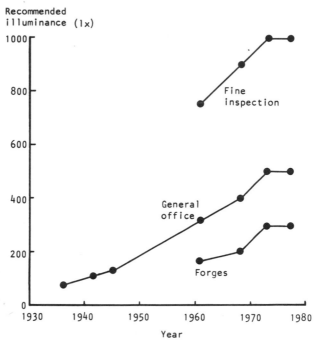

Fig. 9.1.   Illuminances recommended for fine inspection work, general offices, and forges, in successive IES codes.

criteria for all codes of lighting practice are illuminance recommendations. These were the first criteria thought necessary. They are the basis of the simple lumen method of design. They are also the values over which much argument arises, the reason for which can be seen from Fig. 9.1. This shows the changes in illuminances recommended by the IES for office work, fine inspection work and work in forges, over the years since the IES code was first published. The steady upward trend in recommended illuminances is apparent. This trend is viewed with grave suspicion by some people who see it as being due to commercial influence. The question they ask is 'if about 100 lx was sufficient for offices in 1936 why is 500 lx regarded as necessary today?' The answer to this question is that today is not 1936. The nature of work in offices has changed since 1936, the means of providing the lighting has changed, the form of the recommendations has changed and people's expectations have changed. This last point is particularly important. Similar trends to that shown in Fig. 9.1 can be found in many other environmental criteria, e.g. temperatures recommended for thermal comfort.

The changing standards reveal a fact about all lighting recommendations that should always be remembered. Lighting recommendations are not immutable. They are not like the laws of physics, nor are they written on tablets of stone. Rather they represent the best efforts of groups of people to decide on reasonable lighting recommendations in the prevailing conditions. To reach this decision for any particular application, a number of different factors have to be considered. First, there are what might be called the aims of the lighting. For each application it is necessary to consider the relative weights to be given to task visibility, task performance, observer comfort and perceived impression. Once the aims of the lighting have been specified the necessary lighting conditions can be derived from any available experimental evidence and from practical experience. Second, there is the extent to which the lighting desired can be achieved with available equipment. There is little point in recommending lighting which cannot be achieved. Third, the economics need to be assessed. What does it cost to produce the recommended lighting from the available equipment? These three factors: the desired lighting, its technical possibility and its cost are all considered before making a decision about lighting criteria. To appreciate the relevance of this it is only necessary to remember that whilst the recommended illuminances have increased over the years (Fig. 9.1) there have also been changes in the type and luminous efficacy of the light sources used to provide the illuminances [11]. For example, in 1936 offices were generally lit by incandescent lamps with a luminous efficacy of about $10 \, lm \, W^{-1}$. Today, tubular fluorescent lamps are used with luminous efficacies of about $65 \, lm \, W^{-1}$. Thus the increase in recommended illuminances has been matched by an increase in luminous efficacy of the light sources used.

The point to grasp from all this is that recommended illuminances and levels of all other lighting criteria are matters of judgement, involving the balancing of several factors. They represent the consensus view of what is reasonable for the conditions prevailing when they are written. The problem this process creates is that different groups may reach different conclusions, as will be evident from the following review of interior and exterior lighting documents.

9.3 INTERIOR LIGHTING CODES

The aim here is to examine the contents of documents concerning interior lighting which are used by lighting engineers and architects. To include every such document would be impossible in the space available so only those which are widely used in the English speaking world will be considered. There is sufficient variety in this group of publications to indicate most of the problems that the authors and readers of such documents have to face. This examination will not simply be a list of what each document contains. Rather it will attempt to contrast the approaches used in the documents.

9.3.1 CIE Publication 29, Guide on Interior Lighting [12]

This document was published by the Commission International de l'Eclairage (CIE) in 1975. It results from the efforts of an international committee, and, as such, could be said to represent the most basic consensus about lighting. Its basic nature arises because all international committees, and many national committees, have to operate by seeking out those areas that people can agree on and edging round those areas where there is passionate argument. This process tends to produce documents which are definite in some parts but vague in others. The CIE guide is no exception. Table 9.2 shows one area of general agreement, namely a scale of illuminances recommended for use in working interiors. The scale is divided into three parts. Illuminances in the range 20–200 lx are recommended for infrequently occupied rooms; illuminances from 200–2000 lx are suggested for general lighting in working interiors, whilst illuminances covering a range of 2000–20 000 lx, which may be needed for very exacting visual tasks, are suggested as best produced by local lighting. Whilst there is general agreement on this scale, there is no suggestion as to where any specific type of work should lie on the scale. There is some indication of the region of the scale that would be appropriate but this region covers about three steps of the scale and a ratio of about 2:1 in illuminance. This is to be expected given the technical and economic considerations that are involved in selecting recommended illuminances. Different countries will have different lighting costs and different attitudes to that cost. They will have access to technology at different levels and different expectations, so it is not too surprising that different countries will recommend different illuminances for similar tasks.

An area where different national viewpoints have prevented a common recommendation being presented is in limiting discomfort glare. As discussed in Chapter 7, there are a number of different approaches to predicting discomfort glare sensation and hence to framing recommendations that will limit its occurrence. Some of these views are held with theological fervour. From the CIE guide it would appear that the only way

Table 9.2: Recommended illuminances for interiors from the CIE guide on interior lighting 1975 [12]

Range	Recommended illuminance (lx)	Type of activity
A. General lighting for areas used infrequently or having simple visual demands	20 30 50	public areas with dark surroundings
	75	simple orientation for short temporary visits only
	100 150 200	rooms not used continuously for working purposes, e.g. storage areas, entrance halls
B. General lighting for working interiors	300	tasks with limited visual requirements, e.g. rough machining, lecture theatres
	500 750 1 000	tasks with normal visual requirements, e.g. medium machining, offices
	1 500 2 000	tasks with special visual requirements, e.g. hand engraving, clothing factory inspection
C. Additional lighting for visually exacting tasks	3 000 5 000 7 500	very prolonged and exacting visual task, e.g. minute electronic and watch assembly
		exceptionally exacting visual tasks, e.g. micro electronic assembly
	10 000 15 000 20 000	very special visual tasks, e.g. surgical operations

agreement could be reached on something to say about discomfort glare was to devise yet another system and call it a 'simplified interim method'. The method involves two parts. The first is based on the luminance limit method developed by Fischer [13] but in a simplified form. The second is a series of shielding angles applicable to open bottomed, cut-off luminaires, the method being developed in Australia [14]. Between them these two methods are regarded as providing 'a simple safeguard against inadequate glare control in many common situations. They should not be regarded as a substitute for the more comprehensive evaluation methods used in some countries.' That says it all.

Recommendations are also made on the range of reflectances suitable for various room surfaces and the limits to ratios between the illuminances on these surfaces and on the horizontal working plane. These values largely follow the recommendations of the 1973 IES code. This agreement with one

Table 9.3: Lamp colour rendering groups from the CIE guide on interior lighting 1975 [12]

Colour rendering group	Range of general Colour Rendering Index (R_a)	Colour appearance	Examples of use
1	$R_a \geq 85$	Cool	Textile industries, paint and printing industries
		Intermediate	Shops, hospitals
		Warm	Homes, hotels, restaurants
2	$70 \leq R_a < 85$	Cool	Offices, schools, department stores, fine industrial work (in hot climates)
		Intermediate	Offices, schools, department stores, fine industrial work (in temperate climates)
		Warm	Offices, schools, department stores, fine industrial work (in cold climates)
3	Lamps with $R_a < 70$ but with sufficiently acceptable colour rendering properties for use in general working interiors		Interiors where colour rendering is of comparatively minor importance
S (Special)	Lamps with unusual colour rendering properties		Special applications

country's recommendations probably occurs because, with the wide range of values permitted, all countries' views could be accommodated. Also, unlike the glare control system, there is little commercial benefit to be gained by any country's equipment manufacturers in limiting surface reflectances and illuminance ratios.

An area where countries differ considerably in their opinions is in the applications for which lamps with different colour properties are appropriate. The CIE guide does not make any specific recommendations on this topic. Rather it provides a system of classifying light sources by Colour Rendering Index and colour appearance, and gives some examples of where each class may be used (Table 9.3).

The CIE guide also gives some advice on design problems. It discusses the interaction between daylight and electric lighting and the integration of lighting and air conditioning. The depreciation and maintenance of lighting equipment and the factors which affect its energy consumption are also considered. Unfortunately most of this is done in very general terms which indicate the possibilities to the reader without giving him much detail. In fact this could serve as a description of the whole CIE guide. It indicates what is possible and the ranges of some variables that are appropriate but it leaves the final decision as to what is right for any particular country and any specific design to the inhabitants of that country and the designer. This is probably the best that can be done on an international scale and it is no doubt very useful to countries which have no national recommendations. However, for those countries which do have such recommendations, its only purpose is to enable them to check the positions of their recommendations relative to an international consensus.

9.3.2 IES Code for Interior Lighting [4]

The British Illuminating Engineering Society (IES) *Code for interior lighting* has been published intermittently since 1936. The latest edition was published in 1977 and contains much advice, both quantitative and qualitative, on criteria and design approaches. The IES code, as it is known, is divided into three parts. Part 1 is headed 'General Considerations'. This deals with the principal factors, including the need for economy in the use of energy, which affect the design of interior lighting. There are numerous sections on various lighting criteria, at the end of each of which are given a series of recommendations, both qualitative and quantitative. Part 2 is headed 'Lighting Systems'. This covers design methods, lighting equipment and maintenance problems before concluding with a checklist of factors that need to be considered in the design of electric lighting. The checklist is given in Table 9.4. The final part, part 3, is the most widely read. It contains

Table 9.4:   Check list for design of electric lighting systems from the IES code 1977 [4]

General requirements
1. Purpose of the interior (activities may be diverse and more than one lighting scheme may be needed); probable layout of plant, furniture or equipment, where this is known.
2. Availability of daylight and depth of daylight penetration; need for a combined electric lighting/daylighting system. Need for automatic control of electric lighting.
3. Visual difficulty of the task; risk of veiling reflections.
4. Task and general surround illuminances; whether on horizontal, inclined, or vertical plane; local, localised or general lighting; need for optical aids.
5. Strength of modelling required; provision of directional lighting.
6. Colour appearance and colour rendering required.
7. Limiting Glare Index; other requirements for glare control.
8. Statutory requirements or official recommendations.
9. Need for emergency lighting.
Environmental conditions and requirements
10. Presence or absence of hostile environment; need for special luminaires.
11. Unusually high or low ambient temperatures (e.g. foundries, cold stores); effect on control gear and luminaire components and on light output of fluorescent lamps.
12. Possibility of high ambient temperatures near luminaires.

Table 9.4—*contd.*

13. Possible effect of radiant heat from furnaces or other industrial equipment on luminaires and control gear.
14. Effect of heat from luminaires on air temperature in interior; use of lighting heat to supply part of building heating.

Effect of structural features
15. Dimensions of interior: length, width, height.
16. Luminaire mounting height.
17. Luminaire spacing/mounting height ratio.
18. Reflectances of ceiling, walls and floor, including influence of furnishings, windows and glazed partitions.
19. Co-ordination of lighting equipment with other building services.
20. Limitations on luminaire mounting position (e.g. roof structure, number of bays, modular construction of building, space available in ceiling void).
21. Effects of obstruction by parts of structure (e.g. beams), other services (e.g. ventilation trunking, pipework), heavy plant and furniture.

Lamps
22. Lamp types which meet colour appearance and colour rendering requirements for interior or activity.
23. Lamp light output related to source size and mounting height.
24. Rationalisation of lamp type, colour and wattage, particularly with existing installations; availability of replacements.
25. Run-up time to full light output; need for standby system to cover interruption of power supply.
26. Need to reduce flicker and stroboscopic effects.
27. Economics—capital and running costs.

Luminaires
28. Suitability for purpose.
29. Appearance.
30. Luminous intensity distribution required.
31. Authenticated photometric data.
32. Need for special equipment.
33. Ease of maintenance.
34. Availability—present and future—of spare parts.
35. Weight; fixing arrangements.
36. Use of trunking systems or lighting tracks.

Maintenance
37. Maintenance factors.
38. Accessibility of luminaires—need for special equipment.
39. Acceptability of proposed luminaires, lamps, etc. to client's maintenance staff.

Use of energy
40. Check wattage loading for proposed scheme with suggested target loading.
41. Check switching or control systems with object of minimising use of energy.

the detailed recommendations of lighting criteria for a large number of applications. The recommendations appear in two schedules, the first being a daylighting schedule, the second the general schedule for electric lighting. Table 9.5 shows the entry for the electric lighting of milk bottling plants. This table contains recommendations on the illuminance that should be provided, the position where it should be provided, the limiting Glare Index

Table 9.5: Sample of the artificial lighting schedule from the IES code 1977 [4]

Area of application	Standard service illuminance (lx)	Position of measurement	Limiting Glare Index	Colour appearance of light source	Suitable light source†	Notes
Milk Bottling Plants (See SI 1172, *Food hygiene (general) regulations*, 1970)						
General working areas	300	Working plane	25	Intermediate or Warm	C, D, E, F, H, I, K, L, Q	Local lighting required for instruments and sight glasses.
Bottle filling	750	Working plane	25	Intermediate or Warm	C, D, E, F, H, I, K, L, Q	Proof luminaires may be required.
Bottle inspection	Special lighting					

† C, D, E, F, H, I = fluorescent lamps; K, L = high pressure mercury discharge lamps; Q = incandescent lamps.

necessary to control discomfort glare, an appropriate colour appearance for the light source and a list of light sources which are suitable for each application. Also given are any statutory requirements, in this case the *Food hygiene (general) regulations* 1970 and some notes on particular points. Here the notes mention the possibility of local lighting being required in some parts of the plant and the fact that special luminaires may be necessary. Finally, it should be noted that the entry suggests some areas where the recommendations are not applicable. For example, bottle inspection will call for special lighting which cannot be described by illuminances on working planes and Glare Indices. This should explain the attraction of the IES code. In one table are given all the criteria normally used for a wide range of practical applications. The information on more general lighting requirements is also available and references are given for those interested in more detailed consideration of any particular point.

One particular point about the IES Code which is worth considering in more detail is the form of the illuminance recommendations. These are termed standard service illuminances. The measure of illuminance is the usual one, but what do 'standard' and 'service' mean? 'Service' is a term used to reflect the reality of lighting practice. Electric lighting systems, even when luminaires are arranged in regular arrays, produce different illuminances at different points. Further, as the installation ages the illuminances change. Thus any lighting installation will produce illuminances which vary with time and space. The service illuminance is meant to take account of this. It is the mean illuminance throughout the maintenance cycle of an installation averaged over the relevant area. As for the other term, the 'standard' referred to is the standard condition to be expected for the particular application. For the milk bottling plant the conventional conditions expected in a milk bottling plant are the standard conditions. When non-standard conditions for a particular application are experienced, then the service illuminance recommendations should be modified according to the flow chart shown in Table 9.6. This forms a useful summary of the illuminance recommendations in the IES code. The first column classifies work according to the frequency of occurrence and the degree of detail involved and indicates the appropriate standard service illuminances. Then a series of modifying factors are introduced. If reflections or contrasts are unusually low, if errors will have serious consequences or if the area is windowless and lit to less than 500 lx, the standard service illuminance is increased by one step on the illuminance scale. If a task is of short duration the standard service illuminance is reduced by one step. The modified standard service illuminance is called a final service illuminance.

The inclusion of this flow chart in the IES code is valuable. It encourages people to think about the particular situation they are concerned with. This is always necessary when considering lighting recommendations because often they represent simply an informed opinion as to what is suitable. This is certainly the case with the lighting recommendations in the IES code. No attempt is made to justify each and every recommendation. Rather the recommendations are simply declared to be representative of good practice. If people want to depart from the recommended conditions then they are free to do so.

Although most of the above has been about lighting criteria there is other advice contained in the IES code. There is a large section on energy consumption by lighting. This contains some wattage loading targets in watts per square metre. The idea behind this is to encourage designers to consider the energy consumption of their designs. If the electrical loading for a particular installation is much greater than the target given, then the designer should think again about what he is trying to achieve and how he is trying to achieve it.

There are also appendices which contain details of calculation methods for obtaining illuminances at points and on planes from point and extended sources. Other appendices provide information on some lighting criteria which are thought to be of value but that cannot be said to be established as part of the lighting engineer's craft. Examples include a demonstration of the directional effects of lighting, statements on Contrast Rendering Factor and the CIE Colour Rendering Index, and a link between the Glare Index system and the method given in the CIE guide [12].

By now it should be apparent that the IES code contains much of interest. Grouped around the schedules is advice on the general principles of lighting design, the equipment available and the methods of calculation. Therefore it can be considered as both a source of lighting standards and design guidance. The only drawback is that the advice is rather general. As long as the proposed installation is going to use conventional luminaires in regular arrays then the IES code contains all the designer needs, but if some unusual features are involved then additional advice is necessary. This additional advice is provided by a series of documents in the form of application guides or technical reports on particular aspects. The *IES lighting guide: sports* [15] is an example of the former, whilst the technical report on the calculation of utilisation factors [16] is an example of the latter. Taken together, the IES code, the ancillary application guides and the technical reports, provide a complete framework of guidance on lighting for a wide variety of applications.

Table 9.6:   Flow chart for modifying standard service illuminances for unusual conditions, from the IES code 1977 [4]

Task group and typical task or interior	Standard service illuminance (lx)	Are reflectances or contrasts unusually low?	Will errors have serious consequences?	Is task of short duration?	Is area windowless?	Final service illuminance (lx)
Storage areas and plant rooms with no continuous work	150					150
Casual work	200				no → 200 / yes	200
Rough work Rough machining and assembly	300	no → 300 / yes	no → 300 / yes	300 / yes	300 — no → 300 / yes	300
Routine work Offices, control rooms, medium machining and assembly	500	no → 500 / yes	no → 500 / yes	500 — no / yes	500 → 500 / yes	500

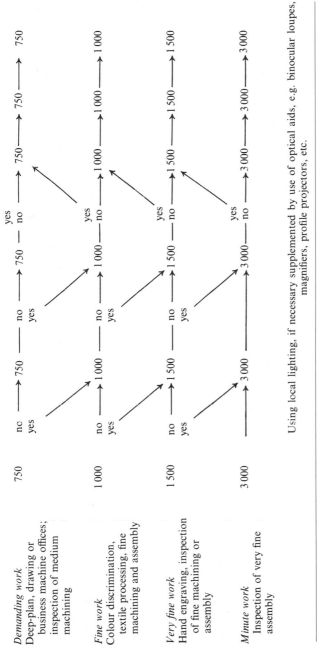

Using local lighting, if necessary supplemented by use of optical aids, e.g. binocular loupes, magnifiers, profile projectors, etc.

It is worth noting that the IES code is not alone. Other European countries have similar publications which include similar recommendations [17, 18]. In general, the illuminance recommendations for particular jobs are similar throughout Europe as are the recommendations on illuminance ratios, surface reflectances and the light sources thought suitable for particular applications. The most obvious difference between the European recommendations concerns discomfort glare control. However, this difference is not so much a matter of the level of discomfort glare that is acceptable but rather the system in which the criterion should be expressed. The Glare Index system is used by the UK, Belgium and Scandinavia, whilst the luminance limit system is used by Austria, France, Germany, Italy and the Netherlands. These two systems provide control of discomfort glare by different methods, the former by expressing the effect of the luminaire in a particular room, the latter by considering only the luminance distribution of the luminaire. This fundamental division means that it is not possible to reconcile the two systems. However, as discussed in Chapter 7, discomfort glare is a rather nebulous concept and either system is capable of providing the control necessary. In which case the lighting codes of Europe can be considered to be in reasonable agreement [19].

9.3.3 Illuminating Engineering Society of North America. IES Lighting Handbook [9]

The *IES lighting handbook* contains all there is to know about lighting and a bit more. First published in 1949, the latest edition was produced in 1972. Its aim is to provide information on light and lighting in a simple condensed style. In pursuit of this aim it provides chapters on the physics of light, light and vision, measurement of radiant energy, colour, luminaire design, light sources, lighting calculations and daylighting. Individual chapters give detailed advice on lighting offices, schools, banks, shops, industries and homes. Other chapters deal with outdoor lighting for sports, roadways, aviation, transportation and even underwater. It will be apparent that this document is nothing if not comprehensive. It is also unique and unusual. It is unique in two ways. The first is that the illuminances recommended for different tasks are implied to be those necessary to produce a Visibility Level of 8·0 for each task under reference lighting conditions (see Chapter 3). Thus the recommendations are linked directly to a well established curve of threshold contrast against luminance [20]. This is not as unrealistic as might be first thought. It is perfectly reasonable to base illuminance recommendations on task visibility, after all one of the main functions of lighting is to make tasks visible. However, it should always be remembered that there is no unique link between Visibility Level and task performance. Different tasks with the same Visibility Level will allow different levels of performance. This Visibility Level approach to specifying illuminance has an interesting consequence. The illuminances recommended in the IES handbook [9] are generally somewhat higher than those recommended anywhere else. Table 9.7 shows the illuminances recommended for a range

Table 9.7: Illuminance recommendations of different countries for similar applications (lx)

Application	UK (IES code, 1977) [4]	West Germany (DIN 5035, 1978) [7]	USA (IES lighting handbook, 1972) [9]
Clothing factories—sewing	1 000	750	5 400
Leather manufacture—grading	1 500	750–1 500	3 200
Foundries—moulding	500	300	1 100
Glass works—furnace rooms	150	200	320
Engineering—medium machining	500	300	1 100
Textiles—carding cotton	300	300	540
Woodwork—rough bench work	300	300	320
General office	500	500	750–1 600
Drawing office	750	1 000	2 200
Corridors	150	50	220

of applications in the American [9], British [4] and German [7] Codes. The similarity between the European recommendations and their difference from the American recommendations is apparent. The question that needs to be asked about this difference is not which approach is correct, since there is no single correct answer, but rather whether the recommendations are reasonable or unreasonable for the time and place in which they were made.

The second way in which the IES handbook is unique is that some of the illuminance recommendations, usually those involving flat, written materials such as occur in offices, are expressed in terms of equivalent sphere illuminance. This quantity is not an illuminance that you can measure. It is the illuminance on the task in a photometric sphere that gives the same level of contrast sensitivity as does the actual installation. This measure allows for the effects on task visibility of veiling reflections, disability glare and temporary changes in adaptation, as well as illuminance. Unfortunately veiling reflections vary greatly, depending on the nature of the task. This means that when comparing installations for an interior in which a number of different tasks occur, misleading conclusions as to which is the better installation are likely to occur [21].

Two other features of the IES handbook are unusual. The first is the detail in which each application, particularly for industrial areas, is considered and the type of lighting installation described qualitatively. The second is that although there are quantitative recommendations other than illuminances, these are given much less emphasis than in other lighting documents. For example, little mention is made of controlling discomfort glare by the visual comfort probability measure (see Chapter 7). In addition, frequent reference is made to the problems caused by veiling reflections without any quantitative advice being given, and only rarely are light sources suitable for different applications suggested.

Thus the *IES lighting handbook* is a strange mixture. On the one hand, it contains much that is useful for specific applications. It is comprehensive in its applications and equipment coverage and in the discussion of the physics of light. On the other hand, its Visibility Level approach to prescribing illuminances has led to recommendations which are out of line with those of other countries. This fact is recognised in a recent development. The Illuminating Engineering Society of North America is proposing to change the basis on which it makes recommendations for illuminances in rooms [22], from the visibility of tasks to the illuminance scale given in the CIE guide [12], the choice of the appropriate illuminance in any situation being left to the designer. This is a radical move but one which confirms that lighting recommendations are ultimately reached by consensus.

To summarise, recommendations for interior lighting can take many forms; from those that merely specify what conditions should be achieved, to those that do this but also offer extensive advice as to the means by which it can be done. No matter whether the recommendations appear in publications entitled guides, codes or handbooks, they all have one thing in common. They represent the considered judgements of groups of people as to what constitutes good lighting at the time they were written. None of them has a magic formula for deriving the lighting conditions necessary to achieve a required level of task performance or to provide a pleasant appearance. Given that the recommendations are derived by consensus it is to be expected that disagreement will occur between different documents. In arriving at a consensus view, the basic information available, the technical capabilities of the equipment available, and the cost of lighting equipment and power all have to be considered. The knowledge that the balancing of these factors is involved should encourage people to treat interior lighting recommendations for what they are, guidance not regulations. If this in turn encourages designers to think more carefully about what they are trying to achieve then so much the better.

9.4   ROADWAY LIGHTING

Documents containing recommendations for lighting of roadways tend to be more concise and in closer agreement than is the case for interior lighting. The reason for this is simply that countries which can afford roadway lighting are likely to have similar types of road. Thus the variety of conditions likely to be met is much reduced for roadway lighting in comparison with interior lighting. This conformity of conditions has enabled the CIE to produce a document, Publication 12.2, *Recommendations for the lighting of roads for motorised traffic* [23], which is not subject to the diplomatic vagueness identified in the *CIE guide on interior lighting* [12]. In fact CIE Publication 12.2 is only one of a number of CIE documents on roadway lighting which together justify the recommendations and demonstrate how they can be achieved in practical design [24, 25].

CIE Publication 12.2 starts by defining the function of roadway lighting. It is to provide good visual conditions at night so that traffic can move safely, quickly and comfortably. To achieve these conditions a number of lighting criteria are given, some concerned mainly with visibility, others mainly with driver comfort. The criteria are derived from an examination of laboratory and field research and practical experience. In other words, just as for interior lighting, the recommendations for roadway lighting represent a consensus of opinion. Once again there is no magic formula. Some of the evidence on which the recommendations are based is given in Chapters 4 and 7.

The quantitative lighting criteria given in CIE Publication 12.2 are shown in Table 9.8. The first point to note is that different criteria are specified for what are called different road classes. The basis of this classification is the volume, speed and composition of traffic which uses the roads and whether the surroundings to the road at night are bright or dark. The actual classes of roads are shown in Table 9.9. Five criteria are given for each class of road. The first is the average maintained road surface luminance. This is the average value being maintained throughout the life of the lighting installation. Values for different classes of roads vary from 0·5 to 2·0 cd m^{-2}, the higher values being allocated to the more difficult traffic conditions, particularly when the road has bright surroundings. The

Table 9.8:　Criteria for roadway lighting from the CIE recommendations for the lighting of roads for motorised traffic 1977 [23]

Class of road	Surrounds	Luminance level:† average maintained road surface luminance ($cd\,m^{-2}$)	Uniformity ratios — Overall uniformity ratio	Uniformity ratios — Lengthwise uniformity ratio	Glare restriction — Glare control mark	Glare restriction — Threshold increment (%)
A	any	2	0·4	0·7	6	10‡
B 1	bright	2	0·4	0·7	5	10
B 2	dark	1			6	10‡
C 1	bright	2		0·5	5	20‡
C 2	dark	1	0·4		6	10
D	bright	2	0·4	0·5	5	20
E 1	bright	1	0·4	0·5	4	20
E 2	dark	0·5	0·4		5	20‡

† The luminance level recommended is the maintained value of the average road surface luminance. In order to maintain the level, a depreciation factor of at most 0·8, depending on the lantern type and on the local degree of air pollution, should be considered.
‡ In view of the rather restricted experience with the application of the threshold increment concept a value of about $\frac{2}{3}$ of the value indicated should preferably not be exceeded.

Table 9.9: The classification of roads from the CIE recommendations for the lighting of roads for motorised traffic 1977 [23]

	Class of road	Type and density of traffic†	Types of road	Examples
Motorised traffic	A	Heavy and high speed motorised traffic	Road with separated carriageways, completely free of crossings	Motorways, express roads
	B	Heavy and high speed motorised traffic	Important traffic roads for motorised traffic only, possibly separate carriageways for slow traffic and/or pedestrians	Trunk roads, major roads
	C	Heavy and moderate speed motorised traffic‡ or heavy mixed traffic of moderate speed	Important, all purpose, rural or urban roads	Ring roads, radial roads
Mixed traffic	D	Fairly heavy mixed traffic of which a major part may be slow traffic or pedestrians	Roads in city or shopping centres, approach roads to official buildings and areas, where motorised traffic meets heavy slow traffic or pedestrians	Trunk roads, commercial streets, shopping streets, etc.
	E	Mixed traffic of limited speed and moderate traffic density	Collector roads between residential areas (residential streets) and A to D type roads	Collector roads, local streets, etc.

† In cases where the road lay-out is below standard for the considered type and density of traffic, it is advised to install a lighting quality higher than the recommended one. In cases, however, where the road layout is considered to be above standard for the expected traffic density, a slight decrease of the lighting quality may be economically justified.

‡ Speed limit approximately 70 km h^{-1}.

aim of this is to give the carriageway some prominence but not too much. For dark surroundings a lower carriageway luminance is acceptable, but it is suggested that care be taken to distribute some of the light onto the surroundings to reveal any movement at the edge of the road. As a rule of thumb, it is suggested that the 5 m wide strips beside the carriageway should be illuminated to not less than half the level of the nearest 5 m wide strip of the carriageway.

The next criteria of interest concern uniformity. There are two of these. The first is simply the overall uniformity ratio, which is the ratio of the minimum luminance to the average luminance of the carriageway as viewed from a position one-quarter of the width of the carriageway from the kerb. The second uniformity criterion concerns lengthwise uniformity. This is the ratio of the minimum to maximum luminance along the centre line of each lane as seen from a point on that line. These two uniformity ratios are said to be different in what they effect. It is believed that the overall uniformity is more closely related to visual performance than driver comfort. The reverse is believed to be true of lengthwise uniformity.

The last two criteria can be divided on the same basis as uniformity. The glare control mark is a measure which quantifies the extent of discomfort glare that will be experienced under the road lighting installation. The threshold increment measures the effect of disability glare. It should be noted that both glare control mark and threshold increment criteria refer to glare produced by the roadway lighting alone. The glare produced by oncoming headlights is ignored. Both the glare control mark and the threshold increment are considered to be minimum conditions. If economics permit, higher levels of glare control mark, and lower threshold increments, should be provided. It can be seen from Table 9.8 that the glare criteria adopted vary with the class of road and with whether the surrounds are bright or dark. When the surrounds are bright, worse values of glare control mark are allowed than when the surroundings are dark.

The final criterion is qualitative. It involves the visual and optical guidance produced by the roadway lighting system. Visual guidance is given to the driver by the road itself, seen against its surroundings and by such aids as road markings, crash barriers and signs. The roadway lighting can assist visual guidance by enabling the road and these aids to be easily seen. Optical guidance simply means the extent to which the lighting installation itself can give guidance to the driver. For example, the layout of lanterns and columns can indicate the approach of a bend in the road or the presence of an intersection.

There is one serious restriction on the quantitative criteria given in the

CIE guide, although there is little that can be done about it. The criteria are minimum requirements for good lighting on dry roads. Wet roads introduce significant changes in the effect of lighting conditions on object visibility (see Chapter 4); but, in spite of considerable research efforts, there is at present insufficient information to justify international recommendations for wet roads.

Another point to be noted about these criteria is that there is no indication as to the trade-off between them. For example, there is no indication as to whether a higher standard of lengthwise uniformity would compensate for a below standard glare control mark. Although these trade-offs are not given, it is possible to separate the criteria into those that most affect driving performance and those that primarily influence driver comfort. Average road surface luminance, overall uniformity and threshold increment are all said mainly to affect driver performance; and lengthwise uniformity and glare control mark are assumed predominantly to influence driver comfort. It is suggested that where economic factors necessitate, any relaxation of standards should first be applied to the criteria that affect driver comfort. This suggestion ignores the possibility that comfort may affect performance indirectly. Also it seems unlikely that a quantity such as lengthwise uniformity will influence comfort alone. It is much more likely that it will influence both performance and comfort. Certainly a factor, such as average road surface luminance, which affects driver performance, can also affect driver comfort.

The recommendations made in CIE Publication 12.2 have been considered in detail because they form the basis of many of the national codes of practice for roadway lighting. This is not to say that the CIE recommendations have been accepted in their entirety. Rather, many countries have accommodated the CIE recommendations as best they can in their existing framework of practice [26]. An example of one country which has followed CIE recommendations fairly closely is France [27]. The standards recommended for roadway lighting in the latest document are basically those recommended by the CIE with the exception of the threshold increment criterion. This criterion has been sacrificed in order to reduce the number of changes that appeared in the new document as compared with its predecessor. The removal of a criterion for this reason may seem illogical but presumably it is necessary if the complete document is to be accepted. Apart from this omission, the main change from the CIE recommendations is in the method of identifying if a proposed roadway lighting scheme would produce the required conditions. Two methods are given for doing this. One is the point luminance method described in CIE

Publication 30 [25]. The other is a much simpler but less precise method, called the R ratio method. This method allows a non-specialist to design roadway lighting using luminance as a criterion. R is the ratio of average illuminance to average luminance on the roadway. The R ratio method involves first choosing the luminance for the road from the CIE classification. Then the necessary average illuminance is calculated from the luminance criterion accepted and the appropriate R ratio for the class of road. The final stage is to use tabulated utilisation factors for different types of lantern and spacing/mounting height ratios. It is assumed that the CIE criterion for luminance uniformity will be met by limiting the allowable spacing/mounting height ratios and the mounting height/road width ratios. The advantages of the R ratio method are clear. It is a very simple method, which enables a quick estimate of the suitability of any proposed scheme to be obtained and which also allows the practical requirements of a suitable lighting scheme to be identified before the luminaires have been selected. The French document [27] expresses the hope that the CIE point luminance method [25] will be used for motorways, etc. For more modest projects, especially where it is known that luminaires will not be installed exactly as planned, as often occurs in urban situations, the R ratio method is suitable.

Overall the CIE recommendations are generally considered as a standard to aim at, but which need to be modified by the various practical factors existing in each country. In only one country has the CIE approach been seriously questioned. The current roadway lighting standard in the USA [28] does not specify luminances of road surfaces but is based on the maintained illuminance on the surface and its uniformity. True, mention is made of roadway luminance and disability glare, but quantitative advice is restricted to appendices and is not part of the Standard Practice. The concentration on horizontal illuminance is now admitted to be un-satisfactory [29]. The present belief is that ideally the concepts of Visibility Level and/or visibility index discussed in Chapter 4 should form the basis of roadway lighting specifications. Unfortunately, whilst these concepts are believed to be of value, they are not considered sufficiently developed and validated to be incorporated into a Standard Practice yet. However, it has been decided, in principle, to introduce roadway luminance as a criterion of roadway lighting design alongside horizontal and vertical illuminance in future editions of the Standard Practice. In addition to this disagreement with the CIE recommendations there is also disagreement with the luminances considered necessary. The value of average road surface luminance proposed for freeways (motorways) is said to be about $0 \cdot 5 – 0 \cdot 6 \, \mathrm{cd} \, \mathrm{m}^{-2}$ compared with the $2 \cdot 0 \, \mathrm{cd} \, \mathrm{m}^{-2}$ suggested by the CIE [30].

The reason for this difference is not explained but it is suggested that, since at an average road surface luminance of $0.6\,cd\,m^{-2}$ on freeways, such objects as a dog or a dark 20 cm cube can be detected at 100 m [31], $0.6\,cd\,m^{-2}$ is all that is necessary. Obviously the CIE roadway lighting committee thinks differently, and, given the variety of objects that are seen in rapid succession by a driver interpreting the road ahead, they may well be right.

To summarise, in spite of finishing on a discordant note, there is considerable agreement on the factors which must be considered in a roadway lighting design and on the level which those factors should reach. Where disagreement breaks out is in the way in which these criteria are turned into design procedures.

9.5. CONCLUSIONS

The overall impression gained from the above discussions of interior and roadway lighting documents should be that the writer of such documents has an unenviable task. He has to strike the right balance between a number of conflicting aims. He has to make recommendations which are precise and preferably quantitative, but not so precise that they lose credibility. Equally he has to avoid making recommendations which are so vague as to be meaningless. The recommendations have to be technically and economically feasible and simple enough to be implemented, although they should also reflect the complexity of the subject. Having achieved a balance between these factors which is appropriate for the application, and having written the document at a level suitable to the audience to which it is addressed, he may then be called upon to justify his recommendations. Often this can only be done by analogy or by an appeal to practical experience and a general consensus. An unenviable task indeed, yet the document writer's efforts are necessary. Advice is needed on appropriate lighting by people who buy lighting installations and by some of those who design them. To such people it does not matter that every last word in an advisory document can be justified by experimental evidence. They do not have time to wait for that. In any case, if the recommendations produce reasonable results in practice, they will be accepted and the judgements of the writers vindicated. If they do not produce reasonable results in practice, then no amount of experimental evidence will convince anyone that the recommendations are correct. It is this practical test which is the ultimate justification of many lighting standards, codes and guides.

From the wide range of publications offering advice on lighting, it should be possible to find suitable information for any particular application. This is not to say that each document is the same. They all have different strengths and weaknesses, as should be evident from the preceding reviews, but they also have something in common. Ultimately, for any particular application, the recommendations come down to a consensus of opinion. For this reason, if for no other, it is always necessary to think about what lies behind any recommendations before using them. Given this circumspect approach, lighting standards, codes and guides can be of great benefit to the art of lighting.

9.6 SUMMARY

This chapter is concerned with the form and basis of the many documents that offer advice on lighting standards and design. It is not a comprehensive listing of such documents but rather a consideration of their common characteristics, illustrated by example. Such documents can be conveniently classified by whether they are statutory or not and whether they are precise or not. Precise statutory recommendations are usually only made when the consequences of the lighting being poor are grave. The other division between documents is between those that declare what lighting should do and those that define what it should be. In fact most documents do both these things by recommending what lighting conditions should be achieved and offering advice on how this may be done. As for the basis of the recommendations, in virtually all documents the ultimate decision on recommendations occurs by a consensus of opinions. The aims of the lighting and the knowledge of the effects of different variables is considered, as is the technical practicality of meeting imposed criteria and the consequent financial cost. The balance between these factors which is considered appropriate at a particular time and place is what determines the actual recommendations made. It is for this reason that lighting recommendations are always changing.

The current state of this balance and the various approaches to preparing documents are shown in a detailed consideration of three widely used documents on interior lighting and three documents concerning roadway lighting. The documents considered are: the *CIE guide on interior lighting*, the *IES code for interior lighting*, and the IES of North America *Lighting handbook*; the CIE Publication 12.2 on roadway lighting, the French roadway lighting recommendations, and the *American national standard practice for roadway lighting.* Each of these documents has different advantages and disadvantages. They should not be used without thought.

REFERENCES

1. Lam, W. M. C. *Perception and Lighting as Formgivers for Architecture*, McGraw-Hill, New York, 1977.
2. Knopp, L. *The Cinematograph Regulations 1955*, The Cinema Press Ltd, London, 1955.
3. Statutory Instrument 2168, *The slaughterhouses (hygiene) regulations*, HMSO, London, 1958.
4. Illuminating Engineering Society. *IES code for interior lighting*, IES, London, 1977. (In 1978 the IES merged with the Chartered Institution of Building Services, London. IES publications are now available from this body.)
5. Health and Safety Executive. *Lighting in offices, shops and railway premises*, HMSO, London, 1979.
6. Statutory Instrument 890, *The standards for school premises regulations*, HMSO, London, 1959.
7. Deutches Institut Fur Normung. DIN 5035, *Innenraumbeleuchtung mit kunstlichem licht*, DIN, Berlin, 1978.
8. British Standards Institution. BS 5489, *Code of practice for road lighting*, BSI, London, 1974.
9. Illuminating Engineering Society of North America. *IES lighting handbook*, 5th edition, IES, New York, 1972.
10. Illuminating Engineering Society. *IES lighting guide: Building and civil engineering sites*, IES, London, 1975.
11. Collins, J. B. The illuminating engineer in a changing world, *Ltg Res. and Technol.*, **9**, 1, 1977.
12. Commission Internationale de l'Eclairage. *Guide on interior lighting*, CIE Publication 29, 1975.
13. Fischer, D. The European glare limiting method, *Ltg Res. and Technol.*, **4**, 97, 1972.
14. Standards Association of Australia. Australian Standard AS 1680, Sydney, Australia, 1975.
15. Illuminating Engineering Society. *IES lighting guide: Sports*, IES, London, 1974.
16. Chartered Institution of Building Services. Technical Memorandum, *The calculation and use of utilization factors*, CIBS, London, 1980.
17. Netherlands Institution for Illuminating Engineering. Recommendations for interior lighting, *International Lighting Review*, 4/5, 1966.
18. Fordergemeinshaft gutes licht, *Informationen zur Lichtenwendeng*, Vols. 1–8, Frankfurt, West Germany, 1975.
19. Fischer, D. Comparison of some European interior lighting recommendations, *Ltg Res. and Technol.*, **5**, 186, 1973.
20. Blackwell, O. M. and Blackwell, H. R. Visual performance data for 156 normal observers of various ages, *J.I.E.S.*, **1**, 3, 1971.
21. Boyce, P. R. Is equivalent sphere illuminance the future? *Ltg Res. and Technol.*, **10**, 179, 1978.
22. Illuminating Engineering Society of North America. Selection of illuminance values for interior lighting design, *J.I.E.S.*, **9**, 188, 1980.
23. Commission Internationale de l'Eclairage. *Recommendations for the lighting of roads for motorised traffic*, CIE Publication 12/2, 1977.

24. Commission Internationale de l'Eclairage. *Glare and uniformity in road lighting installations*, CIE Publication 31, 1976.
25. Commission Internationale de l'Eclairage. *Calculation and measurement of luminance and illuminance in road lighting*, CIE Publication 30, 1976.
26. Association of Public Lighting Engineers. *CIE TC 4.6 Symposium on road lighting design —present and future*, APLE, London, 1979.
27. Association Francaise de l'Eclairage. *Recommendations relatives a l'eclairage des voies publiques*, Paris, 1978.
28. American National Standards Institution. *American national standard practice for roadway lighting*, RP-8, New York, 1977.
29. Odle, H. A. Roadway lighting specifications of the future, *Lighting Design and Application*, **9**, 7, 18, 1979.
30. Ketvirtis, A. Integration of luminance and illuminance calculation methods in roadway lighting design, in *CIE TC 4.6 Symposium on road lighting design—present and future*, APLE, London, 1979.
31. Ketvirtis, A. *Road illumination and traffic safety*, Transport Canada, Ottawa, 1977.

10

The Way Ahead

10.1 INTRODUCTION

What follows is unlike previous chapters in that it does not deal with definitions or quantities, or experimental results and their interpretation. Rather it gives the author's opinions as to the actions that need to be taken if future lighting research is to be fruitful. Broadly these actions can be classified into (a) avoiding the limitations of the past, (b) developing new approaches to the study of lighting and new measures of the effects of lighting and (c) applying these approaches and measures to areas of practical importance. Each of these classes will be examined in turn.

10.2 THE LIMITATIONS OF THE PAST

Anyone faced with evaluating past lighting research rapidly becomes aware of three aspects which limit the applicability of the results. The first aspect is that by present day standards much past experimentation, although ingenious, has been of moderate quality. Numerous experiments can be found which are characterised by very little control of experimental procedure, a small and/or unrepresentative sample of people and a limited statistical analysis of the data collected; although the control and description of the lighting conditions has often been good. This situation is not unreasonable given that many of these studies were done when psychology and statistics were little-known and developing subjects, but that is not the case today. There are now available many excellent publications giving advice on experimental design and analysis [1, 2, 3, 4]. Future lighting research will need to extend the care and attention that has always been given to the lighting conditions to these features of experimentation as well.

The second limiting aspect is the tendency to treat people as 'human meters' for assessing lighting. This is most evident in the choice of the independent variables used in experiments. In the past the independent variables have frequently been such basic features of lighting as illuminance, different glare conditions and colour properties of lamps. These variables have been chosen because they are important to the equipment manufacturer or are used in lighting design calculations, and not because they necessarily reflect factors that are important to the people using the lighting. It may be of some relevance to the designer's immediate problem if, for example, the Colour Rendering Index of a lamp can be shown to relate to the ease with which different hues can be discriminated; but it is not getting at the root of the problem of the discrimination of colours. To do this it is necessary to gain an understanding of the basis on which people discriminate between colours. Once this has been achieved, a suitable measure of the relevant lamp properties can be developed. To achieve a more fundamental understanding of how people respond to lighting it is necessary to put people first and the lighting conditions second. Further, the people considered have to be representative of actual users in all their variety and complexity. Unless this is done, future research will not be building on the successes of the past, but merely repeating them.

The third limiting aspect is the concentration on the general, inherent in the neglect of context. Quite correctly, a lot of past lighting research has been concerned with establishing general rules for good lighting and/or for the effect of different lighting conditions on work. These general rules can now be said to be established. What deserves attention in the future is the extent to which the general rules need to be modified for different contexts. The point is that the lighting conditions which are most suitable for an interior depend on the context. The effect of lighting conditions on task performance also depends on the context. Until the importance of context is acknowledged there is little likelihood of achieving a finer understanding of how lighting affects impression and work.

10.3 APPROACHES AND MEASURES

Although putting people first, careful experimentation and an awareness of the importance of context are necessary for future research to be fruitful; alone they are not sufficient. For fruitful research to occur, the problems to be investigated have to be approached in the right way. In the broadest sense this means considering the direction in which research should aim to move. For the study of the effect of light on work there is plenty of knowledge on which to base very general rules but little that is applicable to specific tasks. Hence the need in this field of study is for a move from the general to the specific. As for the study of the effect of lighting on impression, the direction of movement here should be from the specific to the general. At present there are a few studies done in realistic conditions which have indicated how to create an impression with light. But these results only apply to the specific contexts in which they were obtained. What is needed in the future is for many more different contexts to be examined. If some consistency was then revealed, general rules about using lighting to create impression could be formulated, although it would always be necessary to identify the ranges and contexts over which such rules applied.

The aims of any research are only one aspect of the approach adopted. Another aspect is the techniques used to study the problems. In this respect, what lighting research would benefit from is an increase in the variety of techniques used—and there are several sources of inspiration. Over recent years there have been advances in psychology and statistics which have considerable relevance for lighting research [5, 6]. Further, other environmental features such as noise [7] and thermal conditions [8] have been studied almost independently of lighting, using a wide variety of techniques. An examination of some of these techniques and the indices developed to characterise the results obtained would be helpful. For example, the Noise and Number Index used to predict the annoyance caused by aircraft noise [7] suggests a similar measure for describing the annoyance caused by intermittent lamp flicker. In general, awareness of developments and the techniques used in other environmental research areas can do nothing but help lighting research.

Although choosing an appropriate approach to a problem is important, new ways of measuring the effects of lighting are also required. For examining the effects of light on work, the most urgent requirement is for some measure of how easy a target is to see, as opposed to how well it is lit. At present the measures used describe the lighting but do not necessarily relate to the ease with which the task can be seen. One of the most interesting developments in this field over the last decade has been the visibility approach, containing as it does the concept of Visibility Level (Chapter 3). This measure has the great advantage of needing a human being to take the measurement, because only a human being can integrate all the aspects that influence how easy something is to see. The measurement also reflects the complexity of the situation. Each measurement for each task is unique to that task, viewed from a particular position, under a particular lighting installation. Such measurements are time-consuming and very variable but they are also meaningful.

Visibility Level is but one possible measure of ease of seeing. No doubt others could be derived which might be simpler to use and less subject to individual differences due to variations in subjective criteria. For example, it might prove possible to have a measure of how easy something is to see which was based on minimum presentation time, instead of contrast reduction as used by the visibility meter. The idea here would be to steadily reduce the time for which the scene was available until the details of interest were no longer visible. This method will still have the problems associated with the subjective nature of the threshold measurements which have bedevilled the contrast reducing visibility meter. Against this it would have the advantage that time intervals can be easily and accurately measured and varied; and it would be easy to introduce an element of visual search into the scene, if that was called for. At the moment this is only a possibility which might not work but there can be little doubt that some simple measure of how easy an object is to see would be of value. It would probably be useful as a component in a system for specifying how effective particular lighting conditions were. It would certainly be useful in establishing a taxonomy of visual task difficulty. At present the best that is widely available is based on the physical measures of size and contrast of critical detail, rather than on a direct visual measure of how easy something is to see. A measure of ease of seeing is a topic well worth pursuing.

But a measure of ease of seeing is not the only requirement. One of the points considered in the discussion on light and work was that for the vast majority of tasks the visual component is only part of the complete task. The visual component can vary from very large, as in driving in a strange town at night, to very small, as in audio typing. If a more precise estimate of the effect of lighting on work is to be gained, then a measure of the magnitude of the visual component in the work is necessary. This might be achieved by some form of time analysis combined with eye movement recording. Basically the idea would be to use eye movement recording devices to identify where the observer was looking and how long for. From these data it would be possible to estimate how much of the total time was spent obtaining the visual information. There would be dangers in this process, such as the need to distinguish between looking, and seeing and identifying what was the important information, but at least it would allow some possibility of quantifying the visual component in a task.

As for the possibilities of lighting research on impression, what is needed is a reorientation and the adoption of new techniques. The reorientation would place people at the centre of the stage rather than the lighting. The subject of interest should be how people interpret the space they are in, as it is revealed to them by the lighting, not simply how satisfactory or pleasant or comfortable the lighting appears to be. To obtain this information some of the multi-dimensional procedures discussed in Chapter 6 could be used. By applying these techniques to a number of different contexts it might be possible to extend the idea of lighting acting as cues to certain impressions. If this could be done, then a number of guidelines for creating particular impressions would be available.

So far the emphasis has been on new measures and new approaches for use with conventional experiments. But it should be realised that there are other ways of gaining knowledge apart from the formal experiment. For example, observation of behaviour can tell us a lot about people's reactions to lighting. The technique calls for careful observation and recording and possibly some imagination in the interpretation. However, it can be done with little interference by the observer and it has the immense advantage of allowing people to 'speak' through their actions. Another approach which has been out of fashion for many years but which may still be of value is introspection. By considering one's own reactions to the lighting conditions it is possible to produce some information on the factors that create certain impressions. Finally, the heretical course of listening to artists, stage lighting experts and photographers, can be advocated. Such people know a lot about creating an impression with light. Introspection, observation of behaviour and discussions with artists are unlikely to be sufficient alone to enable firm conclusions about lighting to be drawn because they are so open to bias. However, they can introduce ideas and concepts which can then be tested by carefully designed experiments. This is their real function—to give insight so that a useful hypothesis can be constructed and then tested.

10.4 AREAS OF APPLICATION

Having considered the possible methods of investigation it is appropriate to discuss the problems to which they might be applied. There are numerous specific problems that need investigation but here consideration will only be given to the general areas that deserve study. In what might be called the pure research area four subjects deserve attention. The first subject is the effect of lighting conditions on peripheral vision. Currently the effects of lighting on vision are almost invariably considered in terms of foveal vision, and yet the descriptions of reading, inspecting and driving in Chapter 4 have all revealed the importance of peripheral vision to these activities. This practical relevance, together with what is known about the way in which increases in task difficulty restrict the efficiency of peripheral vision [9], suggest it would be a fruitful field of study. The second subject is the use of lighting to diminish the limitations imposed by visual disabilities. Slight forms of visual disability are quite common [10] so examining how lighting affects the visual capabilities of people with different degrees of visual disability could be rewarding. The third subject is the effect of lighting as revealed in effort and fatigue. Lack of progress in this area has reduced interest but these phenomena still occur, are still important and deserve study. Finally, there is the question of what underlies the phenomenon of discomfort glare. Limiting discomfort glare tends to reduce the energy efficiency of lighting installations so it would be interesting to know the extent to which it was necessary. The limitations of the existing knowledge were discussed in Chapter 7.

As for applied research, there are a number of areas deserving attention. The first is industrial lighting. Much of existing lighting research has been concerned with offices. Unfortunately factories are not like offices, in that they often involve three-dimensional, as opposed to two-dimensional tasks, and the lighting is frequently subject to severe obstruction. Further, many industrial tasks involve the perception of form rather than the meaning of symbols and hence may call for different lighting conditions. At present most industrial lighting is based on data obtained with office work in mind. It may well be that different approaches to the lighting of factories need to be developed. A second area for investigation is the limitations of non-uniform lighting. For example, local lighting, where each work area is lit independently of the general building lighting, increases the risk of discomfort glare occurring and casts doubt on the need for strict uniformity criteria. How local lighting can be achieved economically, and yet provide good lighting conditions throughout the work area without creating discomfort, needs to be examined. A related problem, but one of increasing urgency, is how to provide a luminous environment suitable for the operation of visual display units. This whole topic of non-uniform lighting is one which is likely to be of considerable importance in the future. The third area that deserves investigation is the balance between natural and artificial light. Combined use of these light sources is likely to increase with the increasing cost of energy for artificial lighting, so it would be useful to make some progress beyond the idea of trying to maintain the balance of brightness in the space. Finally, there is a need to examine the requirements for emergency lighting more closely. Although considerable efforts are being put into providing emergency lighting there is little reliable evidence as to what form this lighting should take. This brief list of areas is by no means comprehensive but it does indicate some of the more pressing needs.

10.5 WHY BOTHER?

All the preceding comments may be interesting but are they necessary? After all, existing lighting seems to be adequate. Work gets done and if interiors are not always pleasant they are not usually uncomfortable. In which case why is it necessary to continue lighting research at all, let alone attempt to improve it? There are two answers. The first is simply that the general standard of lighting could be better. Greater knowledge of how people respond to lighting will aid and encourage better practice. The second answer is that present day lighting is often adequate because it is excessive. Liberal quantities of light are distributed about so that all areas are plentifully lit. In an era of cheap energy this was acceptable. In a world with diminishing energy supplies and consequently increasing costs this practice is not acceptable. Pressure for energy conservation will force lighting to be applied with greater precision than it is at present. Lighting research is necessary to achieve this precision.

10.6 SUMMARY

This chapter is addressed to those who are, or will be concerned with lighting research. It discusses the actions that need to be taken if future research is to be fruitful. On a philosophical level these actions involve putting people before lighting and accepting the importance of context. On a technical level they involve developing new approaches to the study of lighting, and new measures to quantify the magnitude of the visual component and how easy things are to see. On a practical level they involve applying these approaches and measures to areas of interest, a number of which are suggested.

REFERENCES

1. Kirk, R. E. *Experimental Design: Procedures for the Behavioural Sciences*, Brooks/Cole, Belmont, California, 1968.
2. Winer, B. J. *Statistical Principles in Experimental Design*, McGraw-Hill, London, 1962.

3. Siegel, S. S. *Non-parametric Statistics for the Behavioural Sciences*, McGraw-Hill, London, 1956.
4. Harris, R. J. *A Primer of Multivariate Statistics*, Academic Press, London, 1975.
5. Canter, D. A. *Environmental Interaction*, Surrey University Press, London, 1975.
6. Flynn, J. E., Hendrick, C., Spencer, J. and Martyniuck, O. A guide to the methodology procedures for measuring subjective impressions of lighting, *J.I.E.S.*, **8**, 95, 1979.
7. Burns, W. *Noise and Man*, J. Murray, London, 1973.
8. McIntyre, D. A. *Indoor Climate*, Applied Science Publishers, London, 1980.
9. Mackworth, N. H. Visual noise causes tunnel vision, *Psychonom. Sci.*, **3**, 67, 1965.
10. Ferguson, D. A., Major, G. and Keldoulis, T. Vision and work, *Applied Ergonomics*, **5**, 84, 1974.

Index